PERSPECTIVES IN CONDENSED MATTER PHYSICS
A Critical Reprint Series

Condensed Matter Physics is certainly one of the scientific disciplines presently characterized by a high rate of growth, both qualitatively and quantitatively. As a matter of fact, being updated on several topics is getting harder and harder, especially for junior scientists. Thus, the requirement of providing the readers with a reliable guide into the forest of printed matter, while recovering in the original form some fundamental papers suggested us to edit critical selections on appealing subjects.

In particular, the present Series is conceived to fill a cultural and professional gap between University graduate studies and current research frontiers. To this end each volume provides the reader with a critical selection of reprinted papers on a specific topic, preceded by an introduction setting the historical view and the state of art. The choice of reprints and the perspective given in the introduction is left to the expert who edits the volume, under the full responsibility of the Editorial Board of the Series. Thus, even though an organic approach to each subject is pursued, some important papers may be omitted just because they lie outside the editor's goal.

The Editorial Board

PERSPECTIVES IN CONDENSED MATTER PHYSICS
A Critical Reprint Series: Volume 2

QUANTUM HALL EFFECT: A PERSPECTIVE

edited by

A. H. MacDonald
Department of Physics Indiana University, Bloomington

KLUWER ACADEMIC PUBLISHERS

Jaca Book

prima edizione
novembre 1989

copertina e grafica
Ufficio grafico Jaca Book

in copertina:

The quantum Hall effect occurs in two dimensional electron systems wich have particularly stable (incompressible) states at certain electron densities. When the stable density depends on magnetic field, the system must be surrounded by a diamagnetic edge current. The change in edge current per change in chemical potential equals the quantized Hall conductance and is fixed by the field dependence of the stable density. The figure shows an incomprensible fluid in yellow surrounded by an edge region in red. The incompressibility may be caused by the quantization of the electron's cyclotron motion, by a periodic potential or by electron-electron interactions.

Quantum Hall effect / edited by A.H. MacDonald.
p. cm. -- (Perspectives in condensed matter physics ; v. 2)
ISBN 0-7923-0537-X
1. Quantum Hall effect. I. MacDonald, A. H. II. Series.
QC612.H3Q37 1990
537.6--dc20 89-28228

ISBN 0-7923-0537-X (Kluwer: Hardbound)
ISBN 0-7923-0538-8 (Kluwer: Paperbak)
ISBN 88-16-96003-5 (Jaca Book)

per informazioni sulle opere pubblicate e in programma
ci si può rivolgere a Editoriale Jaca Book spa
via A. Saffi 19, 20123 Milano, telefono 4982341

Preface

This volume contains reprints of 30 scientific articles which provide a record of some of the main advances in understanding which have lead us from Edwin Hall's discovery of the effect which bears his name to our current understanding of the quantum Hall effect. In the process of deciding which articles were to be included in the collection and which were to be left out I became aware of the gaps in my knowledge of the relevant history , both very recent and more distant. The final choices are, I have no doubt, imperfect. I know that they reflect to too large a degree the aspects of the subject on which I have worked myself, so that the collection represents *my* perspective on the quantum Hall effect. It is my hope ,however, that the reader will find that these articles taken together have a rather interesting story to tell: a story which reminds us of the unpredictable consequences of scientific enquiry and reflects the impressive achievements of condensed matter physics in this century.

The reprints are accompanied by an Introduction which can stand on its own as a brief explanation of the quantum Hall effect or serve as an primer for the reprinted articles. The point of view taken in the introduction is again my own and is not the only useful one. My perspective has been shaped in ways I no longer recall by conversations with many colleaques who have instructed me on various subtleties of the theory and of the experiments. In this group I would like to mention especially J.P. Eisenstein, F.D.M. Haldane, B.I. Halperin, K. von Klitzing, R.B. Laughlin, , P.M. Platzman, Mark Rasolt, H.L. Störmer, P. Streda and most of all the owner of the voice on the other end of many telephone calls between Ottawa and Gaithersburg, Steve Girvin. These people shared their insights freely and cannot be blamed for any evidence contained here that I have sometimes missed the point.

I am grateful to Michael Johnson for many helpful comments on a draft of the Introduction.

In the Introduction references to the n'th reprinted article are indicated as ^{Rn} while other references are indicated sequentially.

Bloomington, Indiana
June, 1989

Allan Mac Donald

Table of Contents

M.H. MacDonald, Quantum Hall Effect: A Perspective

Introduction

A. Overview

A natural starting point in trying to gain some perspective on the quantum Hall effect is a look at Edwin Hall's discovery[R1] in 1879 of a *New Action of the Magnet on Electric Currents.* At the time of Hall's discovery, the origin of the effect was not at all clearly understood. Even the electron was to wait a further ten years before making its apprearance. Clear understanding had to await the development of quantum mechanics and of its application to solids.

By the time of the development of the Drude theory[1] of metallic transport, it was appreciated that the Hall effect was a consequence of the Lorentz force on moving electrons. Soon after evidence that the carriers in a metal could sometimes have a positive effective charge[2] appeared. Gradually, it became clear that while the Hall effect could be easily be quantitatively described for a gas of free electrons, it was exceedingly complicated for the Bloch electrons in real metals. Progress in sorting out all these complications was reviewed by one of the main contributors, W.G. Chambers[R2] in 1960. In this classic paper Chambers distinguishes several regimes for studying the effect of a magnetic field on the transport properties of a metal. The low, intermediate and (classical) high field regimes are distinguished by the number of times an electron collides with a scatterer while executing a cyclotron orbit in a strong magnetic field. Since the cyclotron frequency is proportional to magnetic field we go from many collisions per orbit to many orbits per collision as the field strength is increased.

Chambers also distinguishes other regimes , the quantum regimes, in which the physics which would be responsible for the Hall effect's quantum cousin appears. In a magnetic field the allowed values for the kinetic energy contribution from the cyclotron motion (in the plane perpendicular to the magnetic field) become quantized. This is, of course, the quantization which is responsible for the quantum Hall effect and it was first discovered theoretically by L. Landau in an article[R3] written in February of 1930, which was to be an important year in the history of the quantum Hall effect. Landau showed that for quantum electrons, unlike classical electrons , the electron's orbital motion gives a (diamagnetic) contribution to the magnetic susceptibility. He also remarks that the kinetic energy quantization gives rise to a contribution to the magnetic susceptibility which is periodic in inverse magnetic field, although he did not believe that the oscillation would be experimentally observable. It is amusing to realize now that the quantization has experimental consequences so robust and so readily observable that they are the basis of a resistance standard. Moreover, the analogous oscillation in the electrical resistivity was discovered in measurements on Bismuth at low temperatures by Schubnikow and de Haas and was reported on in an article[R5] submitted for publication during the same month as Landau's theoretical paper. The oscillation in the magnetization itself were discovered a few months later[R6] by de Haas and van Alphen in work motivated by the discovery of the resistance oscillations. Despite the fact that de Haas and van Alphen were aware of Landau's work , where he even pointed out that his theory could explain the unusually large diamagnetic susceptibility of Bismuth, they did not suspect that the oscillations they observed were related to those in Landau's theory. That understanding had to wait for his theory to be generalized from the case of free electrons to the case of Bloch electrons in a metal. Landau's discovery may have been the only one related to the quantum Hall effect predicted by theory in advance of its experimental discovery.

Shortly after Landau's theoretical paper appeared an article[R4] was published by E. Teller in which he showed that Landau's result for the magnetic moment could be obtained in a different way. Teller considered a gas of electrons enclosed within a hard wall in contrast to the infinite system considered by Landau. In Teller's calculation the diamagnetic moment could be seen as arising from a current circulating around the edge of the system, while in Landau's calculation it appeared to be a bulk effect. In fact, the argument which gave a zero orbital susceptibility for classical electrons suggested that the boundaries would have to be treated correctly in calculating this property and there was initially some skepticism about Landau's derivation for this reason [3]. The agreement between the results obtained by Teller and Landau demonstrated that the orbital moment per electron, which is an extensive (bulk) quantity, is indeed due to currents flowing only at the edge of the system at least for the case of a uniform system. The boundaries *are* important but , as Landau's calculation showed, the result for the susceptibility is independent of the details of the boundary. The physics responsible for the equivalence of these approaches would turn out, many years later, to be intimately related to the physics responsible for the quantum Hall effect.

The voyage of discovery leading to the quantum Hall effect took a new track with the beginning of research into magnetotransport properties of two dimensional electronic systems (i.e., systems with one translational degree of freedom removed. The possibility of realizing such systems at the interface between a semiconductor and an insulator or at the interface between two different semiconductors had been suggested as early as 1957[4] .The pioneering paper[R7] of Fowler et al. was the first to establish experimentally the existence of a two dimensional electron gas (2DEG) at a semiconductor interface[5]. The theory of magnetotransport in the 2DEG in the strong magnetic field limit has special complications because of the reduced dimensionality and because of the kinetic energy quantization. (The set of states with one allowed value for the kinetic energy is said to form one Landau level.) A very satisfactory theory was provided in a classic paper[R8] by Ando. et al. in 1975. This theory did nearly everything *except* predict the quantization of the Hall resistance which was to be discovered five years later. An important element of the quantum Hall effect is localization; electron states near the top or bottom of a disorder broadened Landau levels are localized around extrema of the disorder potential. The importance of this phenomena in magnetotransport was recognized[R9] in 1976 by Kawaji et al. and also by Englert and von Klitzing[6]. These authors found that when the Fermi level moved through a region of localized states the Hall resistance remained constant and the voltage drop along the direction of current flow was zero. The constancy could be qualitatively explained by observing that the localized states did not participate in transport. The precise quantization of the Hall resistance in units of e^2/h was not suspected ,however, until February of 1980, 100 years and a few months after the equally surprising discovery by Edwin Hall. The story of von Klitzing's discovery of the quantum Hall effect[R11] is recounted by Landwehr in $R10$.

Introduction

The challenge this discovery posed for theory was to explain why the Hall conductivity on the plateaus (the constant regions) was precisely equal to the value expected for free electrons with the Fermi level between Landau levels. It was a mystery that this was true despite the evident importance of disorder which, through the localization phenomena, was providing the reservoir of non-current-carrying states which created the plateaus and made the effect visible. Over the next few years an understanding of the origin of this quantization was developed with major contributions from Prange[R12], Laughlin[R13], Halperin[R14], and Streda[R15]. As first pointed out by Halperin[7] the occurrence of the quantum Hall effect made it clear that localization properties in two dimensions must be radically altered by a magnetic field. A theory for localization properties in a magnetic field was put forward by Pruisken el al. [8] Quantitative features of the quantum Hall effect depend on the localization properties and this theory accounts for the main features of experimental observations[R169].

We can conclude from the above theoretical work that the Hall conductivity will be quantized whenever the a system ,at zero temperature, has a discontinuity in the increase of the chemical potential with particle density *and* the chemical potential lies within a gap associated with such a discontinuity. The Hall conductivity in the integer quantum Hall effect is quantized at e^2/h times an integer. We now understand that the integer is related to the way in which the density at which the gap occurs depends on magnetic field. It has been known for some time that a magnetic field introduces gaps into the bands of Bloch electrons in an intricate and fascinating way. These gaps were studied as long ago as 1962 by Pippard[R17] in connection with magnetic breakdown in metals [10]. A detailed numerical study by Hofstader[R17] in 1976 found a beautiful recursive structure[11] in the magnetic field dependence of the energy gaps for a two dimensional tight-binding model. These recursive structures result in a very rich behavior for the quantum Hall effect in the presence of a periodic potential [12][13] which is described here in a review by Thouless[R19] and a brief article by Streda[R20]. . Our current understanding of the quantum Hall effect makes it clear that gaps in the bands can be detected, and the magnetic field dependence of the density at which they occur can be determined, by measuring magnetotransport coefficients. Very pure materials are required in order to prevent these small gaps from being destroyed by disorder. The prospects are excellent for the preparation of systems in which these fascinating effects can be observed are excellent [14,15] and should open a new area for quantum Hall effect studies in the future.

The gap required for the integer quantum Hall effect occurs as a result of the quantization of the kinetic energy of a free electron in a magnetic field or, in the case of electrons in a periodic potential, the quantization of the band energy of a Bloch electron. The discovery of the fractional quantum Hall effect ,in which the Hall conductivity has a plateau at e^2/h times a fraction, by Tsui, Störmer and Gossard[R21] in 1982 required a different explanation.For the fractional effect the occurrence of a gap results from the effect of electron-electron interactions on a group of electrons, all of which are executing cyclotron orbits with identical kinetic energies. The seminal paper which led to the development of this explanation was due to Laughlin[R25], with important contributions being made by Haldane[R26], Yoshioka et al.[R27], Halperin[R28], Girvin et al.[R29,R30] and others. The many-body states whose existence is being manifested by the fractional quantum Hall effect have many exotic properties, including quasiparticles of fractional charge and ,possibly, fractional statistics. The properties of these incompressible quantum fluids are still not completely understood, and many experimental studies[R22,R23,R24] provide a challenge to theory.

The following four sections in this Introduction can be read in two ways. They are intended to be as accessible as possible and should be comprehensible to anyone with a level of expertise comparable to what is expected of a graduate student in condensed matter physics. On one level they can be read as an explanation of the quantum Hall effect. Read in this way, the discussion is sometimes simple-minded. The proverbial alert reader will notice points where subtleties of the arguments are left implicit. Nevertheless the main ideas are present and ,as far as I am aware, there are no outright lies. The sections can also be read as a primer for the reprints, which taken together tell a much more complete story. Section B will prove useful in reading reprints R1–R6, section C in reading reprints R7–R16, section D in reading reprints R17–R20 and section E in reading R21–R30. It is my hope that readers who use the introduction in this second way will find that it makes their meeting with quantum Hall effect literature more convivial.

B. The Basics

In the simplest theory of the electrical conductivity of a metal, an electron is accelerated by the electric field for an average time τ_0, the relaxation or mean free time, before being scattered to a state which has average velocity zero. In this picture the electrons have an average drift velocity,

$$\boldsymbol{v}_d = -e\mathbf{E}\tau_0/m, \tag{1}$$

where $\boldsymbol{E}$ is the electric field and m is the electron mass. The current density in an electric field is then given by

$$\boldsymbol{j} = -en\boldsymbol{v}_d = \sigma_0\boldsymbol{E} \tag{2}$$

where

$$\sigma_0 = ne^2\tau_0/m \tag{3}$$

and n is the electron density. In the presence of a steady magnetic field the Lorentz force must be added to the force from the electric field so that

$$\boldsymbol{v}_d = -e(\boldsymbol{E} + \boldsymbol{v}_d \times \boldsymbol{B}/c)\tau_0/m \tag{4}$$

and the current density is no longer parallel to the magnetic field. In Eq.(4) $\boldsymbol{B}$ is the magnetic field. (We will always assume that the magnetic field is directed along the $\hat{\mathbf{z}}$ direction.) Using Eq.(4) for the drift velocity we find that the resistivity tensor ($\boldsymbol{j} \equiv \boldsymbol{\sigma}\boldsymbol{E}, \boldsymbol{E} \equiv \boldsymbol{\rho}\boldsymbol{j}, \boldsymbol{\rho} = \boldsymbol{\sigma}^{-1}$) is given by,

$$\rho_{xx} = \rho_{yy} = \rho_{zz} = \sigma_0^{-1}, \tag{5}$$

$$\rho_{xy} = -\rho_{yx} = B/nec, \tag{6}$$

with the other components remaining zero. Note that for this simple isotropic free electron model the dissipation , proportional to the voltage drop along the direction of current flow, is unchanged by the magnetic field; i.e., there is no magnetoresistance. The effect of the magnetic field is to introduce an electric field , the Hall field, in the direction perpendicular to the current flow. The corresponding off-diagonal components of the resistivity and concuctivity tensors are known as the Hall resistivity and the Hall conductivity. The Hall voltage produced by the Hall field is what was discovered by Edwin Hall[R1]. The force on the electrons from the Hall field exactly cancels the Lorentz force which the electrons experience as they drift.

For electrons in a metal the magnetoresistance is nonzero because of corrections to the above simple model. The most important corrections are usually associated with the anisotropy of the Bloch electrons in a metal and with the anisotropy of their scattering. However there are also important contributions to the magnetoresistance due to the quantization of the kinetic energy of motion in the plane perpendicular to the magnetic field. It is these corrections which produce the Schubnikow-de Hass[R5] oscillations, mentioned in the overview, and which are relevant to the quantum Hall effect. To go any further toward understanding these oscillations we must first consider the quantum treatment of an electron in a magnetic field.

The Hamiltonian for a free electron in a magnetic field is

$$t = \boldsymbol{\pi}^2/2m, \tag{7}$$

where

$$\boldsymbol{\pi} = -i\hbar\nabla + e\boldsymbol{A}/c \tag{8}$$

and $\boldsymbol{A}$ is the vector potential, related to the magnetic field by

$$\nabla \times \boldsymbol{A} = \boldsymbol{B}. \tag{9}$$

The vector potential can clearly be chosen to lie in the $\hat{\mathbf{x}}\hat{\mathbf{y}}$ plane so that the contributions to t from motion along and perpendicular to the magnetic field separate. As expected from classical mechanics the motion along the magnetic field is unchanged. Since we will ultimately be interested in electrons moving freely in only two directions we will ignore this degree of freedom in the remaining discussion. It follows from Eq.(9) that

$$[\pi_x, \pi_y] = -i\hbar^2\ell^{-2} \tag{10}$$

and hence that

$$t = \hbar\omega_c(a^\dagger a + 1/2) \tag{11}$$

where $\omega_c = eB/mc$,

$$a^\dagger = \frac{\ell}{\sqrt{2}\hbar}(\pi_x + i\pi_y) \tag{12}$$

and $[a^\dagger, a] = 1$. Here $\ell = (\hbar c/eB)^{1/2}$ is the classical cyclotron orbit radius of an orbit with kinetic energy $\hbar\omega_c/2$.

The ladder operator commutation relations imply that the allowed kinetic energies are restricted to the discrete set of values $\hbar\omega_c(n+1/2)$ where n is an integer. Classically, however, it is clear that a cyclotron orbit is not specified completely by its energy. The solution to the equation of motion for a classical electron in two dimensions moving in a magnetic field is

$$z(t) = C - iv(t)/\omega_c, \tag{13}$$

where $z(t) = x + iy$ is the electron position ,$v(t) = v_x(t) + iv_y(t)$ is the electron velocity and C is the cyclotron orbit center all expressed as complex numbers. For a given orbit velocity, and hence a given orbit energy, the center is arbitrary. Quantum mechanically we might also expect that there be many possible states for one of the allowed kinetic energies, corresponding to different orbit centers. Expressing the orbit center as a quantum mechanical operator it is easy to show that

$$[C, C^\dagger] = 2\ell^2 \tag{14}$$

where we have substituted $\boldsymbol{v} = \boldsymbol{\pi}/m$ in Eq.(13) . For the quantum case, unlike the classical case, the orbit center cannot be precisely determined. From Eq.(14) it is clear that we can define a set of ladder operators based on the orbit center operator just as we defined a set of ladder operator based on the orbit velocities. Let

$$b = C/\sqrt{2}\ell = \frac{1}{\sqrt{2}}\left((z + iA/B)/\ell + 2\ell\frac{\partial}{\partial z^*}\right) \tag{15}$$

so that $[b, b^\dagger] = 1$. (Here $A = A_x + iA_y$) Then $[a, b] = [a, b^\dagger] = 0$ and a complete orthonormal set of eigenfunctions of t is given by

$$\phi_{n,m} = \frac{(a^\dagger)^n (b^\dagger)^m \phi_{0,0}}{\sqrt{n!m!}} \tag{16}$$

where $a\phi_{0,0} = b\phi_{0,0} = 0$ and the corresponding eigenvalue is $\hbar\omega_c(n+1/2)$. The set of single-particle states with a given value for the kinetic energy is called a Landau level. The ladder operators $a, a^\dagger$ change the kinetic energy and are therefore inter-Landau-level ladder operators, while the ladder operators $b, b^\dagger$ are closed in a subspace of any given kinetic energy and are called intra-Landau-level ladder operators. The ability to separate operators into a factor associated with the cyclotron orbit velocity and a factor associated with the cyclotron orbit center will be very valuable when it comes to discussing the fractional quantum Hall effect.

Everything we have discussed to this point is independent of the choice of gauge. In the remaining portion of this section we carry the discussion a bit further for the gauge choice which has been most valuable for fractional quantum Hall effect studies, the symmetric gauge,

$$\boldsymbol{A} = (-By/2, Bx/2, 0). \tag{17}$$

For this choice

$$b^\dagger = (z^*/2\ell - 2\ell\frac{\partial}{\partial z})/\sqrt{2} \tag{18}$$

and

$$a^\dagger = i(z/2\ell - 2\ell\frac{\partial}{\partial z^*})/\sqrt{2}. \tag{19}$$

Note that $(\frac{\partial}{\partial z})^\dagger = -\frac{\partial}{\partial z^*}$. Using the algebra of the ladder operators it is easy to show that

$$\phi_{n,m} = \frac{(-i)^n G^{m,n}(iz^*/\ell)\exp(-|z|^2/4\ell^2)}{\sqrt{2\pi}\ell} \tag{20}$$

where

$$G^{m,n}(\alpha) = \sqrt{n!/m!}\left(-i\alpha/\sqrt{2}\right)^{m-n} L_n^{m-n}\left(|\alpha|^2/2)\right) \tag{21}$$

and $L_n^k(x)$ is a generalized Laquerre polynomial. The areal density which can be accommodated by each Landau level is

$$n_1 = \sum_m |\phi_{n,m}|^2 = (2\pi\ell^2)^{-1} \tag{22}$$

Eq.(22) follows from Eq.(20), Eq.(21) and the frequently useful identity,

$$\sum_k G^{n,k}(\alpha_1)G^{k,m}(\alpha_2) = \exp(-\alpha_1^*\alpha_2/2)G^{n,m}(\alpha_1 + \alpha_2) \tag{23}$$

The density measured in units of n_1 is referred to as the Landau level filling factor ,$\nu \equiv 2\pi\ell^2 n$. This quantity plays an essential role in both the integer and fractional quantum Hall effects. i.e. how many Landau levels are filled?

C. The Integer Quantum Hall Effect

In this section we give three derivations of the expression for the quantized Hall conductance of a 2DEG. The derivations are given in order of increasing generality, but we shall see that this does not imply that they will be in order of increasing complexity. In the first place we simply add the contribution of a constant electric field ,E, pointing in the x-direction to the Hamiltonian of a free particle, considered in the previous section.

$$h = t + eEx \tag{24}$$

For this problem it is convenient to choose a gauge, the Landau gauge, in which the vector potential is independent of the y coordinate.

$$\boldsymbol{A} = (0, Bx, 0) \tag{25}$$

This choice allows us to choose a wavefunction which has a plane-wave dependence on the y coordinate.

$$\psi(x,y) = \varphi(x)exp(ik_y y)/\sqrt{L_y} \tag{26}$$

Here L_y is the length of the system in the $\hat{\mathbf{y}}$ direction and, adopting periodic boundary conditions in this direction, allowed values of k_y are separated by $2\pi/L_y$. Substituting Eq.(26) into Eq.(24), we see that

$$h_x(k_y)\varphi_n(x) = \varepsilon_n(k_y)\varphi_n(x) \tag{27}$$

where

$$h_x(k_y) = p_x^2/2m + \tfrac{1}{2}m\omega_c^2(x - X')^2 + eEX' + m(cE/B)^2/2, \tag{28}$$

$$X' = X - \frac{eE}{m\omega_c^2}, \tag{29}$$

and

$$X = -k_y \ell^2 \tag{30}$$

The first two terms on the right hand side of Eq.(28) comprise the Hamiltonian for a one dimensional harmonic oscillator centered at X' with oscillator frequency ω_c while the final two terms are constants. Thus

$$\varphi_n(x) = \psi_n(x - X') \tag{31}$$

where $\psi_n(x)$ is one of the familiar harmonic oscillator eigenstates. It will be convenient to label the eigenstates of h by the harmonic oscillator centers in the absence of the electric field, X, rather than by k_y and by the harmonic oscillator indices.

Before proceeding further we can stop and interpret the results that we have obtained to this point. From Eq.(28) it follows that

$$\varepsilon_n(X) = \hbar\omega_c(n + 1/2) + eEX' + m(cE/B)^2/2 \tag{32}$$

The first two terms are the kinetic energy of the cyclotron orbit and the electric potential energy at the position of the orbit center. To interpret the last term we must notice that the drift velocity of the electrons in the $\hat{\mathbf{y}}$ direction is given by

$$v_d = \langle n, X \mid \pi_y \mid n, X \rangle / m = -cE/B \tag{33}$$

This is the classical drift velocity in the free (no scattering) limit (see Eq.(4)) and the same result may be obtained quantum mechanically using the eigenfunctions discussed above. It is clear then that the last term in Eq.(32) is just the additional contribution to the kinetic energy from the drift motion. Knowing the drift velocity we can calculate the current carried in the $\hat{\mathbf{y}}$ direction.

$$I_y = -nev_d L_x = \nu ecE/(2\pi\ell^2 B) = (\nu e^2/h)(EL_x) \tag{34}$$

Eq.(34) shows that for an ideal system the current flow is perpendicular to the electric field ,so that there is no dissipation, and gives an expression for the ratio the current flow to the voltage drop in the perpendicular direction, i.e. for the Hall conductance. ($G_H = \nu e^2/h$ and $R_H = {G_H}^{-1} = h/(e^2\nu)$) Note that we have expressed the density in Eq.(34) in terms of the Landau level filling factor,ν. We have derived Eq.(34) for an ideal disorder-free system. von Klitzing's discovery[R11] was that Eq.(34) remains true at integer values of the filling factor even with disorder and that the Hall conductance is precisely independent of filling factor for a sample and temperature dependent range of filling factors (the *plateau* region) near integer values. The present derivation makes it clear that without disorder, the conductance always depends linerly on filling factor so the occurrence of plateaus near integer values of the filling factor must require disorder.

To proceed further we must relax our assumption of a disorder free system. An informative first step in that direction may be taken by replacing the potential from the electric field in Eq.(24) by a potential which depends only on the x coordinate but is otherwise arbitrary.

$$h = t + V(x) \tag{35}$$

In this case Eq.(28) becomes,

$$h_x(X) = p_x^2/2m + \tfrac{1}{2}m\omega_c^2(x - X)^2 + V(x) \tag{36}$$

We determine the eigenvalues and eigenfunctions of h_x without specifying $V(x)$. However to calculate the conductance we need only note that the total current flow is given by

$$I_y = -e\sum_{n,X} \langle n, X \mid \pi_y \mid n, X \rangle / mL_y = -e\sum_{n,x} \langle n, X \mid \frac{\partial h_x}{\partial k_y} \mid n, X \rangle / (\hbar L_y) \tag{37}$$

The second form of Eq.(37) is obtained from the first by comparing with Eq.(28). Since adjacent values of X allowed by periodic boundary conditions are separated by $2\pi\ell^2/L_y$ Eq.(37) leads to

$$I_y = \frac{e}{h}\sum_n \int dX \frac{\partial \varepsilon_n(X)}{\partial X} \tag{38}$$

where we have invoked the Hellman-Feynman theorem to evaluate the expectation value. In Eq.(37) and Eq.(38) the sums and integrals are over occupied states. The lower and upper limits of the integration over X in Eq.(38) are at the left and right edge of the sample and in evaluating this integral we are led to introduce the edges explicitly, although we shall see that nothing depends on the specific model used to describe the edges. When an integer number ,N,of Landau levels are full we may assume that except at the edges, all states associated with that Landau level index are occupied. At either edge, we can assume that there is a well defined local chemical potential (equal to the energy of the highest occupied level at that edge) which is what is probed when a voltage measurement is performed. Thus Eq.(38) implies that

$$I_y = eN(\mu_R - \mu_L)/h = -e^2 NV/h \tag{39}$$

where V is the voltage difference from the left hand side to the right hand side of the sample. Note that the Hall conductance at integer filling does *not* depend on the disorder potential. Note also that when $\mu_R = \mu_L$, the total current carried in the y direction is zero. This is true even if $V(x) = eEx$ in the middle of the sample!

As impicitly assumed in the above discussion, the confining potential which defines the edges of the system causes $\varepsilon_n(X)$ to increase near the edges of the sample. The way in which this occurs depends on the confining potential: two simple models for which exact results can be obtained are the hard wall [16] case ($V(x) = \infty$ outside the sample) and the case of a parabolic confinement potential. From Eq.(38) we see that this leads to large current densities flowing in one direction near one edge and in the other direction near the other edge of a 2D system. The large current densities at the edges ($v_d \sim \hbar\omega_c/\ell$) can produce a contribution to I_y from the edge regions which is comparable to the contribution from the bulk of the 2D system ($v_d \sim \hbar\omega_c/L_x$) and allow the total current to be expressed solely in terms of the chemical potentials at the edges. These circulating edge currents are the ones which are responsible for the orbital diamagnetism in Teller's derivation[R4] of the Landau diamagnetic susceptibility[R3]. This fact points us toward our final and most general argument for the quantization of the Hall conductance in which we allow a sample of arbitrary shape and arbitrary disorder. We will find that the same argument is important in understanding the fractional quantum Hall effect and we therefore give the argument in a form which is also appropriate for interacting many-particle systems.

Consider a finite 2D electron system with a confining potential of arbitrary shape. Assume that the chemical potential has a value which would be in a gap for a bulk system in the thermodynamic limit. We want to evaluate the change in the magnetic moment of the system when the chemical potential of the system changes at constant magnetic field. The operator for the magnetic moment in the $\hat{\mathbf{z}}$ direction is given by

$$\hat{M} = \frac{-\partial H}{\partial B} \tag{40}$$

where H is the Hamiltonian,

$$H = \sum_i \frac{\pi_i^2}{2m} + V(\boldsymbol{x}_i) + \sum_{i<j} e^2/|\boldsymbol{x}_i - \boldsymbol{x}_j| \tag{41}$$

Only the kinetic term in H depends on the magnetic field. Using Eq.(8) in Eq.(41) and Eq.(40) we find that

$$\hat{M} = \frac{1}{2c} \int d^2\boldsymbol{x}\boldsymbol{x} \times \hat{\boldsymbol{j}}(\boldsymbol{x}) \tag{42}$$

where $\hat{\boldsymbol{j}}(\boldsymbol{x})$ is the current density operator,

$$\hat{\boldsymbol{j}}(\boldsymbol{x}) = \frac{-e}{2m} \sum_i (\delta(\boldsymbol{x} - \boldsymbol{x}_i)\boldsymbol{\pi}_i + \boldsymbol{\pi}_i \delta(\boldsymbol{x} - \boldsymbol{x}_i)). \tag{43}$$

Note that the integrand in Eq.(42) depends on a gauge choice and we have chosen the symmetric gauge. It is easy to show ,however, using the continuity equation and the fact that the current density vanishes outside the sample that the total magnetization is gauge independent. Because the chemical potential has been chosen so that it would lie in the gap for an infinite bulk system, when it changes the expectation value of the current density operator can change only at the system's edge. The situation is illustrated schematically on the cover where the area in red represents the edge region where the current density changes in response to a chemical potential change while the yellow region represents the interior which is not altered. The change in the expectation value of the magnetic moment ,δM, with a small change in chemical potential is

$$\delta\langle\hat{M}\rangle = \frac{1}{2c} \int d^2\boldsymbol{x}\boldsymbol{x} \times \delta\langle\hat{\boldsymbol{j}}(\boldsymbol{x})\rangle. \tag{44}$$

The integral in Eq.(44) is over the edge (red) region. Taking the position ,$\boldsymbol{x}$,outside the integral when integrating across the edge, and using current conservation when integrating around the edge, this reduces to

$$\delta\langle\hat{M}\rangle = A\delta I/c \tag{45}$$

where A is the area of the finite 2D system and δI is the change in the current circulating around the edge. Eq.(45) thus provides us with a very general expression for the ratio of the change in the edge current to the change in the chemical potential at the edge.

$$\frac{\delta I}{\delta\mu} = \frac{c}{A}\frac{\partial M}{\partial\mu}\Big|_B = \frac{c}{A}\frac{\partial N}{\partial B}\Big|_\mu \tag{46}$$

The second form for the right hand side of Eq.(46) follows from a thermodynamic identity.

The above discussion was for a finite 2D system in equilibrium. When current leads are attached to the system, the edge is divided into two branches, one which carries current from the source to the drain and one which carries current from the drain to the source. The net current carried through the system ,δI, is the difference between the edge currents along the two branches. The Hall conductance is the ratio of the net current to the electrochemical potential difference between the two edges. (Since we incorporate any electric fields which may be present in the Hamiltonian the electrochemical potential difference is just e times the chemical potential difference.) If the chemical potential lies within a gap which occurs at filling factor ν, the derivative appearing in Eq.(46) is known and

$$\frac{\delta I}{\delta\mu} = e\nu/h \tag{47}$$

Since the chemical potential (the energy to add an additional electron at fixed magnetic field) of a 2D electron system jumps whenever the filling factor crosses an integral value Eq.(47) and Eq.(46) explain the quantization of the Hall conductance when the chemical potential lies between Landau levels. (We are ignoring the spin degree of freedom for simplicity; there are two possible spin states for each allowed kinetic energy state so that the quantum Hall effect at odd integral values of the filling factor is actually due to the spin splitting.)

The range of filling factors for which only edge states are at the chemial potential vanishes in the thermodynamic limit since the number of edge states scales with the system perimeter while the Landau level degeneracy scales with the system area. An essential part of von Klitzing's discovery was that the Hall conductance is precisely constant over a finite range of filling factors. To exlain the occurance of these plateaus, which allow the Hall conductance to be measured with great precision, we must invoke the phenomenon of localization. The states in the tails of the disorder broadened Landau levels, will be localized near minima or maxima in the random potential. The *mobility gap* between Landau levels will extend over a finite range of filling factors. As long as the Fermi level lies in a mobility gap, the only current-carrying states at the Fermi level will be the edge states. The constant value of the Hall conductance on the plateaus tells us that the change in edge current for a small change in chemical potential does not change when the edge states become degenerate with *localized* states in the bulk. The edge states do not know or care if localized bulk states are occupied since the their wavefunctions do not overlap at all. We can understand this by noting that the localized holes near the top of a disorder broadened Landau level or the localized electrons near the bottom of a disorder broadened Landau level can be considered to form an uncoupled subsystem. The contributions to the two thermodynamic derivatives in Eq.(46) from these subsystems are identical since the subsystem obeys the same thermodynamic identity. Thus the change in edge current for a given change in chemical potential is the same as when a Landau level is empty if the chemical potential lies in the localized states at the bottom of the Landau level and the same as when a Landau level is full if the chemical potentail lies in the localized states at the top of the Landau level.

From the discussion in this section we see that whenever the Fermi level lies in a gap the Hall conductance will be given by

$$G_H = ec\frac{\partial n(\varepsilon_F)}{\partial B}\Big|_{\mu} \tag{48}$$

This formula was first derived, as far as we are aware, by Streda[R15] using the Kubo formula. The argument given above which leads to this formula explains its origin in a more physical way, which is closely related to other[R13,R14] simple arguments for the quantization of the Hall conductance. As we mentioned in the introduction, and as was emphasized early in the development of the theory by Prange[R12], the mysterious aspect of the quantum Hall effect is the fact that the free gas formula is exactly obeyed between Landau levels, despite the fact that localization reduces the density of current carrying states. The Streda formula shows that the Hall conductance in a gap depends only on the magnetic field dependence of the carrier density when the Fermi level lies within that gap. In the case of the integer quantum Hall effect, for example, the Hall conductance depends only on the density which can be accommodated by each Landau level, something which is clearly preserved as long as the disorder is not so strong that different disorder broadened Landau levels overlap. (In fact it is believed that the integer quantum Hall can occur even when Landau levels overlap.) In the next two sections we discuss the quantum Hall effect which is associated with gaps of two other types of physical origin.

D. The Quantum Hall Effect in a Periodic Potential

The integer quantum Hall effect occurs because of gaps which occur at integer Landau level filling factors. The reader should realize that there is something unusual about a gap which occurs at a magnetic-field dependent density. In this section we discuss a qualitatively different example of the quantum Hall effect in which the gap occurs neither at constant density nor at constant Landau level filling factor. The gaps in this case are those which occur when the Landau level is placed in both a constant magnetic field and a periodic potentail. The effect of the periodic potential is both to broaden a Landau level and to open up an intricate array of gaps within a each Landau level.

The main features of the Quantum Hall effect in a periodic potential can be understood on the basis of qualitative arguments. For a 2D system in a strong magnetic field and a periodic potential, there are two length scales. One is the magnetic length ,ℓ, introduced earlier which was important in the integer quantum Hall effect. The second length scale is the period of the external potential which was not present in the case of a free 2D gas. As we learned above, we can understand the quantum Hall effect if we understand the magnetic-field dependence of the filling factors at which gaps occur in the spectrum. In the case of a free gas, gaps occurr at integer filling factors. The effect of the periodic potential is to allow other gaps to occur.

It is clear from a simple scaling argument that the filling factor at which a gap occurs can depend only on the ratio of the two length scales in the problem. This ratio is conveniently parameterized in terms of the quantity

$$\alpha = 2\pi\ell^2/A_0 \tag{49}$$

where A_0 is the unit cell area of the periodic potential. To determine the spectrum of the system we use translational periodicity ,applying periodic boundary conditions to a finite system whose volume can ultimately be taken to infinity. It will not be necessary for our purposes to describe in detail how this is done in the presence of a magnetic field. We merely note that it is possible to apply periodic boundary conditions to the system, only if the number of units cells of the periodic potential in the system is an integer and if $A/(2\pi\ell^2) \equiv N_L$, is an integer. (Here A is the area of the system.) The first condition is obvious and is required at zero magnetic field as well. The second condition is due to the magnetic field, and is required in the absence of a periodic potential as well; in that case it requires N_L, the number of states per Landau level, to be an integer in a finite system. The main features of the intricate gap structure can be understand from these observation and the scaling argument.

It follows from the above that in the thermodynamic limit,

$$\sigma \equiv \frac{\partial n(\varepsilon_F)}{\partial B}/(2\pi\ell^2/B) \tag{50}$$

and

$$s \equiv -A_0^2\frac{\partial n(\varepsilon_F)}{\partial A_0} \tag{51}$$

are integers when ε_F lies in a gap. (ε_F is the Fermi energy which is the zero temperature limit of the chemical potential) σ is the change in the number of states below the gap when the magnetic field changes by an amount sufficient to change N_L by one, while s is the change in the number of states below the gap when A_0 is changed by an amount sufficient to increase the number of unit cells in the system by one. Note that the Streda formula implies that

$$G_H = (e^2/h)\sigma \tag{52}$$

The change in the the filling factor at which a gap occurs when A_0 and the magnetic field change can be expressed in terms of σ and s. Using Eq.(50) and Eq.(51), in $\nu = 2\pi\ell^2 n$ gives,

$$\delta\nu = [\sigma - \nu]\frac{\delta B}{B} - \alpha s \frac{\delta A_0}{A_0} \tag{53}$$

Similarly the change in α for a change in B and A_0 is

$$\delta\alpha = -\alpha(\delta B/B + \delta A_0/A_0). \tag{54}$$

It follows that the change in the filling factor at which a gap occurs at constant α is given by

$$\delta\nu = [\sigma - \nu + \alpha s]\delta B/B. \tag{55}$$

However, it follows from the scaling argument that that the filling factor at which a gap occurs cannot change if α does not change and hence that the filling factor at which a gap occurs is related to σ and s by

$$\nu = \sigma + \alpha s \tag{56}$$

Eq.(56) is a deceptively simple equation from which many conclusions can be drawn. Consider the situation when α has an arbitrary rational value, $\alpha = q/p$. Then any ν at which a gap occurs, using the fact that σ and s in Eq.(56) are integers, must also occur at a rational filling factor with the same denominator, $\nu = t/p$. Thus each Landau level is split into p subbands when $\alpha = q/p$. Each gap in the spectrum is characterized by two integer quantum numbers σ and s and follows the straight line in ν, α space dictated by Eq.(56). A countably infinite number of such lines intersect with any given rational value of ν and α, characterized by the integers t,q and p, and their values of σ and s are given by

$$(s, \sigma) = (s_0 + kp, \sigma_0 - kq) \tag{57}$$

where k is an integer. The allowed values possible values of s are separated by p and the allowed values of σ are separated by q.

All the above conclusions can be drawn without specifying the periodic potential, which determines which of the possible values of σ is selected at a given value of α and ν. The case of a weak periodic potential of the form

$$V(x) = V_0(\cos(2\pi x/a) + \cos(2\pi y/a)) \tag{58}$$

is equivalent to the model studied recursively by Hofstader[R17] and the complicated recursive structure he discovered in the spectrum, *The Hofstader Butterfly*, is equivalent to the statement[R19] that the correct value of s at any gap for this model is the smallest possible one[11], i.e. the one obeying $|s| \leq p$. It is easy to verify that in this case the values of the integral quantum number related to the Hall conductance, σ are very different from ν and, in particular can have large and even negative values within any Landau level. Thus when a periodic potential is present the quantum Hall effect can occur at fractional filling factors but the filling-factor dependence of the gap is always such that the Hall conductance is an integral multiple of e^2/h. In the next section we discuss gaps which originate from electron-electron interactionsand occur at fractional Landau level fillings at any field, and hence yield a fractional Hall conductance.

We have followed Streda[R20] in discussing the quantum Hall effect in a periodic potential in terms of the Streda formula. Thouless et al.[R19] obtained equivalent results starting directly from the Kubo formula. These authors found that the Hall conductivity is related to the change of the phase of the Bloch wavefunction on moving around the perimeter of the Brillouin zone, which must be an integer (σ) times 2π. The Hall conductance quantum number is thus related to a topological invariant of a mapping of the Brillouin zone by the Bloch wavefunction. It has subsequently been realized that the Hall conductance may be regarded as a topological invariant in more general circumstances [17]. The fact that the Hall conductance in a periodic potential turns out to always be an integral multiple of e^2/h despite gaps at fractional filling is then seen to be a very general feauture of the quantum Hall effect. Fractional values of the quantized Hall conductance can only be produced by electron-electron interactions and, as we will see, they are accompanied by a fractionalization of the electron charge.

E. The Fractional Quantum Hall Effect

We have learned that the quantum Hall effect will occur for a 2D system in a magnetic field whenever there is a gap, i.e. whenever the chemical potential has a discontinuity as a function of density at fixed magnetic field. The integer quantum Hall effect is caused by the quantization of the allowed kinetic energy values for an electron moving in a plane perpendicular to a steady magnetic field. In the case of an electron system moving in a periodic potential we learned in the last section that gaps open up within Landau levels. Even in this case ,however, the Hall conductance is an integer multiple of e^2/h. The fractional quantum Hall effect occurs near a certain set of fractional values of ν and has plateaus in the Hall conductance equal to $\nu e^2/h$. We saw in the last section that a fractionally quantized Hall conductance is not possible for non-interacting electrons, even in an external potential which produces gaps within each Landau level. Our object here will be to explain how electron-electron interactions can give rise to chemical potential discontinuities which are pinned to *fractional* Landau level filling factors.

We will restrict our discussion to the extreme quantum limit in which the Landau level degeneracy is large enough that all electrons can be accommodated within the lowest Landau level and the Landau level separation is large enough that quantum mechanical Landau level mixing by disorder or by interactions can be ignored. Since each electron present has the same kinetic energy, only electron-electron interactions can produce the gap which we require to explain the quantum Hall effect. We will see that the kinetic energy quantization indirectly can produce gaps at fractional filling factors, because the restriction to a single Landau level puts powerful filling-factor-dependent restrictions on the properties of the many-electron wavefunctions. According to Eq.(20), the single-particle wavefunctions in the lowest Landau level in the symmetric gauge are

$$\phi_m(\boldsymbol{x}) = \frac{z^{*m} \exp(-|z|^2/4)}{\sqrt{2^{m+1} m! \pi}} \tag{59}$$

where $z = x + iy$, we have dropped the Landau level index on the wavefunctions and we have adopted the magnetic length, ℓ, as the unit of length for the discussion of the fractional quantum Hall effect. In what follows we will redefine $z \equiv x - iy$ in order to avoid complex conjugating everything in sight. Note that these wavefunctions describe electrons located within one magnetic length of a circle centered on the origin and enclosing an area

$$\pi \langle m | |z|^2 | m \rangle = 2\pi\ell^2(m+1) \tag{60}$$

Any many-electron wavefunction formed entirely within the lowest Landau level must be a sum of products of one-electron orbitals for each coordinate which are of the form given be Eq.(59). It follows that the many electrons wavefunction must take the form

$$\Psi[z] = \left(\prod_{k=1}^{N} \exp(-|z_k|^2/4) \right) P[z] \tag{61}$$

where $P[z]$ is a polynomial, and in the $N \to \infty$ limit an analytic function, in each of the z_k's[18]. N is the number of electrons. At zero temperature the chemical potential equals the change in the ground state energy when one electron is added to the system. Since any state of the form of Eq.(61) has the same kinetic energy, $T = N\hbar\omega_c/2$, the discontinuity in the chemical potential responsible for the fractional quantum Hall effect must come from electron-electron interactions. We know that the electron-electron interaction is increasingly repulsive at short distances, so the ground state will be determined, qualitatively, by minimizing the probability of electrons being close together. For an isotropic system ,of identical particles the pair correlation function, which measures the probability of two electrons being at a certain separation compared to the same probability in an uncorrelated system of the same density, is given by

$$g(|\boldsymbol{x}_1 - \boldsymbol{x}_2|) = n^{-2}N(N-1)\prod_{k=3}^{N}\int d^2z_k |\Psi[z]|^2 \tag{62}$$

Substituting Eq.(61) into Eq.(62) gives for the case of interest,

$$g(|z_1 - z_2|) \sim \exp(-(|z_1|^2 + |z_2|^2)/2)\prod_{k=3}^{N}\int d^2z_k \exp(-|z_k|^2)P^*[z]P[z] \tag{63}$$

It will prove convenient to introduce replace the coordinates z_1 and z_2 by a relative coordinate, $\zeta \equiv z_2 - z_1$, and a mean coordinate ,$\bar{z} \equiv (z_2 + z_1)/2$. As noted in Eq.(62) and Eq.(63) the dependence of g on the mean coordinate is expected to be removed by the integrations over d^2z_k for an isotropic fluid state, such as is generally expected in an electron gas[19]. Since $P[z]$ is an analytic function of z_1 and z_2, it will be an analytic function of ζ, and we can expand

$$P[z] = \sum_p \zeta^{2p+1}F_p(\bar{z}, z_3, \ldots, z_N) \tag{64}$$

Note that because of the antisymmetry requirement on the many-electron wavefunction for electrons only odd powers of ζ appear in Eq.(64). Substituting Eq.(64) into Eq.(63) gives

$$g(|\zeta|) = \exp(-|\zeta|^2/4)\sum_{p,p'} \zeta^{*(2p'+1)}\zeta^{2p+1}f_{p',p} \tag{65}$$

where $f_{p',p}$ is given by an integral over the other coordinates. For an isotropic system g can depend only on the magnitude and not on the orientation of the separation ζ. It follows that $f_{p',p}$ must be zero when p is not equal to p' and hence that

$$g(|\zeta|) = \exp(-|\zeta|^2/4)\sum_{p} |\zeta|^{2(2p+1)}f_{p,p} \tag{66}$$

Eq.(66) tells us that for any isotropic state formed in the lowest Landau level of a 2DEG, the pair correlation function must vanish as an odd power of $|\zeta|^2$ at small $|\zeta|$. The fractional quantum Hall effect is due to a connection between the small-separation behavior of g and the Landau level filling factor, which we now establish. Assume that $g(|\zeta|)$ varies as $|\zeta|^{2m}$ at small $|\zeta|$. It follows from the argument leading to Eq.(66) that $P[z]$ has $(z_1 - z_2)^m$ as a factor and hence, since all particles are identical, that $P[z]$ has

$$P_m[z] \equiv \prod_{i<j}(z_i - z_j)^m \tag{67}$$

as a factor. For a wavefunction representing a large but finite number of electrons, N, the maximum power to which z_1 (or any other coordinate) appears in $P[z]$ is therefore

$$M_1 \geq m(N-1) \tag{68}$$

and hence the area occupied by the wavefunction ,according to Eq.(60), is

$$A \geq 2\pi\ell^2 m(N-1). \tag{69}$$

It follows from Eq.(69) that in the thermodynamic limit $A/2\pi\ell^2 N == \nu^{-1} \geq m$. Thus as the electron density is increased at constant magnetic field so that the filling factor crosses $\nu = 1/m$ we go from a regime where it is possible to form states with $g(|\zeta|) \sim |\zeta|^{2m}$ to a regime where $g(|\zeta|)$ vanishes only as $|\zeta|^{2(m-1)}$. This qualitative change in the ability of electrons to avoid each other causes a jump in the chemical potential when the filling factor crosses $1/m$ and ,invoking the Streda formula, also causes the Hall conductance to be quantized at e^2/mh at filling factor $1/m$.

The wavefunction corresponding to the analytic function $P_m[z]$ (Eq.(67)) was first introduced by Laughlin[R25]. When the electrons are in this state we see by inspection that any pair of electrons are always in a state of relative angular momentum at least m. As first emphasized by Haldane[R26], the electron- electron interaction within the lowest Landau level is completely specified by a set of pseudopotential parameters,$\{V_m\}$, which give the interaction strength in each relative angular momentum channel. (Only odd values of m are relevant if the electrons are completely spin polarized.) An extremely attractive picture of the fractional quantum Hall effect can be given in terms of these pseudopotential parameters. For filling factors less than $1/m$, where we have shown that it is possible to form states in which no pair of electrons has a relative angular momentum less than m, there are zero-energy eigenstates in models which have repulsion only in channels of relative angular momentum less than m. At filling factor $1/m$ our discussion shows that there is only one state, the Laughlin state, which has zero energy for such a model. Thus associated with the chemical potential jump which occurs as the filling factor crosses $1/m$, there is an excitation gap which occurs when the filling factor equals $1/m$. The pseudopotential parameters for realistic electron-electron interactions decrease monotonically with m. Exact diagonalization studies for small systems, pioneered by Yoshioka et. al[R27], have provided convincing evidence that the realistic interactions are sufficiently close to the simple pseudopotential models described above, that the ground states maintain a nearly quantitative similarity.

To explain the plateaus which occur in connection with the fractional quantum Hall effect we must also consider states which occur at filling factors close the $1/m$. The expectation is that some type of localization behavior must occur which is analogous to the localization of solutions to the one-body Hamiltonian responsible for the plateaus in the integer quantum Hall effect. For example, consider the case in which the area of the system is larger than that at filling factor $1/m$ by $2\pi\ell^2$ and the disorder potential has a repulsive peak at some point, which we take to be the origin. Laughlin[R25] suggested that the ground state in this circumstance could be obtained from the ground state at filling factor $1/m$ by increasing the single-particle angular momentum labels of all the occupied states by one. This would give a state in which the $m = 0$ single-particle state is never occupied. It is easy to show that away from the origin this state has the same correlations as the Laughlin fluid state, while near the origin there is a decrease in the charge density with a total deficiency of charge equal to e/m. One can form similar states in which there is an excess of charge equal to e/m at some point in space. The picture of the situation on the plateaus near $\nu = 1/m$, is therefore as follows. The ground state may be thought of as consisting of a Laughlin fluid, plus some number of quasiholes or quasiparticles localized near maxima or minima of the disorder potential. The novel feature in the fractional case is that these quasiparticles have *fractional* charge. The fractional charge also leads to the possibility that the quasiparticles may be most conveniently described as having neither Fermi or Bose statistics, but rather an intermediate *fractional* statistics[R28].

To avoid giving the reader the incorrect impression that everything is now explained we must emphasize that the fractional quantum Hall effect also occurs at filling factors not equal to $1/m$. Some of these fractions can be explained by invoking the particle-hole symmetry which exists in the strong magnetic field limit. For example, a fractional quantum Hall effect occurs at filling factor $2/3$, due to the formation of the $1/3$ Laughlin state in the holes of a full Landau level. This notion can be generalized so that, for example the $2/5$ fractional effect can be explained as being due to the formation of a Laughlin fluid in the *fractionally charged* quasiparticles of the $\nu = 1/3$ Laughlin fluid. It should be recognized, however, that the understanding of these so-called hierarchy states in not yet nearly as complete as our understanding of the Laughlin states.

For a non-interacting metal in the absence of a magnetic field, the excited states which are coupled to the ground state by the one-particle operators relevant to most experiments are particle-hole excitations in which an electron in one of the single particle states inside the Fermi surface is promoted to a single-particle state outside the Fermi surface. For an interacting metal the particle-hole excitations still exist and their excitation energies are, for the most part, only slightly changed. However the excitation energy of one particular linear combination of particle-hole excitations is changed qualitatively, especially at long wavelengths. This linear combination is formed by operating on the ground state with the density operator,

$$\hat{\rho}(\boldsymbol{k}) = \sum_i \exp(-i\boldsymbol{k}.\boldsymbol{x}_i). \tag{70}$$

This state is the plasmon state in which the electrons are collectively oscillating with wavevector $\boldsymbol{k}$. For a 2D electron system in a magnetic field,

$$\hat{\rho}(\boldsymbol{k}) = \sum_i A_i(k)B_i(k) \tag{71}$$

where $k = k_x + ik_y$,

$$A_i(k) = \exp(-k^* a_i^\dagger/\sqrt{2})\exp(ka_i/\sqrt{2}), \tag{72}$$

and

$$B_i(k) = \exp(-ik^* b_i/\sqrt{2})\exp(-ikb_i^\dagger/\sqrt{2}) \tag{73}$$

In Eq.(71) we have separated the density operator into a factor involving inter-Landau-level transitions ($A_i(k)$) and a part involving intra-Landau-level transitions ($B_i(k)$). Eq.(71) is obtained from Eq.(70) by using Eq.(18), Eq.(19) and the fact that the inter-Landau-level and intra-Landau-level ladder operators commute. We see that in a magnetic field, the density operators creates particle hole excitations which involve a change of Landau level index as well as particle hole excitations within a Landau level. In a strong magnetic field the Landau level separation ,$\hbar\omega_c$, is a large energy and the electron-electron interaction will be important in coupling only particle-hole excitations with the same change of Landau level index. Of particular importance to the fractional quantum Hall effect is the coupling of intra-Landau-level excitations since they all have zero excitation energy, in the absence of interactions. We may expect that a collective mode is formed, at least at long wave lengths, from the intra-Landau level particle hole excitations. The excitation energy for these collective modes may be estimated by a generalization of the theory used by Feynman[20] to estimate the collective phonon-roton excitation spectrum in superfluid ^4He. We assume that the collective mode state is given approximately by the projection of $\hat{\rho}(k)|\Psi_0\rangle$ onto the lowest Landau level, i.e. by the part of the density wave excitation which does not produce an increase in kinetic energy;

$$|\Psi(k)\rangle = \frac{\bar{\rho}(k)|\Psi_0\rangle}{\sqrt{\langle\Psi_0\,|\,\Psi_0\rangle}} \tag{74}$$

where the projected density wave operator $\bar{\rho}(k)$ is given by

$$\bar{\rho}(k) = \sum_i B_i(k). \tag{75}$$

and $|\Psi_0\rangle$ is the ground state wavefunction. The excitation energies of these magnetoroton states are given by

$$E_{MR}(k) = \langle\Psi(k)\,|\,\bar{V}\,|\,\Psi(k)\rangle - E_0 \tag{76}$$

where E_0 is the ground state energy and $\bar{V}$ is the projection of the electron-electron interaction onto the lowest Landau level,

$$\bar{V} = \int \frac{d^2\mathbf{q}}{(2\pi)^2} V(q)\bar{\rho}(-q)\bar{\rho}(q) \tag{77}$$

In Eq.(77), V(q) is the Fourier transform of the electron-electron interaction. $E_{MR}(k)$ is readily evaluated by expressing the energy difference in Eq.(76) in terms of commutators, expressing $\bar{V}$ in terms of projected density operators using Eq.(77), and using the following expression for the commutator of projected density operators:

$$[\bar{\rho}(k_1), \bar{\rho}(k_2)] = (\exp(k_1^* k_2/2) - \exp(k_1 k_2^*/2))\bar{\rho}(k_1 + k_2) \tag{78}$$

Eq.(78) follows from the commutation relations of the intra-Landau-level ladder operators. The result for the excitation energies is

$$E_{MR}(k) = \int \frac{d^2q}{(2\pi)^2} V(q) \exp(-|q|^2/2)(\exp((q^*k - k^*q)/2) - 1)(\tilde{s}(q) - \tilde{s}(k+q))/\tilde{s}(k) \qquad (79)$$

where $\tilde{s}(k) = \exp(|k|^2/2)\bar{s}(k)$ and

$$\bar{s}(k) = s(k) - (1 - \exp(|k|^2/2)), \qquad (80)$$

the projected static structure factor, is the difference between the static structure factor of the partly filled state and the structure factor when the Landau level is filled. Small system calculations have verified that collective density modes do indeed exist within a partly filled Landau level *when the ground state is incompressible* and that their dispersion is given extremely accurately by Eq.(79). As expected $\lim_{k\to 0} E_{MR}(k)$ is finite and $E_{MR}(k)$ has a minimum where $\bar{s}(k)$ has a maximum, analogous to the roton minimum in Helium.

In two dimensions it is possible for particles to have statistics intermediate between those of Fermi and Bose particles[21]. In fact the Fermi electrons in a two dimensional electron gas can be *equivalently* considered to be bosons provided that a contribution is added to the Hamiltonian which corresponds to an odd number of magnetic flux quanta piercing the system at the position of each particle. Girvin and MacDonald[R30] showed that when the particles at $\nu = 1/m$ are considered to be bosons they show quasi-long range order at zero temperature if and only if the ground state is incompressible. This property has been verified by numerical calculations[22] and has been exploited to develop Landau-Ginzburg theories[23] for the fractional quantum Hall states. The incompressible states of the fractional quantum Hall effect thus exhibit a kind of Bose condensation in which the objects condensing consist of an electron and its associated flux quanta. Very recently there has been a great deal of interest in the possibility that the ground state of high-temperature superconductors may be related to fractional Hall incompressible states[24]. The superconducting property of these states would then emerge as a consequence of the long-range order discussed above. It may turn out that it is this aspect of the surprising incompressible states which are responsible for the fractional Hall effect which has the greatest impact on other sub-fields of physics. Since the existence of of these unique many-body states was revealed to us fundamental research in semiconductors, we are reminded again of the unpredictable consequences of scientific enquiry.

In this introduction we have discussed only the simplest version of the fractional quantum Hall effect where all the electrons are in the lowest orbital Landau level of the conduction band of a semiconductor. The theory has been generalized to higher orbital Landau levels[25] and to electrons in the valence band[26]. The most intricate and interesting complications , however, come when more than one Landau level has its energy close enough to the chemical potential to be relevant. This happens, for example, when the Zeeman energy is small enough that the spin degree of freedom becomes important. Such a *two component* system can also be created by fabricating a system in which two 2D electron layers are in close proximity. The richness of the fractional quantum Hall effect in multi-component systems was appreciated by theory[27]. Continuing technical advances in material fabrication techniques have recently made it possible to study such systems experimentally[28]. This area of study is likely to bring new surprises in the future.

A.H. MacDonald

Reprinted Articles

Important Early Work in Magnetotransport

R1. "The Discovery of the Hall Effect: Edwin Hall's hitherto unpublished account" Katherine Russell Sopka in *The Hall Effect and its Applications* C.L. Chien and C.R. Westgate, eds., (Plenum, N.Y., 1980), pp. 523– 545.

R2. "Diamagnetismus der Metalle", L. Landau, Zeitschrift fur Physik, **64**, pp. 629–637 (1930).

R3. "Der Diamagnetisimus von frien Elektronen, E. Teller, Zeitschrift fur Physics, **67**, pp. 311–319 (1931).

R4. "The Dependence of the Susceptibility of Diamagnetic Metals upon the Field" W.J. de Haas and P.M. van Alphen, Kon.Akad. van Wefenschappen Proc., Vol. 33 No. 10, pp. 1106–1118 (1930).

R5. "Magnetische Widerstandsvergroserung in Einkristallen von Wismut bei tiefen Termperaturen," L. Schubnikow and W.J. deHass, Kon.Akad. van Wetenschappen Proc. **33**, pp. 130–133 (1930).

R6. "Magnetoresistance," W.G. Chambers in "*The Fermi Surface*," edited by W.A. Harrison and M.B. Webb (Wiley, N.Y., 1960), pp. 100–124.

Magnetotransport in 2D—Pre Quantum Hall Effect

R7. "Magneto-oscillatory Conductance in Silicon MOSFET's," A.B. Fowler, F.F. Fang, W.E. Howard and P.J. Stiles, Phys. Rev. Lett., **16** pp 901–903 (1966).

R8. "Theory of the Hall Effect in a Two-dimensional Electron System," T. Ando, Matsumoto and Uemura, J. Phys. Soc. Jpn, **39** pp. 279–288 (1975).

R9. "Quantum Galvanomagnetic Properties of n-Type Inversion Layers, S. Kawaji and J. Wakabayashi, Surf. Sci., **58** pp 238–245 (1976).

The Integer Quantum Hall Effect

R10. "The Discovery of the Quantum Hall effect," G. Landwehr, Metrologia, **22** pp. 118–125 (1986).

R11. "New Method for High-Accuracy Determination of the Fine Structure Constant based on Quantized Hall Resistance," K. von Klitzing, G. Dorda and M. Pepper, Phys. Rev. Lett., **45** pp. 494–497 (1980).

R12 "Quantized Hall Resistance and the Measurement of the Fine- Structure Constant," R.E. Prange, Phys. Rev. B, **23** pp. 4802–4805 (1981).

R13. "Quantized Hall Conductivity in 2 Dimensions," R.B. Laughlin, Phys. Rev. B, **23** pp. 5632–5633 (1981).

R14. "Quantized Hall Conductance, Current-Carrying Edge States and the Existence of Extended States in a 2-Dimensional Disordered Potential," B.I. Halperin, Phys. Rev. B, **25** pp. 2185–2190 (1982).

R15. Theory of the Quantized Hall Conductivity in 2-Dimensions," P. Streda, J. Phys. C, **15**, L717–21 (1982).

R16. "Localization and Scaling in the Quantum Hall Regime," H.P. Wei, D.C. Tsui and A.M.M. Pruisken, Phys. Rev. B, **33**, pp. 1488–1491 (1986).

The Quantum Hall Effect in a Periodic Potential

R17. "Quantization of Coupled Orbits in Metals," A.B. Pippard, Proc.Roy. Soc. A, **270** pp. 1–13 (1962).

R18. "Energy Levels and wave functions of Bloch electrons in rational and irrational magnetic fields," D. Hofstader, Phys. Rev. B, **14** pp.2239–249 (1976).

R19. "Quantized Hall Effect in a Two-Dimensional Periodic Potential," David Thouless, Phys. Rep., **110** pp. 279–291 (1984).

R20. "Quantized Hall Effect in a Two-Dimensional Periodic Potential," P. Streda, J. Phys. C, **15** pp. L1299–L1303(1982).

The Fractional Quantum Hall Effect (Experiment)

R21 "Two-Dimensional Magnetotransport in the Extreme Quantum Limit," D.C. Tsui, H.L. Stormer and A.C. Gossard, Phys. Rev. Lett. **48** pp. 1559–1562 (1982).

R22. "Odd and Even Fractionally Charged Quantized States in GaAs/GaAsAs Heterojunctions," R.G. Clark, R.J. Nicholas, A. Usher, C.T. Foxon and J.J. Harris, Surf. Sci. (Netherlands) **170** pp. 141–147 (1986).

R23. "Magnetic Field Dependence of Activation Energies in the Fractional Quantum Hall Effect," G.S. Boebinger, A.M. Chang, H.L. Stormer and D.C. Tsui, Phys. Rev. Lett., **55** pp. 1606–1609 (1985).

R24. "Second Activation Energy in the Fractional Quantum Hall Effect," J. Wakabayashi, S. Sudou, S. Kawaji, K. Hirakasw and H. Sakaki, J. Phys.Soc. Jpn., **56** pp. 3005–3008 (1987).

The Fractional Quantum Hall Effect (Theory)

R25. "Anomalous Quantum Hall Effect: An Incompressible Fluid with Fractionally Charged Excitations," R.B. Laughlin, Phys. Rev. Lett., **50** pp. 1395–1398 (1983).

R26. "Fractional Quantization of the Hall Effect: A Hierarchy of Incompressible Quantum Fluid States," F.D.M. Haldane, Phys. Rev. Lett., **51** pp. 605–608 (1983).

R27. "Ground State of the 2-Dimensional Electrons in Strong Magnetic Fields and 1/3 Quantized Hall Effect," D. Yoshioka, B.I. Halperin and P.A. Lee, Phys. Rev. Lett., **50** pp. 1219–1222 (1983).

R28. "Statistics of Quasiparticles and the Hierarchy of Fractional Quantized Hall States," B.I. Halperin, Phys. Rev. Lett., **52** pp.1583–1586 (1984).

R29. "Magnetoroton Theory of Collective Excitations in the Fractional Quantum Hall Effect," S.M. Girvin, A.H. MacDonald and P.M. Platzman, Phys.Rev. B, **33** pp. 2481–2494 (1986).

R30. "Off-Diagonal Long Range Order, Oblique Confinement and the Fractional Quantum Hall Effect," S.M. Girvin and A.H. MacDonald, Phys. Rev.Lett., **58** pp. 1252–1255 (1987).

REFERENCES

1. P. Drude Annalen der Physik **1**, 566 (1900); Annalen der Physik **3**, 369 (1900) .
2. O.M. Corbino Il Nuovo Cimento **1**, 27 (1911); Il Nuovo Cimento **2**, 39 (1911)
3. Rudolf Peierls: *Surprises in Theoretical Physics* (Princeton University Press, Princeton 1979)
4. J.R. Schrieffer in *Semiconductor Surface Physics* R.H. Kinston ed. (University of Pennsylvania Press, Philadelphia,1957) pp.55-69
5. The properties of two dimensioal electronic systems are reviewed by T. Ando, A. B. Fowler and F. Stern Rev. Mod. Phys. **54**, 437 (1982)
6. Th. Englert and K. von Klitzing Surf. Sci. **73**, 70 (1978)
7. B.I. Halperin Helv. Phys. Acta **56** 75 (1983)
8. A.A.M. Pruisken Nuclear Physics b **235** 277 (1984)
9. H.P. Wei D.C. Tsui M.A. Paalanen and A.M.M. Pruisken Phys. Rev. Lett. **61** 1294 (1988)
10. R.W. Stark Phys. Rev. Lett. **9** 482 (1962)
11. A.H. MacDonald Phys. Rev. B **28** 6713 (1983)
12. D.J. Thouless, M. Kohomoto, M.P. Nightingale and M. den Nijs Phys. Rev. Lett. **49** 405 (1982)
13. A.H. MacDonald Phys. Rev. B **29** 3057 (1984)
14. R.R. Gerhardts D. Weiss and K. von Klitzing Phys. Rev. Lett. **62** 1173 (1989)
15. R.W. Winkler J.P. Kotthaus and K. Ploog Phys. Rev. Lett. **62** 1177 (1989)
16. A.H. MacDonald and P. Streda Phys. Rev. B **29** 1616 (1984)
17. B. Simon Phys. Rev. Lett. **51** 2167 (1983); Q. Niu, D. J. Thouless and Wu Y-S Phys. Rev. B **31** 3372 (1985)
18. S.M. Girvin and Terrence Jach, Phys. Rev. B **29**, 5617 (1984) S.M. Girvin Phys. Rev. B **30**, 558 (1984); D. Levesque and J.J. Weis Phys. Rev. B **30**, 1056 (1984)

20. see R.P. Feynman *Statistical Mechanics* (Benjamin, Reading Mass,1972) and references therein.
21. F. Wliczek Phys. Rev. B **49**, 957 (1982); D.P. Arovas, J.R. Schrieffer, F. Wilczek and A. Zee Nucl. Phys. B **251**, 117 (1985)

22. F.D.M. Haldane Phys. Rev. Lett. **61**, 1985 (1988)

23. S.M. Girvin, in *The Quantum Hall Effect*, R.E. Prange and S.M. Girvin eds. (Springer-Verlag, New York,1986), Chapter 10; N. Read, Phys. Rev. Lett. **62**, 86 (1989); S.C. Zhang, T.H. Hansson and S. Kivelson, Phys. Rev. Lett. **62**, 82 (1989)

24. V. Kalmeyer and R.B. Laughlin, Phys. Rev. Lett. **59**, 2095 (1987); R. B. Laughlin, Phys. Rev. Lettt. **60**, 2677 (1988).

25. F.D.M. Haldane, in *The Quantum Hall Effect*, R.E. Prange and S.M. Girvin eds. (Springer-Verlag, New York, 1986), Chapter 8 and references therein.

26. A.H. Mac Donald and U. Ekenberg, Phys. Rev. B **39**, 5959 (1989)

27. B.I. Halperin Helv. Phys. Acta **56**, 75 (1983); Tapash Chakraborty and F.C. Zhang, Phys. Rev. B **29**, 7032 (1984)

28. J.P. Eisenstein, H.L. Störmer, L. Pfeiffer and K.W. West, Phys. Rev. Lett. **62**, 1540 (1989); R.G. Clark, S.R. Haynes, A.M. Suckling, J.R. Mallett, P.A. Wright, J.J. Harris and C.T. Foxon, Phys. Rev. Lett. **62**, 1536 (1989)

REPRINTED ARTICLES

THE DISCOVERY OF THE HALL EFFECT:

EDWIN HALL'S HITHERTO UNPUBLISHED ACCOUNT

Katherine Russell Sopka

Harvard University
Cambridge, Massachusetts 02138

ABSTRACT

An annotated transcript of Edwin Hall's notebook account of his 1879 laboratory investigations of the transverse effect of a magnetic field applied at right angle to the direction of electric current in a conductor.

INTRODUCTORY NOTE

The public account of Edwin Hall's laboratory investigations associated with the discovery of what has since become known as the Hall Effect has been widely available for a century. In late 1879, when Hall was a 24 year old graduate student, he published "On a New Action of the Magnet on Electric Currents" in the _American Journal of Mathematics Pure and Applied_,[1] a quarterly journal published under the auspices of the Johns Hopkins University, edited by J. J. Sylvester and William Story of the Mathematics Department with the cooperation of other faculty members, including Hall's thesis supervisor, Henry A. Rowland. There he gave the preliminary notice of that new effect. In addition, Hall's early findings were brought to wider international attention through the publication a few months later of exactly the same text in _The London, Edinburgh and Dublin Philosophical Magazine and Journal of Science_.[2] Less than a year after these preliminary notices, accounts of Hall's more complete investigations that formed his doctoral dissertation at the Johns Hopkins University were published in the _American Journal of Science_ and in the _Philosophical Magazine_.[3]

523

The hitherto unpublished account of the early part of Edwin Hall's investigations into the effect of imposing a magnetic field perpendicular to a current carried by a conductor which is presented in the following pages comes from his handwritten Notebook of Physics. This item is one of the earliest of some 400 papers that were deposited in the Houghton Library of Harvard University by his daughter, Miss Constance Hall in 1964, more than a quarter century after his death. Hall had joined the faculty of the Harvard Physics Department in 1881, serving actively until his retirement in 1921. As an emeritus professor he continued to work in his laboratory at Harvard until shortly before his death in 1938.

The Notebook itself is in remarkably good condition. The paper is substantial and only slightly discolored. The ink has turned brown but is still dark and easily legible, thanks principally to Hall's good penmanship.

The notebook's 124 ruled pages, approximately 7" by 8", are sewn together and covered with a firm binding of a dark, mottled coloring. The early pages are devoted to data taken by Hall on about a dozen experiments ranging from "Use of Large Spherometer" to "On Sp[ecific] Gr[avity] of Carbonic Anhydride compared with Air". While these experiments are not dated they were presumably done before January 14, 1878, the first date recorded (on page 56). The next several pages contain a narrative description of Hall's studies in the areas of spectrometry, polarized light and photography, followed by a List of Experiments in Electricity and Magnetism done during the calendar year 1878.[4]

Beginning on page 73 and continuing for the remainder of the notebook is Hall's personal account of how he came to perform those experiments which revealed the effect, now bearing his name, that is recognized today as being of great practical as well as theoretical importance in solid-state physics.

For the historian of science a notebook such as this is a valuable document, revealing private aspects of doing science that are not found in the public record of scientific results. While Hall's first publication mentioned above[1] is similar in many ways to his notebook account which follows, it seems worthwhile to make the full text from the notebook available now to scientists and historians of science in this centennial year of his discovery of the Hall Effect. Especially noteworthy are the insights the reader can gain into such areas as Hall's student/mentor relationship with Henry A. Rowland and his association with others in the Johns Hopkins physics department, and, perhaps more importantly, the state of knowledge and understanding in physics a century ago through Hall's candid personal account of how he approached his laboratory investigations.

The page numbers referred to below, and found in the text of the transcript, correspond to those used by Hall, those in [ ] having been added to fill in gaps in the sequence written by him.

As we read these pages today some of the ideas expressed by Hall strike us as decidedly naive or just plain wrong. We must, however, keep in mind that in 1879 scientists had little understanding of the nature of electricity. The electron, for example, was not identified until more than a decade after Hall conducted the experiments he described in this notebook. Furthermore, Hall's expectation that a wire with triangular cross section would be better than a round wire in testing for the proposed magnetic effect on resistance was apparently shared by a physicist of no less stature than Rowland himself. Hall's use of the now long outmoded water stream analogy for electric current nevertheless provided him with a line of reasoning that did put him on the path to making his discovery. On the other hand, his expectation that continuous work could be done by a permanent magnet is clearly at odds with the principle of the conservation of energy which had been established well before 1879.

Hall's account transcribed below has some curious aspects of which the reader should be aware. Despite the existence of dated entries (June 13, 14, November 3, 10, 21 and December 15) it appears that the entire piece was not written sequentially, page by page. The first clue to this comes in the opening sentence on page 73 with his reference to the "history" of his experiment. The next two pages are blank and his narrative continues on page 76. Again, pages 88 and 89 are found to be blank also.

It is my conjecture that Hall purposely left several blank pages after his earlier entries in this notebook and then did in fact record the data on pages 82, 84 and 86 on June 13 and 14, 1879 after discussion with his advisor Rowland and at least one other member of the department. It is likely that he then wrote at least some of the accompanying narrative before leaving for the summer with the belief that he had found an increased resistance in his test coil when it was subjected to a magnetic field, but somewhat perplexed by the <u>smaller</u> effect observed when he used a <u>more</u> sensitive galvanometer (see page 85).

By the time Hall made his next dated entry, November 3 on page 90, he had been back at the Johns Hopkins University several weeks. During this time he had redone the initial experiment more carefully, recording his data elsewhere, and come to the conclusion, stated on page 91, that there was no measurable increase in resistance. In addition, however, he had changed his experimental set-up to one proposed by Rowland (see page 95) and looked for a transverse electromotive force in a piece of gold foil carrying an electric current and subjected to a perpendicular magnetic field.

The fact that with this arrangement he did find clear evidence of such an emf is not recorded in this notebook until November 10 (on page 98) although it had been observed about two weeks earlier.

Hall's results were, of course, of considerable interest to Rowland and also to other younger members of the physics department, some of whom were present while Hall was conducting his experiment and assisted him in carrying it out.

Obviously further work was needed and the remainder of Hall's account discusses such variations as changing the location of the tapping points for detecting the emf and using metals other than gold. Each such variation shed further light on the nature of the effect, but also raised further questions that are not adequately resolved in these pages. The interested reader will find a more thorough probing of such questions in "The Experiments of Edwin Hall 1879 - 1881", Chapter IX of "Henry Augustus Rowland and his Electromagnetic Researches", the doctoral dissertation of John D. Miller submitted to Oregon State University in 1970, an account which draws not only on this notebook of Hall but also on other resources.

In Hall's notebook narrative he has been primarily concerned with describing his conception of the phenomena being studied and explaining the routes by which he arrived at his ideas. While it is regrettable that the associated "Experiment Book" has apparently not survived, we rejoice that we do have such a revealing personal account as is contained in the pages transcribed below.

TRANSCRIPT FROM EDWIN H. HALL'S NOTEBOOK OF PHYSICS OF THOSE ENTRIES RELATED TO THE DISCOVERY OF THE HALL EFFECT

page 73

For the last two or three weeks of the year have been engaged in an experiment of which the history is as follows.

I was surprised to read some months ago a statement in Maxwell,[5] Vol II page [144] that the Electricity itself flowing in a conducting wire was not at all affected by the proximity of a magnet or another current. This seems different from what one would naturally suppose, taking into account the fact that the wire alone was certainly not affected and also the fact that in Static Electricity it is plainly the Electricity itself that is attracted by Electricity.

(Note: page 74 and 75 are blank)

[page 76]

Soon after reading the above statement in Maxwell I read an article by Prof. Edlund entitled Unipolar Induction (Phil. Mag and Journal de Chemie et Physique) in which the author evidentally assumed that an electric current was acted on by a magnet in the same way as a wire bearing a current.

As these two authorities seemed to disagree I asked Prof. Rowland about the matter. He told me he doubted the truth of Maxwell's statement and had been thinking of testing it by some experiment though no good method of so doing had yet presented itself to him. I now began to give more attention to the matter and thought of a plan which seemed to promise well. As Prof. Rowland was too much occupied with other matters to undertake this investigation at present I proposed my scheme

page 77

to him and asked whether he had any objection to my making the experiment. He approved of my method in the main, though suggesting some very important changes.

My method was formulated on the following theory. If the Current of Electricity is itself attracted by a magnet while the wire bearing it remains fixed in position, the current should be drawn to one side of the wire and therefore the resistance experienced should be increased. I thought this effect might be magnified if the wire were made of this (▽) section and so wound that the tendency of the magnetic force would be to draw the current into the thin part of the wire. This seemed plausible to Prof. Rowland and he put

[page 78]

me in the way of carrying out the experiment. Dr. Hastings[6] urged that there seemed to be no reason for expecting any advantage in using wire of the propose[d] section and thought round wire ought to show the effect, if there were any, as well. I could not indeed give any very definite reason for using the wire of triangular section, though it seemed to me that the effect would certainly not be diminished and might possibly be increased. Prof. Rowland, however, said that some effect of this kind had already been observed with round wire though not explained on above theory. He thought there would be an advantage in using the triangular wire.

The wire selected was of German

page 79

Silver about 1/50 in. thick. Short pieces of this wire were drawn through a triangular hole punched in a steel ribbon and the pieces were then soldered together making a length of about three feet and having a resistance probably of about two ohms. This wire was wound in a spiral groove turned in the flat surface of a disk of hard rubber. It was held in place as wound by a soft cement and then covered by another disk of hard rubber laid upon the first. The coil thus prepared was pressed between the poles of an electromagnet and made one arm of a Wheatstone Bridge.

The magnet was operated by a battery of 20 Bunsen's cells[7] placed 4 in series, 5 broad.

My plan was simply to send

[page 80]

a current through the wire and observe whether the resistance was varied by operating the magnet.

The disturbing cause might be
1st Current in wires from the magnet battery might directly affect the Galvanometer.
2nd Magnet might directly affect Galvanometer.
3rd Induction currents would probably arise on closing circuit though Galvanometer.
4th Thermo-Electric currents might exist.

In order to avoid the second difficulty magnet was placed some thirty feet distant from the Galvanometer. It was found that the first and second of the above causes affected the Galv. only slightly and in the way of a permanent

page 81

deflection whenever the magnet circuit was closed. These causes then do not affect the experiment.

The combined effect of the 3^d and 4th causes was observed as follows. I replaced the small battery used with the Bridge by a short wire; then with the magnet circuit alternately open and closed, I would complete the circuit through the Galvanometer pressing down the key until the needle reached its maximum deviation. I usually found slight deviations in this way and have recorded them when about to try the main experiment.

Observations made June 13th and 14th are recorded on succeeding pages.

[page 82]

Preliminary, Small Battery replaced by wire.

mag. F East	mag. F —	Mag. F West
0	0	0

Main Experiment

mag. F East	(M. F. 0)	M. F. West
83½ r	72 r	72½ r
	65 r	
69	65½	67
	64	
65	64	64½
	60	
63	61½	64
	60½	
63	62½	64
	61½	
63½	63	63
	56½	
60½	62	63½
	61½	
63½	61	63
	62½	
61½	61½	57
	61	
63	53½	62
	61½	
61	58	63
	52½	
60½	61	59
	62	
64	62½	61½
	61½	
58	61	64
	62	

The observation as made run horizontally to the right, diagonally to the left. It is not therefore allowable to compare the mean of 1st + 3d columns with mean of 2d.

I proceeded as follows.

Mean of 1st col. ; rejecting 1st read.

= 900 ÷ 14 = 64.29 [I]   816½ ÷ 13 = 62.81 [II]

1st readings 2d col.

869 ÷ 14 = 62.07 [III]

2d readings 2d col ; reject last read

852 ÷ 14 = 60.86 [IV]   790 ÷ 13 = 60.77 [V]

Mean of 3d col. ; reject last

888 ÷ 14 = 63.43 [VI]   824 ÷ 13 = 63.38 [VII]

$\frac{I + VI}{2} - III = 1.79$ } 2.055

$\frac{II + VII}{2} - V = 2.32$ }

Small Bat. Discon.

m. m.	m. in	m. F.
¾ r	1¼ r	1¼ r
1¼ r	1¼ r	

This effect evidently

page 83

This effect evidentally cannot be accounted for by thermal currents or induction. It is such an effect as would be caused by a slight increase of resistance in the wire tested.

[page 84]

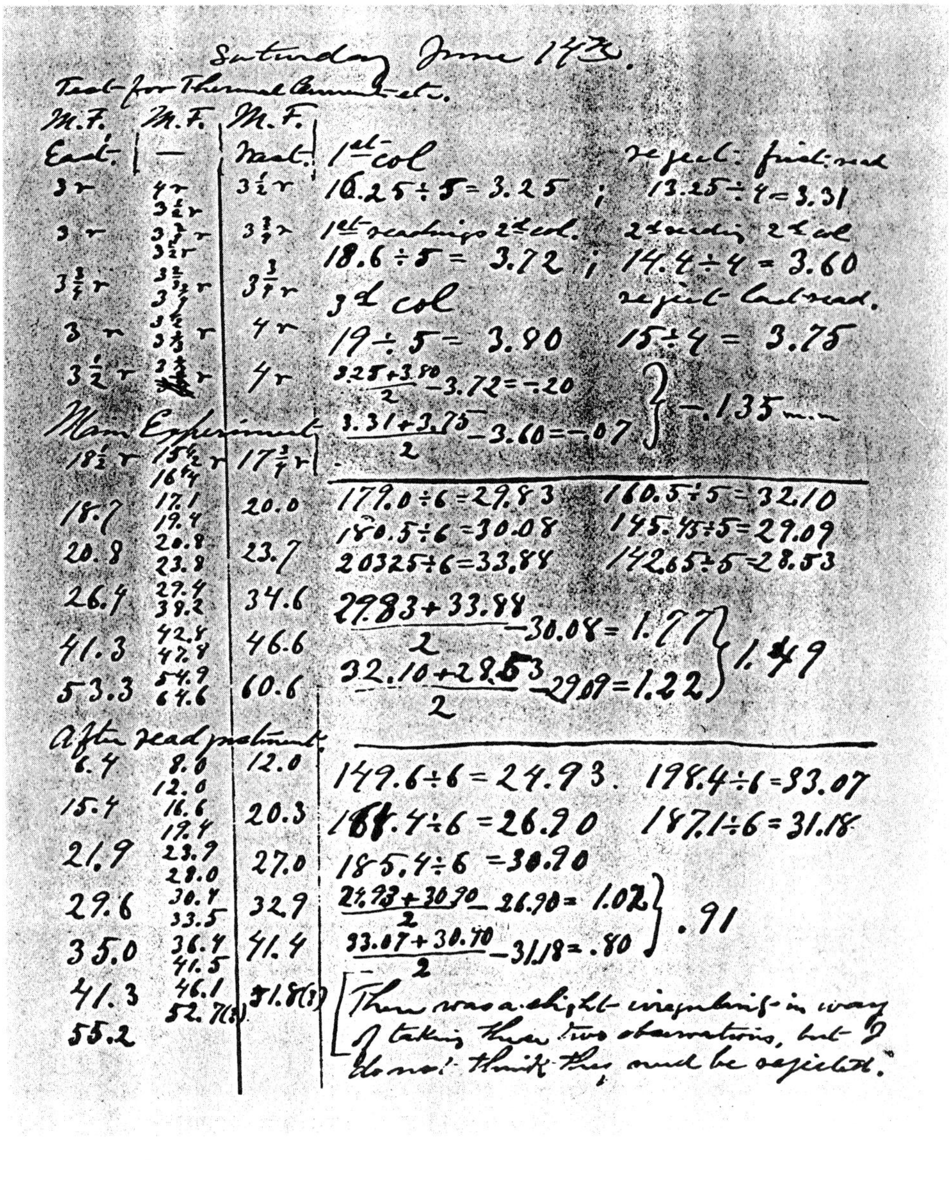

Saturday June 14/73.

Test for Thermal Current etc.

M.F. East	M.F. —	M.F. West
3 r	4 r, 3½ r	3½ r
3 r	3¾ r, 3½ r	3¾ r
3¾ r	3¾ r, 3¾ r	3¾ r
3 r	3½ r, 3⅓ r	4 r
3½ r	3⅓ r	4 r

1st col
16.25 ÷ 5 = 3.25 ; reject first read 13.25 ÷ 4 = 3.31

1st readings 2d col. 2d reading 2d col
18.6 ÷ 5 = 3.72 ; 14.4 ÷ 4 = 3.60

3d col reject last read.
19 ÷ 5 = 3.80 15 ÷ 4 = 3.75

$\frac{3.25+3.80}{2} - 3.72 = -.20$
$\frac{3.31+3.75}{2} - 3.60 = -.07$ } −.135 mean

Main Experiment

18½ r	15½ r, 16¼	17¾ r
18.7	17.1, 19.4	20.0
20.8	20.8, 23.8	23.7
26.4	27.4, 38.2	34.6
41.3	42.8, 47.8	46.6
53.3	54.9, 64.6	60.6

179.0 ÷ 6 = 29.83 160.5 ÷ 5 = 32.10
180.5 ÷ 6 = 30.08 145.45 ÷ 5 = 29.09
203.25 ÷ 6 = 33.88 142.65 ÷ 5 = 28.53

$\frac{29.83+33.88}{2} - 30.08 = 1.77$
$\frac{32.10+28.53}{2} - 29.09 = 1.22$ } 1.49

After readjustment.

6.4	8.0, 12.0	12.0
15.4	16.6, 19.4	20.3
21.9	23.9, 28.0	27.0
29.6	30.4, 33.5	32.9
35.0	36.4, 41.5	41.4
41.3	46.1, 52.7(?)	51.8(?)
55.2		

149.6 ÷ 6 = 24.93. 198.4 ÷ 6 = 33.07
161.4 ÷ 6 = 26.90 187.1 ÷ 6 = 31.18
185.4 ÷ 6 = 30.90

$\frac{24.93+30.90}{2} - 26.90 = 1.02$
$\frac{33.07+30.90}{2} - 31.18 = .80$ } .91

[There was a slight irregularity in way of taking these two observations, but I do not think they need be rejected.

page 85

In making observations recorded on the preceding page and the following a more sensitive Galvanometer has been substituted for the one previously employed. It is to be noticed however that the effect seems to be less marked with the sensitive Galvanometer than with the other. The continued and rapid increase in the deflection is probably due to the heating of some part of the circuit by the repeated currents.

[page 86]

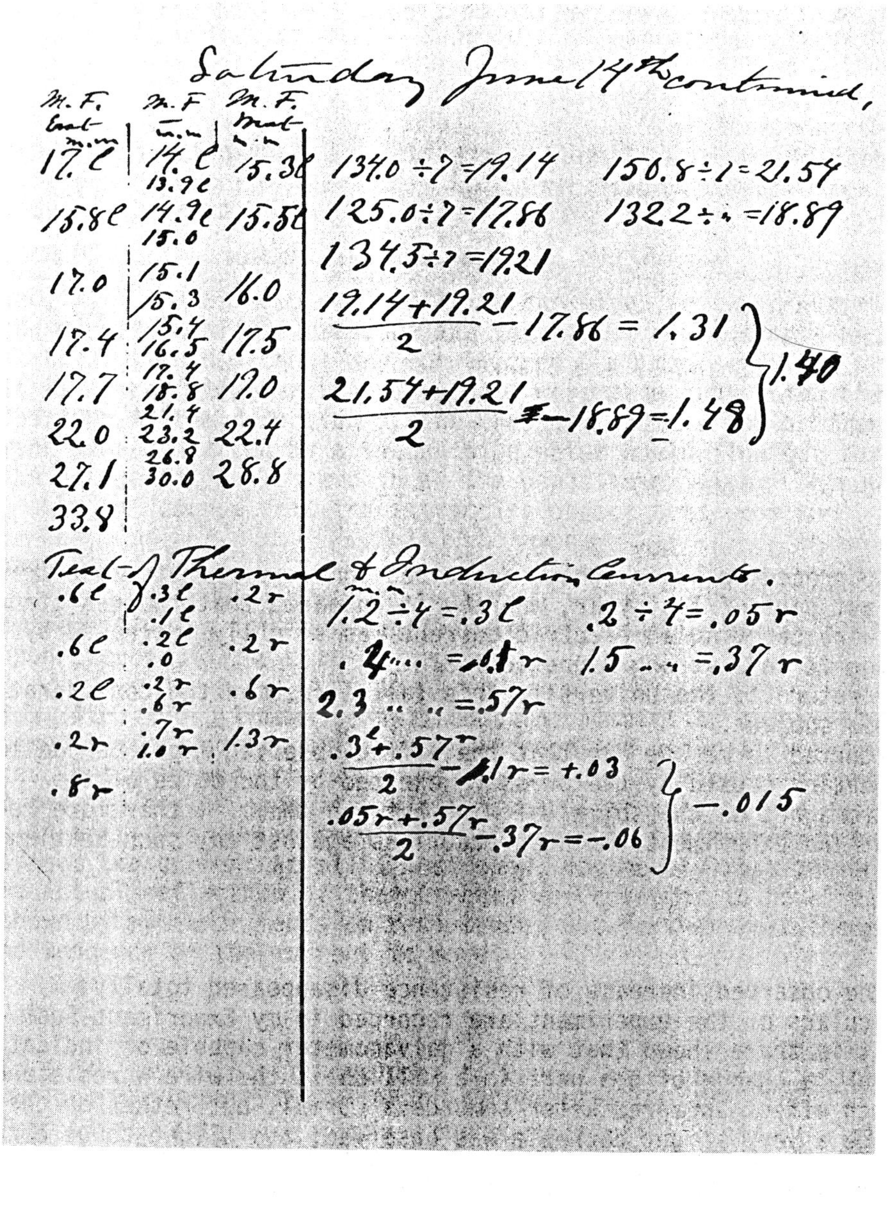
Saturday June 14th continued,

17.2l 14 l 15.3l
13.9l
15.8l 14.9l 15.5l
15.0
17.0 15.1 16.0
15.3
17.4 15.4 17.5
16.5
17.7 17.4 19.0
18.8
22.0 21.4 22.4
23.2
27.1 26.8 28.8
30.0
33.8

134.0 ÷ 7 = 19.14 150.8 ÷ 7 = 21.54
125.0 ÷ 7 = 17.86 132.2 ÷ " = 18.89
134.5 ÷ 7 = 19.21
(19.14 + 19.21)/2 − 17.86 = 1.31
(21.54 + 19.21)/2 − 18.89 = 1.48
} 1.40

Test of Thermal & Induction Currents.
.6l .3l .2r
.1l
.6l .2l .2r
.0
.2l .2r .6r
.6r
.2r .7r 1.3r
1.0r
.8r

1.2 ÷ 4 = .3l .2 ÷ 4 = .05r
.4 " " = .1r 1.5 " " = .37r
2.3 " " = .57r
(.3l + .57r)/2 − .1r = +.03
(.05r + .57r)/2 − .37r = −.06
} −.015

page 87

Deflections recorded on preceding page are mainly to the left while those previously recorded were to the right.

This was caused by reversing current rhrough the Bridge so that the current from the zinc pole entered the test coil by the circumference instead of the center as previously it had done. The last arrangement made current in the test coil flow from north to south above

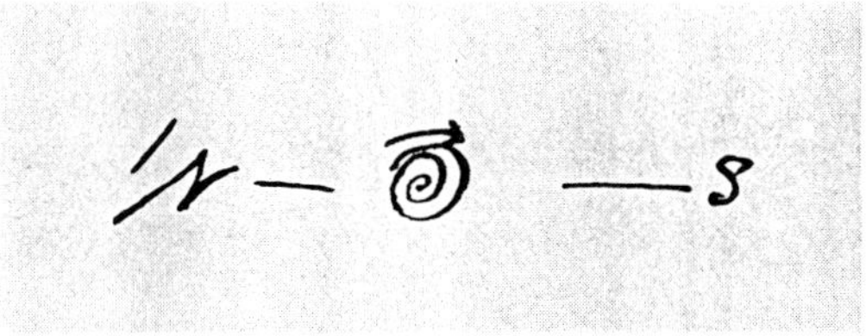

(Note: pages 88 and 89 are blank)

[page 90]

Record for '79 and '80

Nov. 3 '79

In the preceding pages I have described some experiments made in the last part of last year, whereby it appeared that the resistance of a wire carrying an electric current was slightly increased by placing it in a strong magnetic field.

On my return to the University this fall I resumed the consideration of this subject.

It occurred to me one day that the effect observed might be due to the heating caused by the pressure exerted by the poles of the electromagnet on the spiral of wire between them. I therefore repeated the experiment taking precautions against any such heating effect

page 91

and the observed increase of resistance disappeared totally. Particulars of the experiment are recorded in my Experiment Book and it is there shown that with a galvanometer capable of indicating, I think, a change of one part in a million in the wire's resistance, no such effect appeared after some days' trial, but rather on the average a very slight decrease was observed, too slight to be considered anything more than an error of the experiment. It seemed therefore very probable that no measurable change in resistance was caused by the magnet. This might have seemed to settle the question of the action or non action of the magnet on the electric current as such, but the whole

[page 92]

ground was not yet covered. I reasoned thus:

If electricity were an incompressible fluid it might be acted on in a particular direction without moving in that direction. I took an example about like this. Suppose a stream of water flowing in a perfectly smooth pipe which is however loosely filled with gravel. The water will meet with resistance from the gravel but none from the pipe, at least no frictional resistance. Suppose now somebody brought near the pipe [something] which has the power of attracting a stream of water. The water would evidently be pressed against the side of the pipe but being incompressible

page 93

and, with gravel completely filling the pipe, it could not move in the direction of the pressure and the result would simply be a state of stress without any actual change of course by the stream. It is evident however that if a hole were made traversing through the pipe in the direction of the pressure and the two orifices thus made were connected by a second pipe water would flow out toward the attracting object and in at the opposite orifice. This suppose[s] of course that the attracting object acts upon the

[page 94]

current flowing in one direction without acting, equally at least, upon current in the other direction.

Nov. 4th '79 I mean by this that the attracting object is supposed to act, not upon the water at rest and under all circumstances, but only when the water is flowing and flowing in a certain direction or the opposite. In this way I arrived at the conclusion that in order to show conclusively that the magnet does not affect the current at all I must show, not merely that there was no actual deflection of the current, which seemed to be already shown by my experiments on resistance, but further that there was no tendency of the current to move.

page 95

In order to do this I concluded to repeat an experiment which Prof. Rowland had once tried without any positive results.
This was to determine whether the equipotential lines in a disk of metal carrying a current would be affected by placing the disk between the poles of the electro-magnet. I set something of the above reasoning before Prof. Rowland and he advised me to try the

experiment as he had not made the trial very carefully himself.

He advised me to use gold leaf mounted on a glass plate and I did so. The plan was this

[page 96]

The theory was this

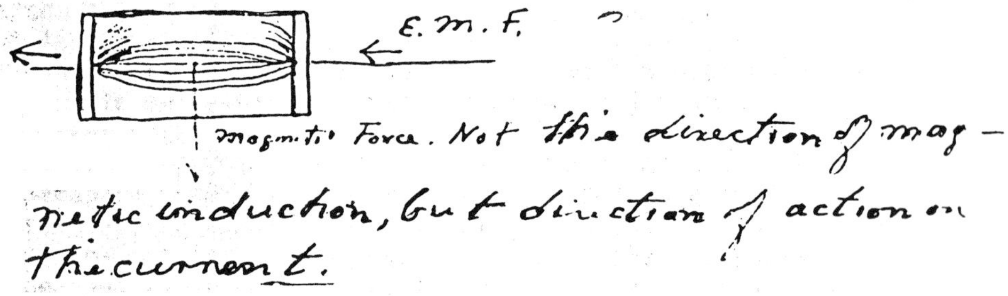

Prof. Rowland advised me to place my tapping points (a,b) near the end of the disk on the ground that the equipotential lines crossing the disk in the center would not be deflected by a deflection of the current. His theory was

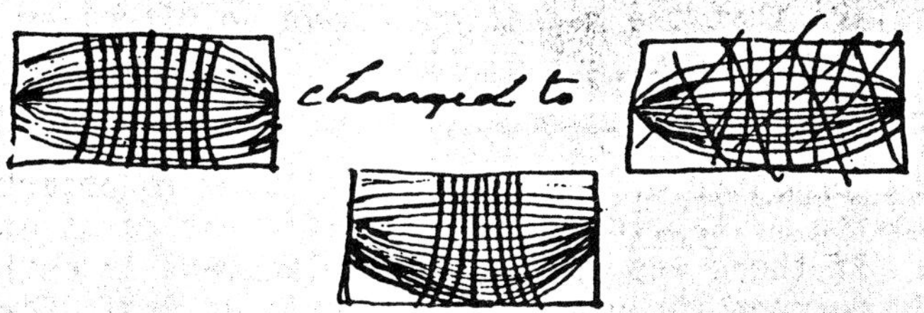

I followed his advice in placing my points but I pointed out to

page 97

him before trying the experiment that in case there was a tendency to deflect the current without any actual deflection, the current would be in a state of stress and the equipotential lines would not be simply perpendicular to the lines of flow but would be oblique to them and that the whole length of the disk, in the middle as elsewhere. Experiment afterward proved the correctness of my theory.

Nov. 10th '79
In trying this experiment I used a Thomson galvanometer which was perceptibly affected by an EMF of 1/1,000,000 of a Bunsen cell, the resistance of the galvanometer being 6491 ohms. With this instrument, having in its circuit an additional resistance of 100,000 ohms,

[page 98]

there was detected on the evening of, I believe, Friday the 24th of October, a somewhat doubtful indication of the action of the magnet. A trial the next day however seemed to fail in confirming this indication. Means were taken however to make the test more delicate and finally we were able to do away with all additional resistance in the galvanometer circuit. On the evening of, I believe, Tuesday Oct. 28th, very marked and seemingly unmistakable evidence of the looked for effect was observed. Mr. Freeman[8] was observing the galvanometer. The deflection observed was a permanent one of two or three centimeters. I was myself at the magnet. It seemed hardly safe even then to believe that a new phenomenon had

page 99

been discovered but now after nearly a fortnight has elasped and the experiment has been many times and under various conditions successfully repeated, meeting at the same time without harm the criticism of fellow students and professors, it is not perhaps too early to declare that the magnet does have an effect on the electric current or at least an effect on the circuit never before expressly observed or proved.

It now became a matter of great interest to me to ascertain the effect of moving my tapping points to different parts of the gold leaf plate. If there was an actual displacement or deflection of the electric current through the gold leaf, as Prof. Rowland had

[page 100]

thought possible, the effect should be reversed on moving the tapping points to the other end of the gold leaf and in the transverse diameter the effect should disappear. If the effect of the magnet was simply to cause a tending of the current to change its course without effecting any deflection, which I thought to be the only view consistent with my failure to discover an increase of resistance in the spiral of wire previously tried, then the observed effect would not be reversed in direction when the points were moved to the other end of the gold leaf and furthermore the effect on the galvanometer should not disappear when the points were placed

page 101

opposite the center of the gold foil. I stated both these views to Mr. Jacques[9] and Mr. Ayres[10] immediately before making the test and then asked Mr. Ayres to assist me in the trial telling him just what I expected to find, viz. that the effect would not be reversed.

The points were first moved to the opposite end of the gold leaf and then to the middle.

The result was what I expected. Prof. Rowland came in while the experiment was in progress. On learning the result he after a little thought proposed an explanation of the phenomenon which seemed to me at the time very like my own already given. It was discovered in these experiments just described or soon after that the effect was

[page 102]

in one respect the reverse of what had been anticipated by both Prof. Rowland and myself. That is, the current seemed to tend to move in the opposite direction to that in which the disk itself would tend to move under the action of the magnet.

At first this seemed to be easily accounted for on the supposition that the electric current really flows from negative to positive instead of from positive to negative as usually assumed for convenience. A little reflection has however convinced me that this explanation is not sufficient. If we make the suggested change in our conception of the direction of the current through the gold leaf,

page 103

we must make the same change in our conception of the direction of the current in the electromagnet and the two changes will annul each other leaving us still face-to-face with [a point] of difficulty. To me at present it seems probable that two parallel currents of electricity flowing in the same direction tend to repel each other, just as two quantities of static electricity do. Further experiments will probably be necessary to ascertain the truth of the matter.

Note by EHH: see p. 114

I have been for a week past making ready, by Prof. Rowland's advice and under his direction to make a series of observations which will be capable of reduction to absolute measure, thus giving the quantitative laws of the phe

[page 104]

nomenon observed. This is for the purpose of getting something definite to publish.

More than a week ago I suggested an experiment which Prof. Rowland, not getting my idea fully I think, did not regard favorably. I have since been able to discuss the matter more fully with him and think it altogether probable that he will ultimately approve of it. The experiment proposed is suggested by one described by Edlund in his article on "Unipolar Induction" (Journal de Chemie et Physique).

Edlund states that if a magnet be placed vertical and surrounded with a metal cylinder, this cylinder tends to revolve about the magnet when a current of electricity is

page 105

made to flow from one end to the middle of the cylinder, the wires from a battery being applied at these points. Some months ago as I was considering this phenomenon I wondered what could make the cylinder move if, as Edlund seemed to suppose, an electric current could be acted upon and moved by a magnet just as a wire bearing a current is acted upon and moved. Prof. Rowland to whom I mentioned my difficulty suggested that the current might be made by the magnet to move around the cylinder and by the metallic resistance drag the cylinder about. Some time after this however I approached Prof. Rowland again on this subject.

If, I argued, the current is ac

[page 106]

tually made to pursue a spiral course along the cylinder more work must be done in the cylinder than if the current flowed in its natural course. This must be true even when the cylinder is at rest. Now this additional work must be done either by the battery maintaining its current, in which case the work must show itself in an increased resistance of the cylinder, or the extra work must be done by the magnet. On one supposition we have the resistance of a circuit increased by the presence of a magnet, a thing hitherto unobserved; on the other supposition we nave a permanent magnet doing continuous

page 107

work, a thing equally improbable.

Prof. Rowland admitted the full force of my reasoning and remarked at the time that he didn't see why this argument was not conclusive. Conclusive against Edlund's theory I suppose he meant. He observed however at the same time that if there was no action at all on the current itself he didn't see how his experiment on Electric Convection[11] could succeed.

When I began the experiments with the gold leaf I expected to obtain a negative result as I had in the experiment on resistance and I expected to publish those negative results together with a criticism of Edlund's theory as applied to this phe

[page 108]

nomenon of the revolving cylinder. Naturally after discovering the action of the magnet on the gold leaf I recurred to this phenomenon and asked myself how I could reconcile it with the results of my own experiments. So I took up again the suggestion of Prof. Rowland that the current went in a spiral course along the cylinder and concluded to boldly face the two horns of the dilemma and test them in turn.

Who knows that the proximity of a magnet may not increase the resistance of an electric current? Who knows that a permanent magnet cannot do continuous work?

At present I reason thus. The deflection of the current in the cylinder cannot be the cause of the

page 109

cylinder's motion, for according to my experiment thus far the current would have a tendency to drag the cylinder in a direction contrary to the one it follows. Moreover the effect of the deflected current would be, so far as I can judge now, altogether inadequate to produce such a result as I presume the motion of of the cylinder is affected in one way while the current flowing in it is affected in the opposite way. It is not difficult to suppose this possible, though I have not now any clear conception of the way in which it is possible.

Note by EHH: see p. 114

The explanation, as I conceive, of the

[page 110]

non deflection of the current in a disk is that the circuit in the transverse direction is not completed. When we complete it through the galvanometer a current is set up. Suppose now we bend our disk into a cylinder and so dispose it in the magnetic field that many lines of force pass out through the wall of the cylinder thus giving the current flowing therein a tendency to turn or slide around the cylinder. Why will it not pursue a spiral course? If it does take a spiral course, and I think we can ascertain whether it does so or not, we will try the two horns of our dilemma in succession.

My opinion is that the experiment will succeed and that the magnet will be found to do the

page 111

extra work.

Nov. 21st '79

Since writing the preceding notes I have extended my experiments on the newly discovered phenomenon and the numerical results of my work are given in my experiment notebook where the observations made are recorded. It was my idea before performing my later experiments, and I expressed my view to Dr. Nichols[12] at least before making the trial, that with the apparatus used the effect on the Thomson galvanometer would be directly proportional to the product of the strength of the magnetic field

page 112

and the strength of the current through the gold leaf.

That this is the law is made extremely probable by the experiments made Nov. 12th and recorded in the notebook above referred to.

I had formed and expressed moreover the further opinion that the new phenomenon would not with our instruments appear when a strip of moderately thick copper was substituted for the gold leaf.

Prof. Rowland however thought the copper would serve quite as well as the gold leaf.

He happened to come in just as I was about to try the strip of copper which was of dimensions somewhat similar to those of the gold leaf except that the

page 113

thickness was perhaps 1/4 mm.

Immediately before making the trial which he witnessed, I told Prof. Rowland I thought the copper would not show the effect.

He declared his opinion that it would show it and further remarked that we would now see who had the right idea of the phenomenon. Upon making the trial he saw and at once admitted that I was right.[13]

In this experiment with the copper the magnetic field, as well as the current through the strip, was considerably stronger than had been the case when the measurements with the gold leaf were made as mentioned above. By Prof. Rowland's advice

[page 114]

I have written a short abstract of what has been recorded in the preceding pages, adding thereto the numerical results arrived at, and this article will, I presume, appear in the next number of the Mathematical Journal.[14]

Nov. 25th '79

On pages 102 and 103 I have expressed my opinion that there was no advantage to be gained by supposing the current of electricity to flow from the negative to the positive pole. Prof. Rowland however was inclined to think there would be an advantage, or at least a difference [in] conception of the direction of the newly discovered effect, by assuming that the current was from negative

page 115

to positive. Upon further consideration I have come to the same conclusion. I have said above that reversing our conception of the current in the gold leaf made it necessary to revse our conception of the current through the magnet circuit, and consequently these two changes would neutralize each other. I forgot however that we must also change our ideas of the course of the current in the Thomson galvanometer. It seems then that if we conceive of the current as flowing from the zinc pole of the battery through the gold leaf to the carbon pole of the battery, we find that the new phenomena are in accordance with the supposition that two parallel currents attract each other. I tried yesterday the

[page 116]

experiment of placing between the poles of the magnet a strip of gold leaf about 1 cm. wide lying in a horizontal position, flat side up and testing it for a change of relative potential of points on opposite edges of the strip when the magnet was put in operation. No effect was discerned beyond that ordinarily caused by the direct action of the magnet and its circuit on the galvanometer needle.

The current in [the] gold leaf and the strength of the magnetic field were in this case probably not so large as in some previous experiments, but were sufficient to have shown a marked effect had the gold leaf strip been placed in the usual position.

page 117

Dec. 15th '79

Since making my last entry in this book I have been at work quite steadily upon my experiments but have not reached any very important results as yet. I made a silver strip by depositing the metal on glass. This showed the same effect as the gold leaf though no absolute measurements have as yet been [made?] with the silver. I have a strip of tin foil prepared but have not yet tried this. Prof. Rowland wants me to try iron and thinks the effect in this case will be reversed.

I referred some time ago to an experiment I wanted to make with a cylinder. Prof. Rowland did not approve of this experiment at first but finally admitted

[page 118]

it would be worth trying. He suggested using instead of the cylinder a disk of metal with one pole at the centre. The radiating currents would of course be affected by the magnetic force. This was an obvious improvement over the cylinder and I adopted his plan.

Neither of us saw at first though Prof. R. has since pointed out that any increase of resistance would be very small indeed. Thus where I had found a transverse electromotive force equal to about 1/3000 of the direct electromotive force in the strip, we could look for a change of resistance equal to the square of 1/3000 or 1/9000000. This slight change is of course

page 119

very difficult to detect and I doubt whether it can be discovered with our instruments even if it exists which I somewhat doubt.

I have heretofore expressed my opinion that the magnet would be found to do the extra work, but here too the change of resistance (in the magnet circuit) would be extremely small; too small probably to detect even if it exists. At present however I see no absurdity in the way of expecting continuous work from a magnet. It is quite evident that a magnet does work while it is attracting any body toward itself. The difficulty appears to be that ordinarily in removing the attracted body just as much work has to be as it were returned to the magnet as has been done

[page 120]

by the magnet in attracting the body. That is, the body attracted has a Potential with respect to the magnet. I conceive the magnet to act like a stretched spring which is capable of doing work and losing energy by contraction but which recovers the same amount of energy when it is again stretched.

If now we could allow a body to approach a magnet in such a way as to be attracted during the approach and then remove the body by some path in which it was no longer affected by the magnet, it seems to me that we could have the magnet doing work and losing energy. Obviously this is impossible in ordinary cases. But suppose now that the magnet

[page 121]

acts upon the electric currents radiating from the center of the disk mentioned above and continues to act upon them until they reach the ring surrounding the disk. Beyond this there seems to be no reason for attributing to the magnet any considerable action on the current. Now if in this case a new electromotive force is set up causing a current around the disk at right angles to the original radiating currents, these radiating currents remaining unchanged meanwhile, it seems to me that in this particular part of the circuit the magnet does work and I do no see how the energy thus lost to the magnet can be made up in any other part of the circuit.

A week or two ago I expressed those views to Dr. Nichols who him

[page 122]

self had formed somewhat similar ones, though hardly so definite as mine. He immediately set about trying to find some way of making a magnet do work. A few days afterward he suggested the experiment of allowing a magnet to attract a stream of iron filings which after being attracted to the vicinity of the magnet were to be dissolved or otherwise changed in such a way as to lose their magnetic property. I suggested, though, he may have thought of it before, that in this case there might be a retardation or enfeeblement of the chemical reactions owing to the influence of the magnet on the filings. It appears that this must be true or that the magnet must be able to do con

[page 123]

tinuous work. I have been myself thinking of attacking the problem in a different way.

It seems to me that a magnet ought to do work when as in Faraday's experiment (Ganot p. 712)[15] a part of an electric circuit revolves about a magnet.

The difficulty is that if the magnet does do work in this case it ought to have been discovered long ago. There may be some consideration which I have overlooked and which will show the absurdity of my ideas at once. Nichols and I are thinking of making some experiments on the thing some time, in his way or mine or both, unless someone shows us our folly before we have a chance to test out theories practically.

NOTE, added by transcriber: There were no more entries after the above although 3 more pages were available before the end of the notebook.

ACKNOWLEDGEMENTS

I gratefully acknowledge the encouragement and helpful comments made to me during the preparation of this manuscript by Professors Gerald Holton and Edward M. Purcell of the Harvard Physics Department. I am indebted to Ms. Julia Morgan of the Ferdinand Hamburger Jr. Archives of the Johns Hopkins University for information on the identities of persons mentioned by Edwin H. Hall in his Notebook of Physics. The above pages were transcribed by permission of the Houghton Library, Harvard University.

NOTES

1. Amer. Jour. Math. 2:287 (1879). The exact date when this publication reached the hands of readers is not clear since it was in the "September" issue, but Hall's paper and appended note are dated November 19 and 22, respectively.
2. Phil. Mag.(Series 5) 9:225 (1880). This was in the March issue and bore the notation "From a separate impression from the 'American Journal of Mathematics' 1879, communicated by the Author."
3. On the new action of magnetism on a permanent electric current, Amer. Jour. Sci. 20:161 (1880) and Phil. Mag.(Series 5) 10:301 (1880).
4. The notebook described here is the only one presently known to exist from among those used by Hall during his years at the Johns Hopkins University. It is clear, however, from references made by Hall in this notebook that he did keep at least one other "Experiment Book".
5. J. Clerk Maxwell Treatise on Electricity and Magnetism was published for the first time in 1873 by the Clarendon Press of Oxford University.
6. Charles Sheldon Hastings, Associate in Physics at the Johns Hopkins University 1876-1883.
7. Bunsen's cells were zinc carbon batteries, invented by Robert Wilhelm Bunsen in 1843. These double fluid cells (sulfuric and nitric acids) yield an emf of 1.9 volts.
8. Spencer Hedden Freeman was a fellow graduate student of Hall.
9. William White Jacques received his Ph.D. in physics at the Johns Hopkins University in 1879 and was a "Fellow by Courtesy" in the Physics Department during the academic year 1879-80.
10. Brown Ayres was a Fellow in Mathematics, 1879-80, who also studied physics with H. A. Rowland.
11. While working in Helmholtz' laboratory in Berlin during the year 1875-6 H. A. Rowland successfully demonstrated the magnetic effect of a rotating charged disk. For further discussion of this work see J. D. Miller, op. cit. pp. 114-135.
12. Edward Leamington Nichols was a Fellow in the Physics Department in 1879-81. Later, 1887-1919, he taught at Cornell University and was a founding editor of The Physical Review.
13. The significance of the thickness of the metallic strip is now understood with the recognition that a crucial element in the Hall Effect is the current density rather than the total current.
14. See Note 1.
15. "Ganot" is Hall's shortened version of the title Elementary Treatise on Physics Experimental and Applied: For the Use of Colleges and Schools, translated and edited from Ganot's

Elements de Physique by E. Atkinson and published in New York by William Wood and Company. This work went through many editions in French and English and was widely used as a college text in the United States. Presumably the Faraday experiment referred to by Hall is one in which a metallic conductor shaped in a double loop and carrying a current can be mounted in such a way that it rotates in the presence of a suitably oriented magnetic field. Such an apparatus is pictured and described on page 817 of the 1886 edition of Ganot's Physics.

Diamagnetismus der Metalle.

Von **L. Landau**, zurzeit in Cambridge (England).

(Eingegangen am 25. Juli 1930.)

Es wird gezeigt, daß schon freie Elektronen in der Quantentheorie, außer dem Spin-Paramagnetismus, einen von den Bahnen herrührenden, von Null verschiedenen Diamagnetismus haben, welcher in der Teilendlichkeit der Elektronenbahnen im Magnetfeld seinen Ursprung hat. Einige weitere mögliche Folgerungen dieser Bahnenendlichkeit werden angedeutet.

§ 1. Es wurde bis jetzt mehr oder weniger stillschweigend angenommen, daß die magnetischen Eigenschaften der Elektronen außer dem Spin ausschließlich von der Bindung der Elektronen in Atomen herrühren. Für freie Elektronen übernahm man für den Bahneffekt das klassische Nullresultat mit der Begründung, daß auch das Fermische Integral von der entsprechenden Hamiltonfunktion wie das Boltzmannsche vom magnetischen Felde unabhängig ist. Dabei wird aber eine Quantenerscheinung unberücksichtigt gelassen. Bei Vorhandensein eines Magnetfeldes wird nämlich die Elektronenbewegung in der zum Felde senkrechten Ebene finit. Das führt notwendigerweise zu einer Teildiskretheit (entsprechend der Bewegung in der genannten Ebene) der Eigenwerte des Systems, was, wie im folgenden gezeigt wird, zu einem von Null verschiedenen Bahnenmagnetismus Anlaß gibt.

Die Hamiltonfunktion eines freien Elektrons im Magnetfeld schreibt sich, wie bekannt, in der Form

$$E = \frac{m v_1^2}{2} + \frac{m v_2^2}{2} + \frac{m v_3^2}{2}, \tag{1}$$

wo

$$v_1 = \frac{1}{m}\left(p_1 - \frac{eH}{2c} y\right), \qquad v_2 = \frac{1}{m}\left(p_2 + \frac{eH}{2c} x\right), \qquad v_3 = \frac{1}{m} p_3, \tag{2}$$

die Geschwindigkeiten des Systems sind (H ist der Absolutwert des in die Richtung der z-Achse gerichteten Magnetfeldes). Die Bewegung in der Richtung des Feldes ist vom Felde und anderen Bewegungskomponenten unabhängig und kann abgesondert werden, indem man einfach p_3 gleich einer Konstanten setzt, was der Schrödingerfunktion

$$\psi(x, y, z) = f(x, y)\, e^{\frac{i}{h} p_3 z} \tag{3}$$

entspricht. Die Energiewerte des Systems werden sich dann als Summe zweier unabhängiger Glieder darstellen. Anstatt nun die entsprechende

Schrödingergleichung für die xy-Bewegung zu lösen, können wir zur Aufstellung der Energiewerte eine künstliche Methode benutzen, indem wir die Vertauschungsrelationen der Geschwindigkeitskomponenten v_1 und v_2 aufstellen. Aus (2) ergibt sich unmittelbar:

$$[v_1 v_2] = v_1 v_2 - v_2 v_1 = \frac{h}{i} \frac{eH}{cm^2}, \tag{4}$$

da bekannterweise $[xy] = [p_1 p_2] = 0$, $[p_1 x] = [p_2 y] = h/i$ ist. Die Konstanz der rechten Seite von (4) erinnert an die gewöhnlichen p, q-Vertauschungsrelationen. Um zu diesem Falle überzugehen, können wir nun einen Augenblick die Koordinaten P und Q mittels

$$v_1 = \frac{P}{\sqrt{m}}, \quad v_2 = \frac{eH}{cm\sqrt{m}} Q \tag{5}$$

einführen. Die Vertauschungsrelation geht dann in die gewöhnliche $[PQ] = h/i$ über. Was die Energie betrifft, so schreibt sie sich nun in der Form

$$E = \frac{P^2 + \left(\frac{eH}{mc}\right)^2 Q^2}{2}. \tag{6}$$

Das ist aber nichts anderes als die Hamiltonfunktion eines linearen Oszillators mit der Masse m und der Frequenz $\omega = eH/mc$. Die Eigenwerte eines solchen Systems sind, wie bekannt, gleich

$$E = \left(n + \frac{1}{2}\right) h\omega = \left(n + \frac{1}{2}\right) \frac{eh}{mc} H, \tag{7}$$

wo n alle positiven ganzzahligen Werte annehmen kann. Zusammen mit der z-Bewegung ergibt das

$$E = \left(n + \frac{1}{2}\right) \frac{eh}{mc} H + \frac{p_3}{2m}, \tag{8}$$

als Eigenwerte der Translationsbewegung des Elektrons.

In einfacher Weise können auch die Eigenfunktionen bestimmt werden. Zu diesem Zwecke eliminieren wir aus den Geschwindigkeitsoperatoren (und somit auch aus dem Energieoperator) eine der Koordinaten, beispielsweise ψ, indem wir

$$\psi = e^{-\frac{ieH}{2hc} xy} \chi \tag{9}$$

setzen. Das ergibt

$$\left.\begin{aligned} v_1 \psi &= \frac{h}{i} \frac{\partial \psi}{\partial x} - \frac{eH}{2c} y \psi = e^{-\frac{ieH}{2hc} xy} \left(\frac{h}{i} \frac{\partial \chi}{\partial x} - \frac{eH}{c} y \chi\right), \\ v_2 \psi &= \frac{h}{i} \frac{\partial \psi}{\partial y} + \frac{eH}{2c} x \psi = e^{-\frac{ieH}{2hc} xy} \frac{h}{i} \frac{\partial \chi}{\partial y}. \end{aligned}\right\} \tag{10}$$

Dementsprechend schreibt sich die Schrödingergleichung:

$$\left\{\left(\frac{h}{i}\frac{\partial}{\partial y}\right)^2 + \left(\frac{h}{i}\frac{\partial}{\partial x} - \frac{eH}{hc}\,y\right)^2 - 2\,mE\right\}\chi = 0. \qquad (11)$$

Diese Gleichung enthält x nicht explizite; somit können ihre Lösungen in der exponentiellen Form

$$\chi = e^{\frac{i}{h}\sigma x}\,\varphi(y) \qquad (12)$$

geschrieben werden, wobei σ eine Konstante ist und φ nicht mehr von x abhängt. Einsetzen von (12) in (11) ergibt ohne weiteres für φ eine Oszillatorgleichung

$$\frac{d^2\varphi}{dy^2} + \frac{2m}{h^2}\left[E - \frac{m}{2}\left(\frac{eH}{mc}\right)^2\left(y - \frac{c}{eH}\sigma\right)^2\right]\varphi = 0, \qquad (13)$$

was auch nach dem Vorhergehenden wohl zu erwarten war. Der „Ruhepunkt" dieses Oszillators befindet sich im Punkt $\eta = c\sigma/eH$. Somit erhalten wir endgültig für die vollständige Eigenfunktion des Systems

$$\psi = e^{\frac{i}{h}\left(p_3 z + \sigma x - \frac{eH}{2c}xy\right)}\,\varphi_n\left[\sqrt{\frac{eH}{hc}}\left(y - \frac{c}{eH}\sigma\right)\right], \qquad (14)$$

wobei φ_n die Eigenfunktionen der Gleichung

$$\frac{d^2\varphi_n}{du^2} + (2\,n + 1 - u^2)\,\varphi_n = 0 \qquad (15)$$

bezeichnet.

Die Größe σ geht in die Eigenwerte nicht ein. Da sie beliebige Werte annehmen kann, so ist unser Problem noch in kontinuierlicher Weise entartet. Um die Dichte der Eigenwerte zu bestimmen, ersetzen wir, wie üblich, den unendlichen Raum durch ein endliches Gefäß mit den Lineardimensionen A, B und C in den x-, y- und z-Richtungen. In der z-Richtung ist die Zahl der möglichen p_3-Werte im Intervall Δp, wie bekannt, gleich

$$R_{\Delta p} = \frac{C}{2\pi h}\Delta p. \qquad (16)$$

In ganz analoger Weise erhalten wir für die x-Richtung

$$R_{\Delta\lambda} = \frac{A}{2\pi h}\cdot\Delta\sigma. \qquad (17)$$

In der y-Richtung müssen wir fordern, daß die Bahn im Kasten immer in genügender Entfernung von den Wänden liegt. Dann brauchen wir wegen des schnellen Abklingens von φ_n mit der Entfernung den Einfluß der „y"-Wände nicht zu berücksichtigen. Da die Zahl der an die Wände stoßenden Bahnen bei genügend großen Gefäßen evidenterweise als klein betrachtet werden kann, so können wir annehmen, daß diese Forderung uns praktisch alle existierenden Bahnen ergibt. Wegen der großen Gefäßdimensionen können wir dabei auch den Radius der Bahn vernachlässigen und einfach schreiben:

$$0 < \frac{c}{eH}\sigma < B$$

oder

$$0 < \sigma < \frac{eB}{c}H. \tag{18}$$

Wollen wir nun die gesamte Zahl der der gegebenen nicht entarteten Quantenzahl n entsprechenden Eigenwerte erhalten, so haben wir in (17) $\Delta\sigma = \frac{eB}{c}H$ einzusetzen. Das ergibt

$$R_n = \frac{eH}{2\pi hc}AB = \frac{eH}{2\pi hc}S,$$

wo S die Fläche der Kastenseite ist. Zusammen haben wir

$$R_{\Delta p, n} = R_{\Delta p}R_n = \frac{eH}{4\pi^2 h^2 c}V\Delta p, \tag{19}$$

also, wie zu erwarten war, dem Volumen proportional. Wie leicht nachzurechnen ist, geht (19) beim Grenzübergang $H \longrightarrow 0$ in die gewöhnliche Eigenwerteverteilung der freien Bewegung über. Mit dem Spin zusammen haben wir:

$$E' = E \pm \frac{eh}{2mc}H, \tag{20}$$

das heißt

$$E = \frac{ehH}{mc}n + \frac{p_3^2}{2m}, \tag{21}$$

wobei jedem $n > 0$ die doppelte Entartung

$$R_{n,\Delta p} = \frac{eH}{2\pi^2 h^2 c}V\Delta p \tag{22a}$$

entspricht, und bei $n = 0$

$$R_{0,\Delta p} = \frac{eH}{4\pi^2 h^2 c}V\Delta p \tag{22b}$$

ist.

§ 2. Um die magnetischen Eigenschaften des Körpers zu erhalten, brauchen wir, wie bekannt, nur die Summe

$$\Omega = -kT\sum \lg\left(1 + e^{\frac{\omega - E}{kT}}\right) \tag{23}$$

über alle Energiewerte zu ermitteln. ω bezeichnet dabei das sogenannte chemische Potential. Die Teilchenzahl N ist mit ω durch die Beziehung

$$N = -\frac{\partial \Omega}{\partial \omega}, \tag{24}$$

und das magnetische Moment durch

$$M = -\frac{\partial \Omega}{\partial H} \tag{25}$$

verknüpft.

In unserem Falle haben wir einen kontinuierlichen und einen diskreten Parameter, so daß die Summe (23) sich als Summe von Integralen darstellen läßt. Dabei werden wir, um die Effekte klarer zu trennen, von den Bahnenenergien (8) ausgehen und den Spin zunächst nur in der Multiplizität berücksichtigen. Setzen wir

$$\frac{eH}{hmc} = \mu, \tag{26}$$

so ist

$$\Omega = -kT \sum_{n=0}^{\infty} \int \lg \left[1 + e^{\frac{\omega - (n + \frac{1}{2}) \mu H}{kT} - \frac{p_3^2}{2mkT}} \right] \frac{eH}{2\pi^2 h^2 c} V \, d p_3. \tag{27}$$

Bezeichnen wir nun der Kürze wegen

$$-kT \int \lg \left(1 + e^{\frac{\omega}{kT} - \frac{p_3^2}{2mkT}} \right) \frac{m}{2\pi^2 h} d p_3 = f(\omega), \tag{28}$$

so nimmt Ω die Form

$$\Omega = \mu H \sum_{0}^{\infty} f[\omega - (n + \tfrac{1}{2}) \mu H] \tag{29}$$

an. Zur Ermittlung dieser Summe können wir die bekannte Reihenentwicklung

$$\sum_{a}^{b} f(x + \tfrac{1}{2}) = \int_{a}^{b} f(x) \, dx - \tfrac{1}{24} | f'(x) |_{a}^{b} \cdots \tag{30}$$

anwenden. Ihre Zulässigkeit fordert im allgemeinen

$$\frac{f_{x+1} - f_x}{f_x} \ll 1. \tag{31}$$

In unserem Falle entspricht das, wie leicht einzusehen ist,

$$\mu H \ll kT. \tag{32}$$

Diese Bedingung ist bei sehr niedrigen Temperaturen und in starken Feldern nicht mehr erfüllt. Dieser letzte Fall würde deswegen zu einer komplizierten, nicht mehr linearen Abhängigkeit des magnetischen Momentes von H führen, welche eine sehr starke Periodizität im Felde haben würde. Wegen dieser Periodizität dürfte es aber kaum möglich sein, diese Erscheinung experimentell zu beobachten, da wegen der Inhomogenität des vorhandenen Feldes immer eine Mittelung auftreten wird. Mitteln wir aber die Reihe (29) über ein Intervall ΔH, so wird die Bedingung (31) wieder erfüllt, wenn im „gefährlichen" Teil neben $\omega - (n + \frac{1}{2}) \mu H = 0$ die Änderung des Arguments wesentlich größer als die Differenz zweier einander folgender Argumente wird, d. h.

$$n\mu \Delta H \gg \mu H,$$

$$\omega \frac{\Delta H}{H} \gg \mu H,$$

woraus

$$\frac{\Delta H}{H} \gg \frac{\mu H}{\omega}. \tag{33}$$

Sogar bei den stärksten jetzt möglichen Feldern ($H = 3 \cdot 10^5$ Gauß) ergibt die rechte Seite bei $\omega = 3$ Volt nur 0,1%.

Wenden wir nur die Summationsformel (30) explizite an, so ergibt sich:

$$\Omega = \mu H \int_0^\infty f(\omega - n\mu H)\, dn + \frac{1}{24} \mu^2 H^2 \left| \frac{\partial f(\omega - n\mu H)}{\partial \omega} \right|_0^\infty$$

$$= \int_{-\infty}^{\omega} f(x)\, dx - \frac{\mu^2 H^2}{24} \frac{\partial}{\partial \omega} f(\omega) \tag{34}$$

$[f(\infty) = 0]$. Das erste Glied dieser Summe hängt vom Magnetfeld nicht ab. Es stellt den Wert der Summe im feldfreien Zustande dar, so daß wir an Stelle von (34)

$$\Omega = \Omega_0 - \frac{\mu^2 H^2}{24} \frac{\partial^2 \Omega_0}{\partial \omega^2}$$

schreiben können. Daraus folgt:

$$M = -\frac{\partial \Omega}{\partial n} = \frac{\mu^2}{12} \frac{\partial^2 \Omega_0}{\partial \omega^2} H. \tag{35}$$

Setzten wir nun:

$$\frac{\partial \Omega}{\partial \omega} = -N, \quad \omega = \frac{\partial F}{\partial N},$$

wo $F = \Omega - \omega \frac{\partial \Omega}{\partial \omega}$ die freie Energie des Systems ist, so geht (35) über in

$$M = -\frac{\mu^2 H}{12 \frac{\partial \omega}{\partial N}} = -\frac{\mu^2 H}{12 \frac{\partial^2 F}{\partial N^2}}. \tag{36}$$

Wir haben also wirklich einen Diamagnetismus, welcher exakt gleich einem Drittel des Paulischen* Spinparamagnetismus ist, für welchen wir bekannterweise

$$\Omega = \frac{1}{2}\Omega_0\left(\omega + \frac{\mu H}{2}\right) + \frac{1}{2}\Omega_0\left(\omega - \frac{\mu H}{2}\right) = \Omega_0 + \frac{\mu^2 H^2}{8}\frac{\partial^2 \Omega_0}{\partial \omega^2} + \cdots \quad (37)$$

haben. Insgesamt sind also freie Elektronen doch paramagnetisch.

Befinden sich die Elektronen im periodischen Felde eines Gitters, so kann bekanntlich** ihre Bewegung in gewissem Sinne doch als frei betrachtet werden. Der prinzipielle Charakter der Wirkung des Magnetfeldes bleibt deswegen ungeändert, obgleich die obige Rechnung natürlich nicht mehr quantitativ anwendbar ist. Insbesondere ändert sich das Verhältnis vom Para- und Diamagnetismus, und es ist wohl möglich, daß in gewissen Fällen das letzte auch das erste übertreffen kann, so daß wir eine diamagnetische Substanz wie Wismut erhalten. Das ist aber wohl nur bei stärkerem Gittereinfluß möglich, so daß eine quantitative Theorie dieser Erscheinung kaum möglich sein dürfte. Ein anderer Einfluß der Wechselwirkung besteht darin, daß der Diamagnetismus seine Symmetrie verliert und nunmehr in verschiedener Richtung verschieden wird, eine Eigenschaft, die diese Art des Diamagnetismus vom gewöhnlichen Atomdiamagnetismus sowie vom notwendig symmetrischen Spinparamagnetismus unterscheidet.

Eine analoge Erscheinung kann auch bei nicht leitenden Substanzen, und zwar *paramagnetischen* stattfinden, wo wir ja auch ein kontinuierliches Eigenwertspektrum haben. Auch hier bekommen wir diskrete Eigenwerte im Magnetfeld und infolgedessen einen Diamagnetismus. Dieser Diamagnetismus ist zwar klein gegen den vorhandenen Paramagnetismus, unterscheidet sich aber von ihm durch seine Asymmetrie, so daß er vielleicht den Hauptgrund (ein anderer Grund ist die sogenannte magnetische oder relativistische Wechselwirkung der Spins miteinander) der beobachteten Asymmetrie in paramagnetischen Kristallen bildet. Es ist deswegen von Interesse, die Größenordnung des Effekts abzuschätzen. Das geschieht am einfachsten aus Dimensionsgründen. Die Suszeptibilität ist erstens proportional mit $(e/c)^2$, da die Wirkung des Magnetfeldes immer durch eH/c eingeführt wird. Die Elektronenmasse m tritt in diesem Falle in die Rechnungen nicht explizite ein. Ihre Rolle spielt sie in dem Austauschintegral, welches die Austauscherscheinungen im Gitter charakterisiert.

* W. Pauli, ZS. f. Phys. **41**, 81, 1927.

** F. Bloch, ebenda **52**, 555, 1928.

Außerdem können nur noch h und die Dichte N/V eintreten. Das führt eindeutig zum Ausdruck

$$\chi \sim \frac{e^2}{h^2 c^2} \left(\frac{V}{N}\right)^{1/3} J. \tag{38}$$

Das Austauschintegral J bestimmt, wie bekannt, die Curietemperatur, wobei $k\Theta$ von der Größenordnung J ist, so daß wir an Stelle von (38)

$$\chi \sim \frac{e^2}{h^2 c^2} \left(\frac{V}{N}\right)^{1/3} k\Theta \tag{39}$$

schreiben können.

Ganz anders gestalten sich die Erscheinungen, wenn die äußeren Einwirkungen nicht periodischer Natur sind. Solche Einwirkungen zerstören die Richtungsentartung der Bewegung und somit, wenn sie nicht als klein angenommen werden können, die Möglichkeit eines Einflusses des Feldes der untersuchten Art. Dazu genügt, daß die diesen Einflüssen entsprechende „freie Weglänge" klein wird gegen den Durchmesser der Elektronenbahnen im Magnetfeld. Da dieser Durchmesser in gewöhnlichen Feldern von der Größenordnung eines Zehntelmillimeters ist, so können dazu schon sehr kleine Verunreinigungen oder sogar das Zerpulvern der Substanz genügen. Solche Änderungen der Suszeptibilität sind bei Wismut und für den ersten Fall bei einer ganzen Reihe von Substanzen nachgewiesen worden. Es wäre von großem Interesse, in diesen Fällen eine Änderung der Suszeptibilität mit dem Felde beobachten zu können, welche nach der angeführten Theorie beim Übergang von $r_H \gg \lambda$ (r_H Radius der Kreisbahn im Magnetfeld, λ die freie Weglänge bzw. Dimensionen der Kristalle) zu $r_H \lesssim \lambda$ stattfinden müßte.

Zum Schluß möchte ich noch die Vermutung aufstellen, daß die untersuchte Erscheinung auch den Kapitzaeffekt der linearen Widerstandsänderungen im Magnetfeld erklären dürfte. Für die Zulässigkeit der vorausgesetzten Näherung freier Elektronen im Magnetfeld ist dabei nicht notwendig, das r_H kleiner als die dem Gitter entsprechende freie Weglänge ist (was bei gewöhnlichen Temperaturen unmöglich wäre), weil die Wechselwirkung mit den Gitterschwingungen außer der Impulsabgabe auch Energieabgabe hervorruft. Es ist aber nach der vorhergehenden Bemerkung wohl notwendig, daß r_H wesentlich kleiner als die freie Weglänge der Gitterstörungen wird, was nach kurzen Rechnungen zur Beziehung

$$H \gg ec \frac{N}{V} R \tag{40}$$

führt, wo R den spezifischen Restwiderstand (in elektrostatischen Einheiten) des betreffenden Kristalls bezeichnet. Ist die Beziehung (40) nicht erfüllt, so ist die betrachtete Methode nicht anwendbar und man kann wohl einsehen, daß alle Einwirkungen des Feldes unbedingt quadratisch werden müssen. Das Feld (40) steht in gutem Einklang mit dem kritischen Felde der Kapitzaschen Versuche, was wohl als eine Stütze der Theorie angesehen werden könnte. Eine quantitative Ausbildung der Theorie ist mir bis jetzt nicht gelungen.

Ich möchte auch an dieser Stelle Herrn P. Kapitza für Diskussionen über Ergebnisse der Versuche und Mitteilung einiger noch nicht veröffentlichter Daten herzlichst danken.

Cambridge, Cavendish Laboratory, Mai 1930.

Der Diamagnetismus von freien Elektronen.

Von **E. Teller** in Leipzig.

(Eingegangen am 15. November 1930.)

Es wird das diamagnetische Bahnmoment von freien, in einen Kasten gesperrten Elektronen aus den Strömen berechnet, welche unter der Wirkung eines Magnetfeldes entlang den Kastenwänden fließen. Das Ergebnis ist dasselbe wie bei L. Landau, der den Diamagnetismus mit Hilfe der Zustandssumme unter Vernachlässigung der Randelektronen berechnete.

Bekanntlich ist der Diamagnetismus von freien, in einen Kasten gesperrten Elektronen nach der klassischen Theorie Null. In einer kürzlich erschienenen Arbeit hat nun L. Landau* gezeigt, daß das diamagnetische Bahnmoment nach der Quantenmechanik nicht verschwindet. Landau führte seine Rechnung mit Hilfe der Zustandssumme aus. Er vernachlässigte dabei den Beitrag derjenigen Elektronen, die den Kastenwänden nahe kommen. Obzwar diese Vernachlässigung für den Fall, daß der Kasten genügend groß ist, durchaus gerechtfertigt erscheint, war es doch wünschenswert, dasselbe Problem in einer solchen Weise zu behandeln, welche die Rolle der Randelektronen deutlich hervortreten läßt, da bei den üblichen Herleitungen des klassischen Ergebnisses die Randelektronen eine wesentliche Rolle spielen. Dies wird erreicht, indem man das magnetische Moment aus den Strömen berechnet, die unter dem Einfluß eines Magnetfeldes den Kastenwänden entlang fließen. Während man nämlich den Beitrag, den die Randelektronen zur Zustandssumme liefern, wegen ihrer geringen Anzahl vernachlässigen darf, ist ihr Einfluß auf den Strom ganz wesentlich, da der Strom überall verschwinden wird, außer in der Nähe der Kastenwände.

Es soll also das magnetische Moment von freien Elektronen in einem homogenen Magnetfeld unter folgenden Annahmen berechnet werden: Die Elektronen befinden sich in einem Kasten, der durch zu dem Magnetfeld senkrechte und parallele Wände begrenzt ist. In großer Entfernung von der Wand sollen sich die Elektronen bis auf das Magnetfeld kräftefrei bewegen. Wir werden also das elektrostatische Potential so ansetzen, daß es sich im Innern des Kastens in großer Entfernung von der Wand einer Konstanten nähert und unendlich wird, wenn man sich der Wand des Kastens

* L. Landau, ZS. f. Phys. **64**, 629, 1930.

annähert. Die Abmessungen des Kastens sollen groß sein im Verhältnis zu den „Kreisbahnen", welche die Elektronen im Magnetfeld beschreiben. Es wird von der Wechselwirkung der Elektronen sowie von der Wirkung des Spins abgesehen. Es sei übrigens bemerkt, daß die ganze Überlegung ohne alle Schwierigkeit auch für die Diracgleichungen durchgeführt werden kann. Es wird ferner angenommen, daß die Anzahl der Elektronen, die sich in einem bestimmten Zustande befinden, nur von der a priori-Wahrscheinlichkeit dieses Zustandes, vom Verhältnis der Energie zu kT und von der Dichte der Elektronen* abhängt. Man nimmt also Temperaturgleichgewicht an; über die spezielle Art der anzuwendenden Statistik braucht nichts vorausgesetzt zu werden.

Das Prinzip der Rechnung ist, daß man erst den elektrischen Strom und dann aus diesem das magnetische Moment berechnet. Man sieht sofort, daß in großer Entfernung von der Wand die Ströme sich aufheben müssen. Nur entlang der zum Magnetfeld parallelen Wand bleibt ein Strom übrig. Wir zählen ihn positiv, wenn er ein paramagnetisches Moment erzeugt, d. h. wenn Stromrichtung, Innennormale der Wand und Magnetfeld ein Rechtssystem bilden. Das gesamte magnetische Moment der Elektronen wird dann gleich dem Gesamtstrom multipliziert mit dem zum magnetischen Feld senkrechten (konstanten) Querschnitt des Kastens. Dividiert man noch mit der Zahl der Elektronen, so erhält man das mittlere Moment pro Elektron. Dasselbe kann man im Falle, daß der Kasten genügend groß ist, auch so erhalten, daß man den in einer zum Magnetfeld senkrechten Schicht von Einheitsdicke an der Wand entlang fließenden Strom (wir wollen diese Größe Stromdichte nennen) mit der Dichte der Elektronen dividiert. Dabei muß man für die Dichte der Elektronen den Wert nehmen, dem sie sich bei genügend großer Entfernung von der Wand annähert. Aus dem vorhergehenden folgt, daß unser Problem, falls der Kasten genügend groß ist (vgl. mit der Dimension der „Bahn" der Elektronen) und falls auch die Krümmung der Wände genügend klein bleibt, durch ein einfacheres ersetzbar ist. Man betrachte nämlich eine unendlich ausgedehnte ebene Wand**, die den Raum in zwei Teile teilt.

---

* Die Dichte der Elektronen ist in großer Entfernung von der Wand zu nehmen, denn nur da nähert sie sich einem konstanten Wert.

** Wir werden dann später feststellen müssen, welche Zustände in diesem System a priori gleichwahrscheinlich sind. Solange wir einen geschlossenen endlichen Kasten haben, sind offenbar die Quantenzustände a priori gleichwahrscheinlich. Bei dem Übergang zum unendlich ausgedehnten System wird diese Bestimmung unsinnig, doch wird die Feststellung der gleichwahrscheinlichen Zustände sich leicht ergeben.

Es sei ein Potential gegeben, welches nur von der Entfernung von der Wand abhängt, und zwar sei es auf der einen Seite unendlich, auf der anderen Seite beliebig, mit der einzigen Beschränkung, daß es sich in genügender Entfernung von der Wand einem konstanten endlichen Werte nähert. Wir betrachten nur Elektronen, die sich in diesem Potential unter der Einwirkung eines zur Wand parallelen homogenen Magnetfeldes bewegen. Wir berechnen die Stromdichte entlang der Wand, dividieren durch die Dichte der Elektronen und erhalten das gesuchte magnetische Moment pro Elektron. Wir werden nach dieser Methode eine Formel für das magnetische Moment pro Elektron finden, welche mit der nach der Landauschen Methode (mit Hilfe der Zustandssumme) gewonnenen übereinstimmt. Unsere Methode trägt in keinem Punkte weiter als die Landausche, ist ihr sogar im Falle, daß die Wände schief zum Magnetfeld stehen, wesentlich unterlegen (diesen Fall lassen wir hier außer Betrachtung) und ihr einziger Vorteil dürfte der sein, daß sie vielleicht anschaulicher ist.

Die Bewegung der Elektronen im Magnetfeld wird gewöhnlich in Zylinderkoordinaten behandelt. Der vorhin gestellten Aufgabe ist das nicht angepaßt. Es ist aber leicht möglich, die Separation der Differentialgleichung auch in rechtwinkligen Koordinaten durchzuführen, solange das elektrostatische Potential von einer zum Magnetfeld senkrechten Richtung unabhängig ist. Wir legen die z-Achse in die Richtung des magnetischen Feldes, die y-Achse wird die Innennormale der Wand sein (dabei sei für die Stelle, wo das Potential unendlich wird, $y = 0$) und die Richtung, in der wir den Strom positiv zählen, die x-Achse. Für das Vektorpotential nehmen wir an:

$$A_x = -Hy, \quad A_y = A_z = 0. \tag{1}$$

(Das Vektorpotential ist also divergenzfrei, und seine Rotation ist $H_z = H$, $H_x = H_y = 0$.) Die Schrödingergleichung lautet dann:

$$\frac{h^2}{8\pi^2 m} \Delta \psi + \frac{h\varepsilon}{2\pi i m c} H y \frac{\partial \psi}{\partial x} + \left(E - U(y) - \frac{\varepsilon^2}{2mc^2} H^2 y^2\right) \psi = 0. \tag{2}$$

Für $U(y)$ gilt $U(y) \longrightarrow \infty$, wenn $y \longrightarrow 0$ und $U(y) \longrightarrow 0$, wenn $y \longrightarrow \infty$. Die allgemeine Lösung von (2) hat die Form:

$$\psi = e^{\frac{2\pi i}{h} k z} e^{\frac{2\pi i}{h} w x} v(y). \tag{3}$$

(2) läßt sich also tatsächlich in rechtwinkligen Koordinaten separieren. In (3) können k und w beliebige reelle Werte annehmen.

Bis jetzt haben wir mit einem fest vorgegebenem Vektorpotential gerechnet. Die Rechnung läßt sich etwas übersichtlicher durchführen, wenn man für jeden Wert von w ein anderes Vektorpotential benutzt. Wir führen also eine Umeichung durch, indem wir

$$\psi = e^{\frac{2\pi i}{h} k z} e^{\frac{2\pi i}{h} w x} v(y) \quad \text{durch} \quad e^{\frac{2\pi i}{h} k z} v(y), \tag{4}$$

$$A_x = -H y \quad \text{durch} \quad -H\left(y - \frac{c}{\varepsilon H} w\right)$$

ersetzen*. Wir bekommen also für $v(y)$ die Gleichung

$$\frac{h^2}{8\pi^2 m}\frac{d^2 v}{d y^2} + \left[E - \frac{k^2}{2m} - U(y) - \frac{\varepsilon^2}{2 m c^2} H^2 \left(y - \frac{c}{\varepsilon H} w\right)^2\right] v = 0. \tag{5}$$

Im Falle, daß $U(y) = 0$ wird, bekommen wir die Gleichung eines harmonischen Oszillators, dessen Gleichgewichtslage durch den Wert von w bestimmt wird. Für späteren Gebrauch wollen wir die Lösung von (5) mit $U(y)$ überall gleich Null (d. h. die Wellenfunktion für ganz freie, nicht in einen Kasten gesperrte Elektronen) mit $v_0(y)$ bezeichnen. Es gilt $v \longrightarrow v_0$, wenn $y \longrightarrow \infty$. Man beachte, daß für negative y die Funktion v, aber nicht v_0 verschwindet.

Es soll nun das magnetische Moment eines Elektrons berechnet werden**. Zu diesem Zwecke ändern wir H um δH und berechnen die Änderung der Energie δE. Nach dem Störungsverfahren erhält man mit Rücksicht auf (4) und (5)

$$\begin{aligned} \delta E &= \delta H \int \frac{\varepsilon^2}{m c^2} H \left(y - \frac{c}{\varepsilon H} w\right)^2 v \bar{v}\, d y \\ &= -\delta H \frac{\varepsilon^2}{m c^2} \int A_x \left(y - \frac{c}{\varepsilon H} w\right) v \bar{v}\, d y. \end{aligned} \tag{6}$$

Die rechte Seite von (6) ist das mit $-\delta H$ multiplizierte magnetische Moment des Elektrons***. Es ist also

$$\frac{\partial E}{\partial H} = -M. \tag{7}$$

---

* Durch diese Umeichung erreicht man, daß die Ausdrücke von x unabhängig werden und bringt die wesentliche Rolle von w (die darin besteht, daß sie die mittlere Lage des Elektrons in der y-Richtung regelt) besser zum Ausdruck. Für ψ und A_x setzen wir von nun an immer diese umgeeichten Ausdrücke.

** Diese Rechnung wird hier nur angegeben, um die Analogie zur Berechnung des Stromes besser hervortreten zu lassen.

*** Das magnetische Moment ist dabei auf die Ebene $y = \frac{c}{\varepsilon H} w$ bezogen. Falls der vom Elektron herrührende Gesamtstrom verschwindet, ist es übrigens gleichgültig, auf welche Ebene man das Moment bezieht. Dies ist erfüllt, wenn man v durch v_0 ersetzen darf, wie es in der Anwendung geschehen wird.

Die partielle Differentiation ist dabei so zu verstehen, daß die beiden anderen kontinuierlichen Variablen, von denen E abhängt (nämlich k und w), konstant gehalten werden.

Wir wollen jetzt H festhalten und w (also im wesentlichen die Gleichgewichtslage) um δw variieren. Für δE bekommt man dann

$$\delta E = -\delta w \int \frac{\varepsilon}{m c} H \left(y - \frac{c}{\varepsilon H} w\right) v \bar{v} \, d y = \delta w \frac{\varepsilon}{m c} \int A_x v \bar{v} \, d y \tag{8}$$

und daraus erhalten wir*

$$\frac{\varepsilon}{c} \left(\frac{\partial E}{\partial w}\right) = -s_x, \tag{9}$$

wo s_x den von dem betrachteten Elektron herrührenden Strom in der x-Richtung (in elektromagnetischen Einheiten) bezeichnet. Bei der partiellen Differentiation sind diesmal H und k festzuhalten.

Es muß nun noch angegeben werden, wie im Temperaturgleichgewicht die Elektronen auf die verschiedenen Zustände verteilt sind. Der Zustand eines Elektrons wird durch k, w und n beschrieben. Dabei bedeutet n die um Eins vermehrte Knotenzahl der Funktion $v(y)$. Es ist aus der Definition von k und w (3) zu ersehen, daß die a priori-Wahrscheinlichkeit dessen, daß ein Elektron sich im Zustand k, w, n mit

$$n = n_0, \quad k_0 \leqq k \leqq k_0 + \delta k, \quad w_0 \leqq w \leqq w_0 + \delta w \tag{10}$$

befindet, durch den Ausdruck $\delta k \delta w$ gegeben wird. Die Wahrscheinlichkeit, ein Elektron in irgendeinem Zustand anzutreffen, hängt außer der a priori-Wahrscheinlichkeit noch vom Quotienten E/kT und von D ab, wo E [gemäß (5)] eine Funktion von k, w und n, und D die Dichte der Elektronen für große y bedeutet**. Die Abhängigkeit der Wahrscheinlichkeit von

* Van Vleck hat gezeigt, daß man diese Beziehung, die für das Folgende wesentlich ist, auch durch korrespondenzmäßige Betrachtungen herleiten kann. Er betrachtete dabei einen zylindrischen Kasten mit dem Radius R, dessen Achse parallel zum Magnetfeld gerichtet ist. Die Randelektronen werden dann um den ganzen Zylinder der Wand entlang herumlaufen. Die Bewegung der Elektronen ist also bedingt periodisch und man kann sie nach der alten Quantentheorie quanteln. Zwischen der Umlauffrequenz ν und der Energie E besteht die klassische Beziehung $\nu = \partial E / \partial J_\nu$. J_ν steht dabei mit w in der Beziehung $2 \pi R w = J_\nu$, wie das besonders deutlich aus der nicht umgeeichten Form (3) hervorgeht. Da ferner $-\varepsilon 2 \pi R \nu / c$ der Strom ist, gelangen wir wieder zu (9). Auf diesem Wege weitergehend kann man die ganzen folgenden Betrachtungen schon in der alten Quantentheorie durchführen. Ich bin Herrn Prof. van Vleck für die Erlaubnis, diese Bemerkung mitzuteilen, sowie für viele interessante Diskussionen über den Gegenstand der vorliegenden Arbeit zu Dank verpflichtet.

** Große y bedeuten hier, wie immer, solche Werte von y, für welche $U(y)$ schon Null gesetzt werden kann, und welche außerdem groß sind verglichen

E/kT und D ist in den verschiedenen Statistiken verschieden. Wir brauchen darüber keine Annahme zu machen und setzen die Wahrscheinlichkeit, daß ein Elektron sich im Zustand (10) befindet, für genügend kleine δk und δw

$$F\left(\frac{E(k_0, w_0, n_0, H)}{kT}, D\right)\delta k\,\delta w. \tag{11}$$

Für große Werte von w_0 nähern sich E und also auch F einer von w_0 unabhängigen Konstanten. Wir wollen diese mit E_0 bzw. mit F_0 bezeichnen. E_0 hängt also nur mehr von k_0, n_0 und H ab. Dabei gilt noch $F \longrightarrow 0$, wenn $E \longrightarrow \infty$, und ferner für große Werte von y_0:

$$\int\limits_{x=x_0}^{x_0+1} \int\limits_{y=y_0}^{y_0+1} \int\limits_{z=z_0}^{z_0+1} \iint\limits_{k,\,w=-\infty}^{\infty} \sum_{n=1}^{\infty} F\,\psi\,\overline{\psi}\,dk\,dw\,dx\,dy\,dz = D. \tag{12}$$

Hierbei muß, da y groß ist, auch w groß sein (sonst wird $\psi\overline{\psi}$ klein). Es kann dann F_0 für F und v_0 für v gesetzt werden, und da v_0 nur von $y - \frac{c}{\varepsilon H} w$ abhängt, liegt es nahe, mit der folgenden Substitution w durch ω zu ersetzen:

$$y - \frac{c}{\varepsilon H} w = \omega, \qquad dy\,dw = -\frac{\varepsilon H}{c}\,dy\,d\omega. \tag{13}$$

Man kann dann die Integration nach x, y und z ausführen, und indem man noch die Normierungsbedingung für v_0 in Betracht zieht, erhält man

$$-\frac{\varepsilon H}{c} \int\limits_{\omega=\infty}^{-\infty} \int\limits_{k=-\infty}^{\infty} \sum_{n=1}^{\infty} F_0\, v_0\, \overline{v}_0\, dk\, d\omega = \frac{\varepsilon H}{c} \int\limits_{k=-\infty}^{\infty} \sum_{n=1}^{\infty} F_0\, dk = D. \tag{14}$$

Wir können nun die Stromdichte S_x (also den Strom, der in einer zum Magnetfeld senkrechten Schicht von Einheitsdicke entlang der Wand fließt) berechnen. Da nun für große y die Ströme sich wegheben, genügt es, den Strom zwischen den zwei Ebenen $y = 0$ und $y = Y$ zu berechnen, wenn man nur Y genügend groß wählt. Da ψ von x unabhängig gemacht ist [Umeichung (4)], wird dieser Strom (in elektromagnetischen Einheiten) gegeben durch

$$S_x = -\frac{\varepsilon^2}{m c^2} \int\limits_{w=-\infty}^{\infty} \int\limits_{y=0}^{Y} \int\limits_{k=-\infty}^{\infty} \sum_{n=1}^{\infty} A_x\, v\, \overline{v}\, F\, dk\, dy\, dw. \tag{15}$$

Dieses Integral zerlegt man zweckmäßig in zwei Integrale.

$$S_x = S_1 + S_2 = \int\limits_{w=-\infty}^{w_1} + \int\limits_{w=w_1}^{\infty}, \tag{16}$$

---

mit der Dimension der Elektronenbahn, d. h. wo man v schon mit guter Näherung durch v_0 ersetzen kann.

wobei w_1 so gewählt ist, daß $\frac{c}{\varepsilon H} w_1$, sowie $Y - \frac{c}{\varepsilon H} w_1$ groß sind im Vergleich mit der Bahndimension und $U\left(\frac{c}{\varepsilon H} w_1\right)$ genügend klein wird. In S_1 begeht man nur einen kleinen Fehler*, wenn man für die obere Grenze von y statt Y unendlich setzt. Indem man die Integration über y ausführt, bekommt man mit Rücksicht auf (8) und (9)

$$S_1 = -\frac{\varepsilon}{c} \int\limits_{w=-\infty}^{w_1} \int\limits_{k=-\infty}^{\infty} \sum_{n=1}^{\infty} \frac{\partial E}{\partial w} F \, dk \, dw; \tag{17}$$

für die obere Grenze von w können wir hier noch statt w_1 ∞ setzen und begehen damit nur einen kleinen Fehler. Im Integral S_2 begeht man nur einen kleinen Fehler, wenn man statt v v_0 und statt F F_0 einführt. Wir eliminieren jetzt w durch die Substitution (13) und erhalten

$$S_2 = \frac{\varepsilon^3 H^2}{m c^3} \int\limits_{y=0}^{Y} \left[\int\limits_{\omega = y - \frac{c}{\varepsilon H} w_1}^{-\infty} \int\limits_{k=-\infty}^{\infty} \sum_{n=1}^{\infty} A_x v_0 \bar{v}_0 F_0 \, dk \, d\omega \right] dy. \tag{18}$$

Da der Klammerausdruck von y nur in den Integrationsgrenzen abhängt, liegt es nahe, nach y partiell zu integrieren

$$\left.\begin{aligned} S_2 = {} & \frac{\varepsilon^3 H^2}{m c^3} Y \int\limits_{\omega = Y - \frac{c}{\varepsilon H} w_1}^{-\infty} \int\limits_{k=-\infty}^{\infty} A_x v_0 \bar{v}_0 F_0 \, dk \, d\omega \\ & + \frac{\varepsilon^3 H^2}{m c^3} \int\limits_{y=0}^{Y} \int\limits_{k=-\infty}^{\infty} \sum_{n=1}^{\infty} y A_x v_0 \bar{v}_0 F_0 \, dk \, dy. \end{aligned}\right\} \tag{19}$$

Im ersten Summand von (19) begeht man nur einen kleinen Fehler, wenn man für die Integrationsgrenzen von ω ∞ und $-\infty$ setzt. Der erste Summand wird dann der zur Eigenfunktion v_0 gehörige Gesamtstrom (abgesehen von dem Faktor F_0) und dieser verschwindet aus Symmetriegründen. Im zweiten Summand ist das Argument von A_x und v_0 $y - \frac{c}{\varepsilon H} w_1$.

* In der folgenden Ableitung können die Größen, welche als klein bezeichnet und vernachlässigt werden, immer dadurch zum Verschwinden gebracht werden, daß man in naheliegender Weise zur Grenze übergeht (z. B. im vorliegenden Falle dadurch, daß man Y und $Y - \frac{c}{\varepsilon H} w_1$ gegen Unendlich gehen läßt).

Man sieht wieder, daß man in der Integrationsgrenze Y durch ∞ ersetzen darf, und so gelangen wir unter Benutzung von (6) und (7) zu

$$S_2 = -\frac{\varepsilon H}{c} \int\limits_{k=-\infty}^{\infty} \sum_{n=1}^{\infty} \frac{\partial E_0}{\partial H} F_0 \, d k. \qquad (20)$$

Wir definieren nun

$$\sigma = \frac{\varepsilon H}{c} \int\limits_{k=-\infty}^{\infty} \sum_{n=1}^{\infty} \int\limits_{\eta=\eta_0}^{\infty} F\left(\frac{E_0+\eta}{k T}, D\right) d \eta \, d k. \qquad (21)$$

Man erhält dann mit Berücksichtigung von (14), (17) und (20)

$$\frac{\partial \sigma}{\partial \eta_0} = -D,$$

$$\frac{\partial \sigma}{\partial H} = S_1, \qquad \frac{\partial \sigma}{\partial E_0} \frac{\partial E_0}{\partial H} = S_2,$$

$$\frac{d \sigma}{d H} = \frac{\partial \sigma}{\partial H} + \frac{\partial \sigma}{\partial E_0} \frac{\partial E_0}{\partial H} = S_1 + S_2 = S_x,$$

$$M_{\text{pro Elektron}} = \frac{S_x}{D} = -\frac{\frac{d\sigma}{dH}}{\frac{\partial \sigma}{\partial \eta_0}} \qquad \text{für } \eta_0 = 0,$$

was im wesentlichen mit den Formeln übereinstimmt, welche man aus den Landauschen Überlegungen erhält. Zu beachten ist, daß in der Definition von σ nur E_0 vorkommt, entsprechend der Tatsache, daß bei Landau der Betrag der Randelektronen zur Zustandssumme vernachlässigt wird. In unserer Überlegung sind die Randelektronen mitberücksichtigt. Es sind eigentlich keine Vernachlässigungen gemacht, da die vernachlässigten Größen durch einen Grenzübergang (in welchem wir nur solche Größen ändern, über die wir frei verfügen können, wie etwa Y oder w_1) zum Verschwinden gebracht werden können. Die wesentliche Annahme bei Landau, daß der Kasten genügend groß ist, steht bei uns da, wo wir das Problem durch die Bewegung der Elektronen an einer ebenen Wand ersetzten (und den Kasten dadurch unendlich groß machten). Es folgt also, da wir zum selben Resultat gekommen sind wie Landau, daß die Vernachlässigung des Beitrages der Randelektronen in der Zustandssumme berechtigt ist (was übrigens schon auf anderem Wege gezeigt worden ist). Man bemerkt noch, daß S_2 der Strom ist, welcher erzeugt werden würde, wenn nicht der Ort der Elektronen, sondern der Mittelpunkt ihrer Bahn (also nicht y, sondern w) durch die Kastenwände beschränkt sein würden.

S_1 ist der Teil des Stromes, welcher von der Reflexion (von der Verzerrung der Wellenfunktion durch den Rand) herrührt. In der klassischen Theorie heben sich diese beiden Ströme auf. In der Quantenmechanik ist dies nicht mehr der Fall. Man kann den Strom S_2 allein erhalten, wenn man sehr tiefe Temperaturen und sehr kleine Werte der Dichte D nimmt. In diesem Falle werden sich alle Elektronen im Grundzustand befinden (v wird knotenlos sein) und die Elektronen werden sich in großer Entfernung von der Wand befinden (w wird große Werte haben), da ja kleine Werte von w die Energie erhöhen würden. In diesem Falle sieht man, daß S_1 vernachlässigbar wird, und man erhält das diamagnetische Moment: ein Magneton pro Elektron. Es ist interessant, zu bemerken, daß sich in diesem extremen Falle der diamagnetische Effekt gerade gegen den paramagnetischen Spin-Effekt weghebt. Man versteht das um so leichter, da nach einer Arbeit von Rabi* der Strom der vom Diracelektron im Magnetfeld herrührt, überall verschwindet, falls das Elektron sich im untersten Zustande befindet. Da nun also im ganzen Kasten nirgends ein Strom fließt, muß sich das System magnetisch indifferent verhalten.

Die vorliegende Arbeit wurde durch eine Diskussion über die Landausche Arbeit, an der Herr Prof. Pauli, Herr Prof. van Vleck und Herr Dr. Peierls teilnahmen, angeregt. Ich bin Herrn Prof. Heisenberg für seine fördernden Ratschläge zu Dank verpflichtet.

Leipzig, Physikalisches Institut.

* I. I. Rabi, ZS. f. Phys. **49**, 507, 1928.

Physics. — *The dependence of the susceptibility of diamagnetic metals upon the field.* By W. J. DE HAAS and P. M. VAN ALPHEN. (Communication N⁰. 212*a* from the Physical Laboratory, Leiden.)

(Communicated at the meeting of December 20, 1930).

Introduction. To explain the diamagnetism of metals we have to take into account three factors:

1st The diamagnetism of the electrons bound within the atom.

2nd The influence of the so-called free electrons on the whole diamagnetic effect.

3rd The influence on the bindings of the crystal lattice.

Under 2.

H. A. LORENTZ has shown that the free electrons do not contribute to the diamagnetism. In the classic calculation the conduction electron has been assumed to be absolutely free, which assumption however is not valid.

On the contrary, the great susceptibilities of some metals, bismuth, antimony etc. seem to make it probable that also the conduction electrons contribute to the diamagnetism.

It is possible that the classical theory of the conduction electrons may be able to let them contribute to the diamagnetism; but then these electrons must be assumed to be more or less bound.

In our opinion it seems probable, that a great diamagnetism may be expected there where the binding of the conduction electron is rather strong, so that during some time this electron is strongly under the influence of *two neighbouring atoms.*

Under 3.

The crystal lattice has a very great influence on the phenomena of diamagnetism.

For example this is very evident for tin. White tin is paramagnetic and grey tin diamagnetic. See also the results for copper by HONDA, Nature vol. 126, p. 990, 1930.

In general the problem of the influence of the crystal lattice is closely connected with the question whether the free electrons are bound more or less loosely. For high diamagnetism at least these two questions cannot be separated in our opinion.

Literature.

See for the theoretical treatment of the diamagnetism: LANGEVIN [1]), PAULI [2]), LANDAU [3]), P. EHRENFEST [4]).

---

1) P. LANGEVIN, Ann. de chim. et phys. (8) **5**, p. 70, 1905.

2) W. PAULI, Rapports Solvay 1930.

3) LANDAU. Zeits. f. Phys. Bd. **64**, p, 629, 1930.

4) P. EHRENFEST, Zeit. f. Phys, Bd **58**, p. 719, 1929.

For the explanation of the high diamagnetic values of some metals: P. EHRENFEST [1]); for the group of metals: bismuth, antimony and gallium. See also the summary of H. J. SEEMANN [2]). Detailed experimental investigations have been made by K. HONDA [3]), M. OWEN [4]) and others. The latter determined the temperature dependency of the susceptibility of most elements down to —190°.

§ 1. *Previous considerations.*

We investigated lead, tin and bismuth, the latter also in the form of single-crystals.

The most detailed investigation was that of the bismuth single-crystals.

1st. because we had at our disposal a single crystal of extremely pure bismuth;

2nd because of the desirability of the examination of bismuth single-crystals in connexion with the anomalous results of the resistance measurements by L. SCHUBNIKOW and W. J. DE HAAS [5]) with these crystals. Because of the evident correlation [6]) of the diamagnetic susceptibility with the change of resistance we were inclined to expect a dependance of the susceptibility on the field analogous to that found for the resistance. Further on we shall see that our expectation was fulfilled.

§ 2. *General difficulties in measurements of the diamagnetic susceptibility.*

The general difficulty in the measurement of diamagnetic forces is their smallness. This makes the use either of the torsion balance or of the ordinary balance inevitable. At low temperatures the torsion balance is of no practical use. So we had to choose a balance method, which will be described in § 3.

Besides, a high degree of purity of the specimen used was required. The slightest traces of iron or of other ferromagnetic substances might spoil the results; even the occurrence of paramagnetic substances may be dangerous at low temperatures.

From the phenomena we can conclude however with great surety whether the substances are sufficiently pure. In high fields the ferromagnetic substances certainly become saturated.

Moreover a *strong* diamagnetism is always a sign of purity, both ferro- and paramagnetic substances having the other sign.

---

1) P. EHRENFEST, Zeits. f. Phys. Bd. **58**, p. 719, 1929.

2) H. J. SEEMANN, Zeits. f. Techn. Phys. **10**, p. 399, 1929.

3) K. HONDA, Ann. de Phys. **32**, p. 1027, 1910.

4) M. OWEN, Ann. d. Phys. **37**, p. 657, 1912.

5) L. SCHUBNIKOW and W. J. DE HAAS, These Proc. **33**, pp. 130, 363, 1930, Comm. Leiden N°. 207.

6) W. J. DE HAAS, These Proc. **16** p. 1110, 1914.

§ 3. *Description of the apparatus.*

Principle :

By means of a balance the substance is weighed with and without magnetic field.

In the case of a long rod with transverse section *d*, the extremities of

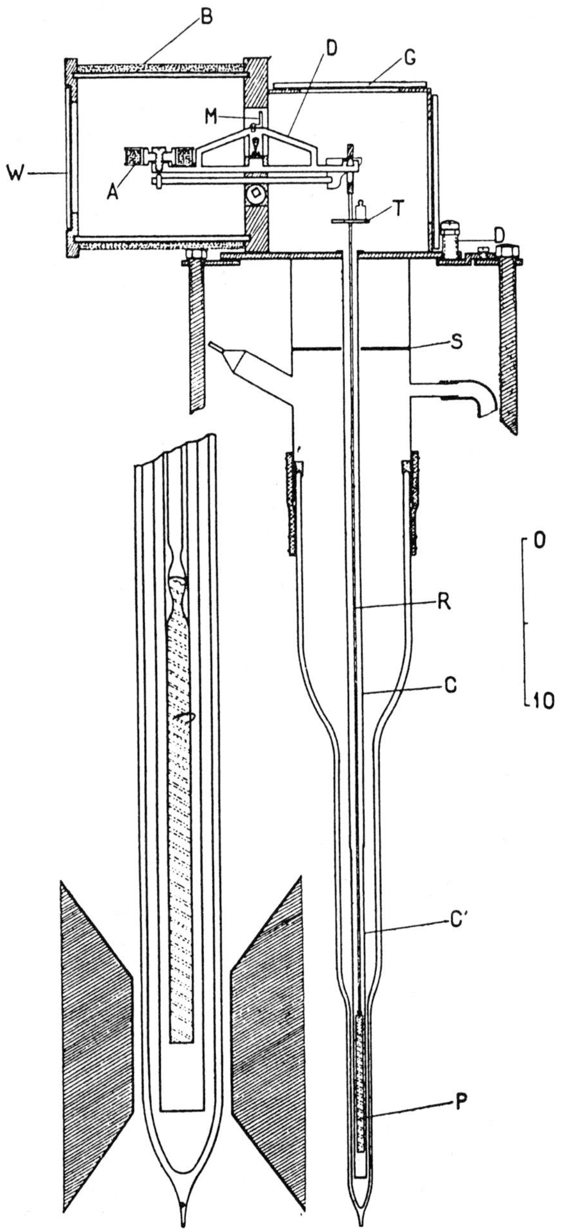

Fig. 1.

which are hanging in the fields H and H' respectively the change in weight is given by

$$p = \frac{2\,Kd}{g}(H^2 - H'^2),$$

where K is the volume susceptibility.

The field changing from H' to H in the space occupied by the substance this method cannot be used when the susceptibility is a function of the intensity of the field.

In that case the Faraday method is preferable, which makes use of a small quantity of matter placed there where $H\frac{dH}{dx}$ has its maximum value

The change in weight is now given by

$$p = \frac{m\chi}{g} H \frac{dH}{dx}$$

or

$$p = \frac{m\sigma}{g} \cdot \frac{dH}{dx},$$

where σ is the magnetic moment belonging to the intensity of the field H.

Performance.

The substance is suspended on a glass rod R at one arm of a balance. The other arm carries a coil A through which an electric current can be passed, by means of two thin springs at the side of the balance. The same current passes through the large coil B, the axis of which is perpendicular to that of A. The couple of forces which acts on the coil A is used to compensate the couple of forces cause by the change in weight of the substance.

By commutating the current in A or in B we can change the sign of the couple of forces. The deflexions of the balance are read with the aid of a telescope, a scale and the mirror M.

The substance P can not hang immediately in the refrigerating liquid. In this case the boiling of the liquid would have too disturbing an influence. Therefore the substance is suspended within a copper tube C' This tube, closed at the bottom, is suspended by a German silver tube in order to reduce the conduction of head as much as possible, and filled with helium gas.

Experiments with a thermo-element showed that the temperatures inside and outside the copper tube differed less than 0.1°, over a range in vertical direction of about 20 cm.

The apparatus was calibrated by means of weights placed on the table T.

After the rod R has been introduced the apparatus is shut with the glass plate G. Through a side-tube the balance space is then evacuated and afterwards filled with helium gas.

The screen S protects the upper part against the cold vapours of the liquid.

The cryostat glass has three narrow places, of which the lowest one is fitted between the pole-pieces, the middle one between the windings of the magnet.

The oscillations of the balance are aperiodic and strongly damped by the FOUCAULT currents. The zero position is however well reproduced.

§ 4. *Measurements with a lead rod.*

To control whether the apparatus works satisfactorily the susceptibilities of lead and tin were determined at low temperatures. For this purpose rods

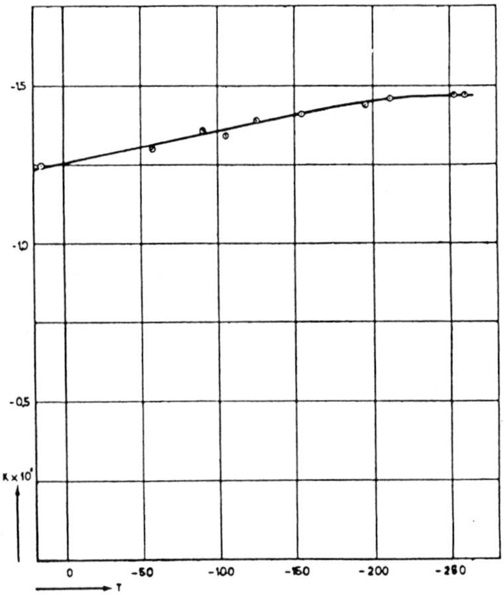

Fig. 2.

TABLE I (lead).

T	$-K \times 10^6$
+ 16° C.	1.25
— 57°.9	1.30
— 90°.5	1.36
—105°.5	1.34
—125°.7	1.39
—154°.8	1.41
—195°.9	1.44
—211°.1	1.46
—252°.7	1.47
—259°.4	1.47

of about 10 cm. length were made from the metal obtained from KAHLBAUM. First the impure surface was removed by etching and the metal was melted in vacuo and poured into a glass tube. After cooling down the lower part of the glass tube could easily be removed. The upper part was left intact, so that it could be used to fuse the rod to the glass tube.

The measurements from 0° to —154° are made with a vapour cryostat, so that the temperatures were not so constant as during the measurements in liquid nitrogen and hydrogen.

The diamagnetism of lead is found to increase about 15 %.

§ 5. *Measurements with a paramagnetic tin rod.*

TABLE II (tin).

T	$-K \times 10^6$
+ 16° C.	0.187
−252°.7	0.173

§ 6. *Measurements with a poly-crystalline bismuth rod.* (Table III.)

Already before the resistance measurements of L. SCHUBNIKOW and W. J. DE HAAS with bismuth single-crystals we examined a bismuth rod obtained from HILGER. This rod was made by melting the metal in vacuo, and letting it cool down in a glass tube. This cooling down had to be slow in order not to spring the glass.

This caused however the formation of rather large crystals the crystal directions of which strongly depend on the temperature gradient.

For most rods investigated by us (lead and bismuth at room temperature) an increase of the diamagnetic susceptibility with the field intensity was caused by traces of iron. For bismuth at hydrogen temperatures this was found to be no longer the case; even the opposite effect was observed.

The results of these measurements are given in table III.

TABLE III (Bismuth).

H_{max}	$T = 14.2$ K. $-K \times 10^6$	$T = 16°$ C. $-K \times 10^6$
6.4 K.G.	14.9	11.7
8.4	15.1	11.8
10.4	15.0	11.8
12.0	15.0	11.8
13.1	15.0	11.8
14.0	14.9	11.8
14.6	14.9	11.8^5
15.1	14.9	11.8^5
15.6	14.8^5	11.9

71*

§ 7. *Measurements with a long single-crystal of bismuth.*

The first research on bismuth single-crystals was made with a long crystal (22 mm.). The principal axis coincided with the longitudinal direction of the crystal.

The results of this investigation have already been published [1]). As was expected the susceptibility proved to depend on the field, so that another method should be applied.

§ 8. *Measurements with a small bismuth single-crystal.*

For the above reason Dr. L. SCHUBNIKOW prepared for us a small single-crystal ($5 \times 5 \times 5$ mm.). With this crystal the measurements were repeated in a known field, in which $H\frac{dH}{dx}$ was constant within a range of 8 mm. The material for this crystal was chemically pure bismuth which was further recrystalised 12 times. The bismuth was therefore extremely pure.

The principal crystal axis was parallel with an edge of the crystal, so that the plane through the binary axes was parallel with a side of the cube.

The direction of the binary axes was derived from the line system on a cleavage plane.

§ 9. *Adjustment of the crystal.*

By means of a very thin strip of copper the crystal was suspended from a ground glass rod in such a way that the principal axis was directed vertically, that is to say perpendicular to the magnetic lines of force.

The correction for the carrier was determined for each temperature separately by taking the crystal each time out of the strip.

The glass becoming paramagnetic at low temperatures the lower part was later replaced by a zinc rod which was diamagnetic. The upper part remained of glass to prevent heat conduction. As this part was not in the magnetic field it had no considerable influence.

§ 10. *The magnets.*

For the low fields a CARPENTIER magnet was used, for the high fields a large type WEISS magnet.

Both magnets could be turned round a vertical axis and the angle of rotation could be read.

In both cases the field intensity was measured at different points and for different currents by means of a small coil and a ballistic galvanometer.

In this way the place of the maximum of $H\frac{dH}{dx}$ was determined as well as its maximum value.

---

[1]) W. J. DE HAAS and P. M. VAN ALPHEN, These Proc. Vol. **33**, N⁰. 7 p. 680, 1930; Comm. Leiden N⁰. 208*d*.

The corresponding value of H was 12.9 K.G. with the first magnet, 20.4 K.G. with the second one.

§ 11. *Results at 16° C.*

As might be expected the susceptibility was found to be independent of the position of the magnet. Nor did the values for the different fields differ more than ± 1 % and these deviations were not systematic.

For low field we found $\chi = 1.47^3 \,.\, 10^{-6}$,
for high fields $\chi = 1.48^2 \,.\, 10^{-6}$,
the mean value of which is $1.48 \,.\, 10^{-6}$,
which is in good agreement with the measurements of FOCKE [1]) and MC LENNAN [2]), who found respectively $\chi = 1.487 \,.\, 10^{-6}$ and $1.50 \,.\, 10^{-6}$.

§ 12. *Measurements at the temperatures of liquid hydrogen.* (20°.4 K. and 14°.2 K.).

Dependency of the magnetisation on the field intensity and on the direction of the magnetic field.

The principal axis of the crystal was again perpendicular to the lines of force. By rotating the magnet about the vertical axis all values could be given to the angles between the field and the binary axes.

In tables IV and V the value of the specific magnetisation has been given, together with the intensity of the field for different field directions.

As might be expected the magnetisation does not only change with the

TABLE IV. (Fig. 3 and 4). $-\sigma \times 10^3$.

$H \times 10^6$ G.	$T = 20°.4$ K.		$T = 14°.2$ K.	
	H // Bin. Axis	$H \perp$ Bin. Axis	H // Bin. Axis	$H \perp$ Bin. Axis
3.4	5.3	5.4	5.8	5.9
5.1	8.7	9.0	9.1	9.5
6.7	12.5	11.8	13.5	11.7
8.3	13.9	15.1	13.4	15 6
9.6	17.7	18.8	18.7	20.2
10.5	20.9	20.8	22.6	22.3
11.3	23.0	21.9	24.9	22.9
11.9	24.5	22.3	26.4	22.9
12.5	24.9	22.3	26.6	22.1
12.9	25.2	22.2	26.6	21.5

1) A. B. FOCKE, Phys. Rev. 36, p. 316, 1930.
2) J. C. MC LENNAN and E. COHEN, Trans. Roy. Soc. Canada 23 sect. 3 p. 159, 1929.

intensity of the field but also with the direction of the latter. This latter change has a period of 60°. Extreme values are reached for those positions of the magnet for which the lines of force are either parallel or perpendicular to the binary axes.

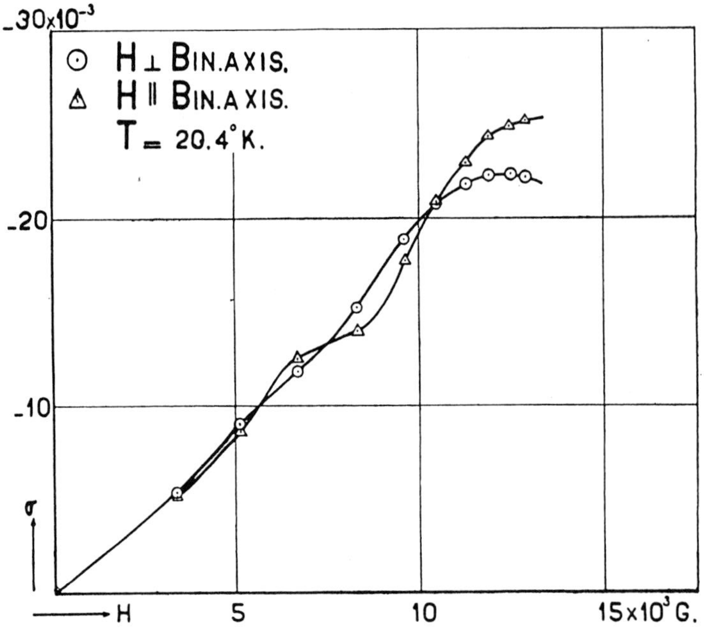

Fig. 3.

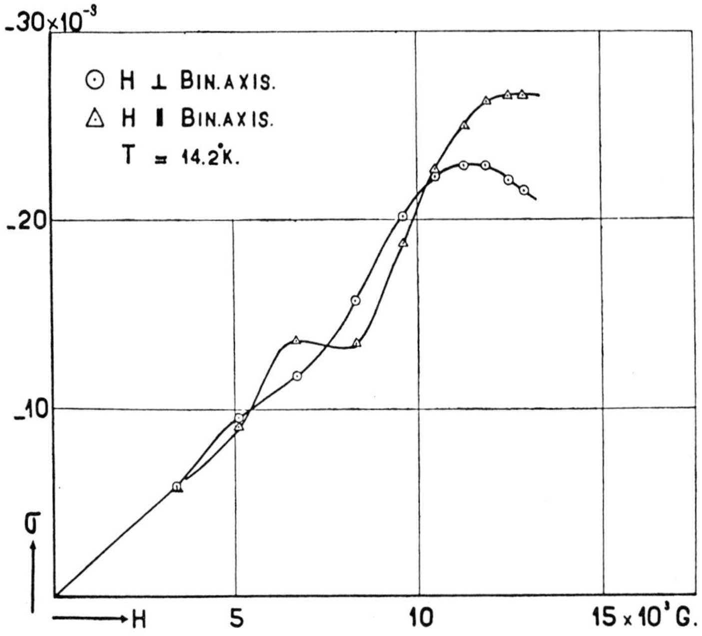

Fig. 4.

TABLE V. (Fig. 5). — $\sigma \cdot 10^3$.

$H \times 10^6$ G.	$T = 20°.4$ K.		$T = 14°.2$ K.	
	$H//$Bin. Axis	$H\perp$ Bin. Axis	$H//$Bin. Axis	$H\perp$ Bin. Axis
2.4	4.0	4.2	4.6	4.8
4.9	8.6	8.5	8.0	9.2
6.2			11.5	10.4
7.4	13.0	12.7	11.9	12.3
8.6			13.7	16.2
9.8	18.8	19.1	18.8	20.0
11.0			22.6	21.4
11.9	23.1	20.6	23.9	20.4
12.8			24.3	19.2
13.7	24.4	21.4	23.9	19.3
15.7	24.6	24.9	22.6	23.6
16.9			23.0	28.0
17.8	26.3	30.7	24.2	30.8
18.8	28.0	32.9	26.6	33.8
19.6			29.4	35.7
20.4	33.5	36.3	32.8	37.6

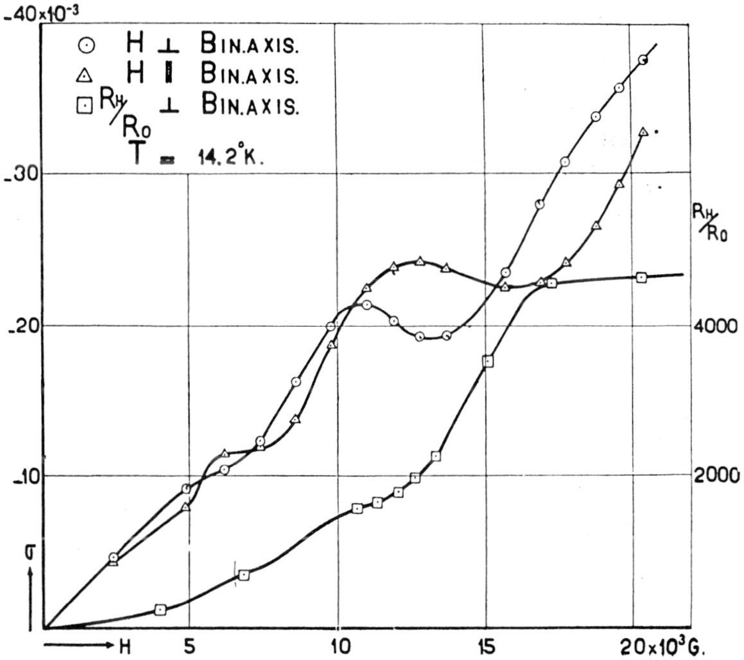

Fig. 5.

From the figures 3, 4, 5 it is evident that between 5 and 7.5 K.G. and between 10 and 15 K.G. the specific magnetisation does not increase much and in the second interval even diminishes with increasing intensity of the field.

TABLE VI. (Fig. 6). $-\chi \times 10^6$.

H	$H_{//}$ Bin. Axis	$H\perp$ Bin. Axis
4.9	1.59	1.82
6.2	1.88	1.70
7.4	1.57	1.63
8.6	1.55	1.82
9.8	1.87	2.00
11.0	2.04	1.94
11.9	1.95	1.66
12.8	1.84	1.46
13.7	1.71	1.38
15.7	1.41	1.47
16.9	1.34	1.63
17.8	1.34	1.70
18.8	1.39	1.77
19.6	1.49	1.81
20.4	1.58	1.80

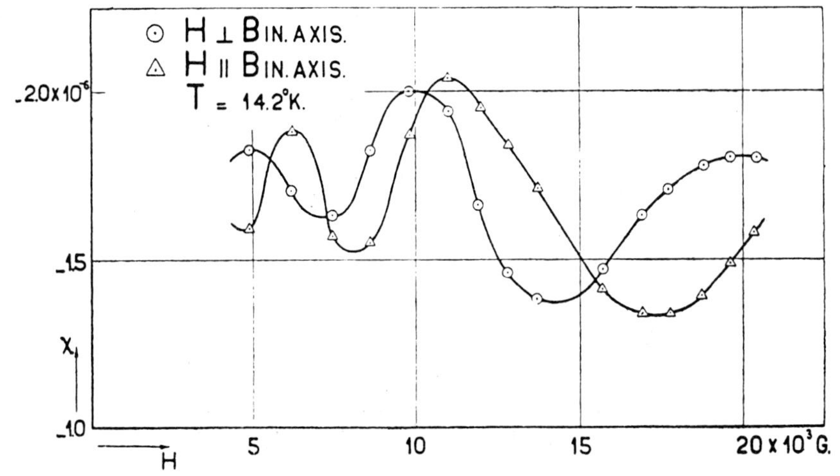

Fig. 6.

$T = 14^\circ.2$ K.

TABLE VII. (Fig. 7). $-\sigma \times 10^3$.

	$H = 11.9$	13.7	15.7	17.4	18.8
0°	19.9	20.6	24.0	28.8	32.3
5°	20.2	19.8	23.5	29.7	34.0
10°	20.0	20.7	24.2	28.8	31.9
15°	20.3	21.6	25.2	27.8	29.2
20°	21.7	23.0	24.6	26.2	28.4
25°	22.9	23.6	23.5	24.9	27.0
30°	23.9	23.7	22.4	23.6	26.1
35°	24.0	23.7	22.2	23.3	26.0
40°	23.6	23.5	22.5	23.6	26.1

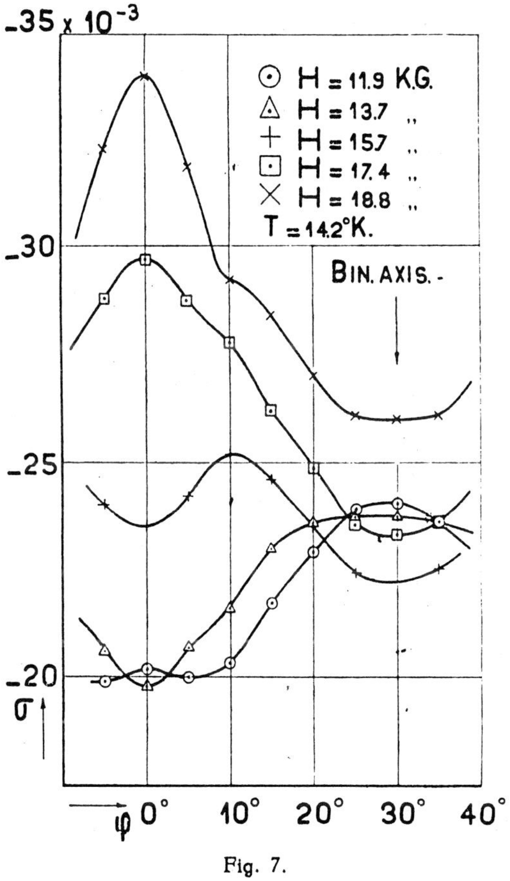

Fig. 7.

These regions of slightly increasing magnetisation are displaced towards the higher fields in the case $H \parallel$ binary axis.

In fig. 5 the change in resistance of *Bi* (for the case binary axis perpendicular to the field, bisectrix parallel and the current vector perpendicular to the field) has been plotted beside the magnetisation. It is evident that a connexion exists between this resistance change and the magnetisation perpendical to it.

The resistance change for the case H parallel with the binary axis shows at 14°.2 a character too weakly pronounced to allow comparison.

In fact the determination of the susceptibility generally offers a more refined means of research than the change of resistance.

Table VII and fig. 7 give the change of the susceptibility with the direction of the field for different field intensities.

It is evident that still other periods occur beside that of 30°. Fourier analysis may reveal the details.

§ 13. *Summary.*

The susceptibility of bismuth at low temperatures (hydrogen) is found to be a periodical function of the field.

The susceptibility in directions perpendicular to the principal axes are complicated functions of the field and of the angle between the binary axes and the field.

The *Bi* atoms that are very near to each other are subjected to their mutual influences.

Finally we wish to express our best thanks to Messrs T. JURRIAANSE, J. W. BLOM, and J. DE BOER for their valuable help during the measurements.

Physics. — *Magnetische Widerstandsvergrösserung in Einkristallen von Wismut bei tiefen Temperaturen.* Von L. SCHUBNIKOW und W. J. DE HAAS. (Mitteilung N°. 207*a* aus dem Physikalischen Institut Leiden).

(Communicated at the meeting of February 22 1930).

An zwei drahtförmigen Einkristallen von Wismut (Querschnit 1.5×0.8 mm. und lange 22 mm.) mit Orientation der Hauptachse || Drahtlänge wurden in Magnetfeldern bis zu 22 Kgauss resp. 31 Kg. $R_H/R_{0^\circ\,C.}$ [1]) Werte gemessen und zwar bei verschiedener Orientierung des Feldes H senkrecht zur Drahtrichtung (ist gleich Hauptachse des Kristalles).

In Fig. 1 und 2 sind $R_H/R_{0^\circ\,C.}$ Werte für beide Kristalle N 622 und N 674 in Abhängigkeit vom Felde H gegeben.

Die Kurven „Max'' und „Min'' entsprechen denjenigen Orientierungen der binären Achsen des Kristalls die extreme R_H Werte liefern.

Der Kristall 622 ist weniger gut; dies erkennt man an der Temperaturabhängigkeit seines Widerstandes ohne Magnetfeld.

Folgende $R/R_{0^\circ\,C.}$ werte wurden für beide Kristalle bei Wasserstoftemperaturen gefunden:

$R/R_{0^\circ\,C.}$

N	20.43° K.	20.37° K.	14.22° K.	14.15° K.
622		0.04738	0.02520	0.02496
674	0.04689	0.04666		0.02435

Man berechnet hieraus extrapolatorisch [2]) für $T = 1{,}5^\circ K$ folgende „Restwiderstände''.

$$(R_{1.5\,K.}/R_{0^\circ\,C.})_{622} = 0{,}0023\ .\ (R_{1.5\,K.}/R_{0^\circ\,C.})_{674} = 0{,}0017.$$

Diese Extrapolation wurde gemacht mit Hilfe der Messungen von $R_{20^\circ\,K.}$, $R_{14^\circ\,K.}$ und $R_{1,5^\circ\,K.}$ die an anderen Exemplaren von Wismut Einkristallen gemacht wurden. Man kann mit Hilfe der angegebenen Zahlen den Verlauf von $R_H/R_{0^\circ\,C.}$ umrechnen in den von R_H/R bei den in der Tabelle angegebenen Temperaturen.

---

[1]) R_H ist Widerstand im Felde H bei tiefer Temperatur, $R_{0^\circ\,C.}$ Widerstand ohne Feld beim Eispunkte.

[2]) Die Kristalle 622 u. 674 konnten im Augenblick noch nicht bei Heliumtemperaturen gemessen werden.

Der Kristall 674 zeigt bei 14,15° K. und $H = 30{,}9$ K.G. den enormen Wert

$$(R_H/R)_{14{,}15\,\mathrm{K.}} = 176000.$$

Der Kristall 622 zeigt eine kleinere Widerstandsänderung. Bemerkenswert ist, dass ein verhältnismässig kleiner Unterschied des Restwiderstandes der beiden Kristalle einen so grossen Unterschied in $R_H/R_{0^\circ\,\mathrm{C.}}$ bei tiefen Temperaturen hervorruft.

N 622

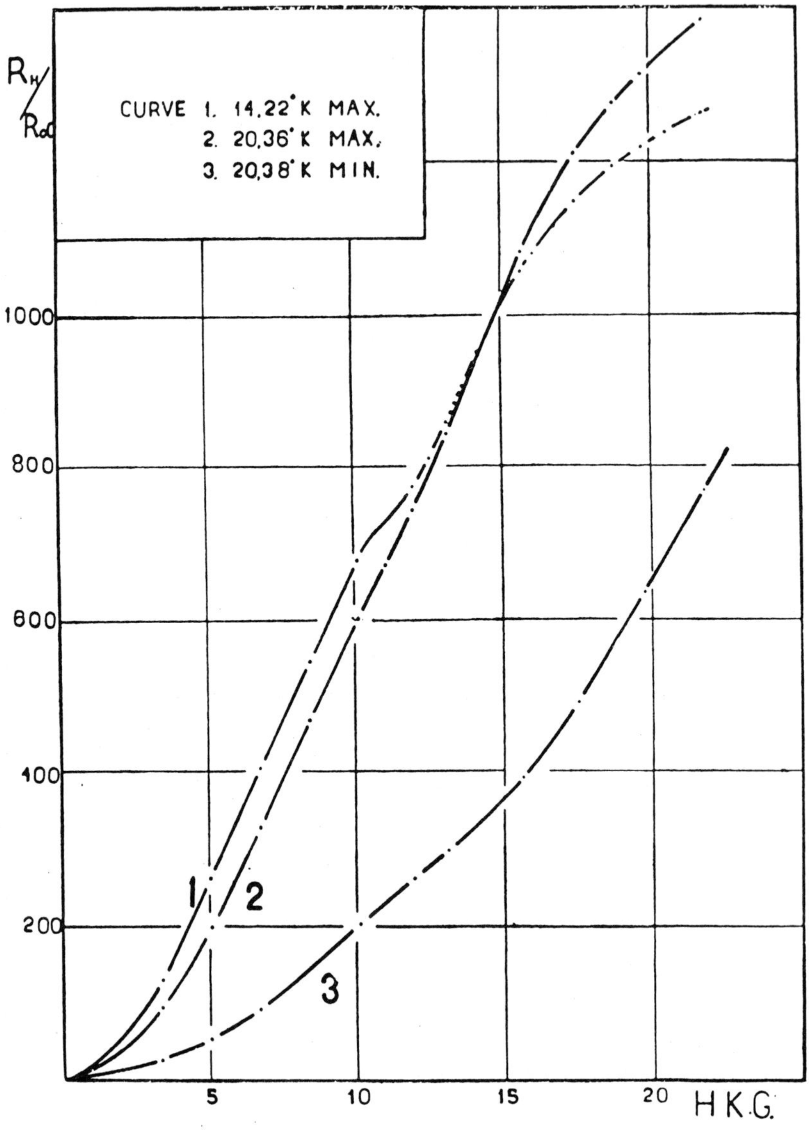

Fig. 1.

Bei den höheren Temperaturen 77.4° K. und 64° K. dagegen wurde innerhalb der Fehlergrenzen für beide Kristalle $R_H/R_{0^\circ\,\mathrm{C.}}$ gleich gross gefunden.

Wie die Fig. 1 und 2 zeigen liefern unsere Messungen besonders bei

14° K. eine überraschend komplizierte Abhängigkeit des R_H von H, welche noch einer näheren Aufklärung bedarf.

Um zu prufen ob es sich hierbei nicht um einen blossen Störungseffekt handelt, haben wir eine Reihe von Kontrollversuchen vorgenommen. Wir konnten vor allem feststellen, dass die Kurven gut reproduzierbar sind und für steigendes und fallendes Feld H exakt zusammenfallen.

N 674

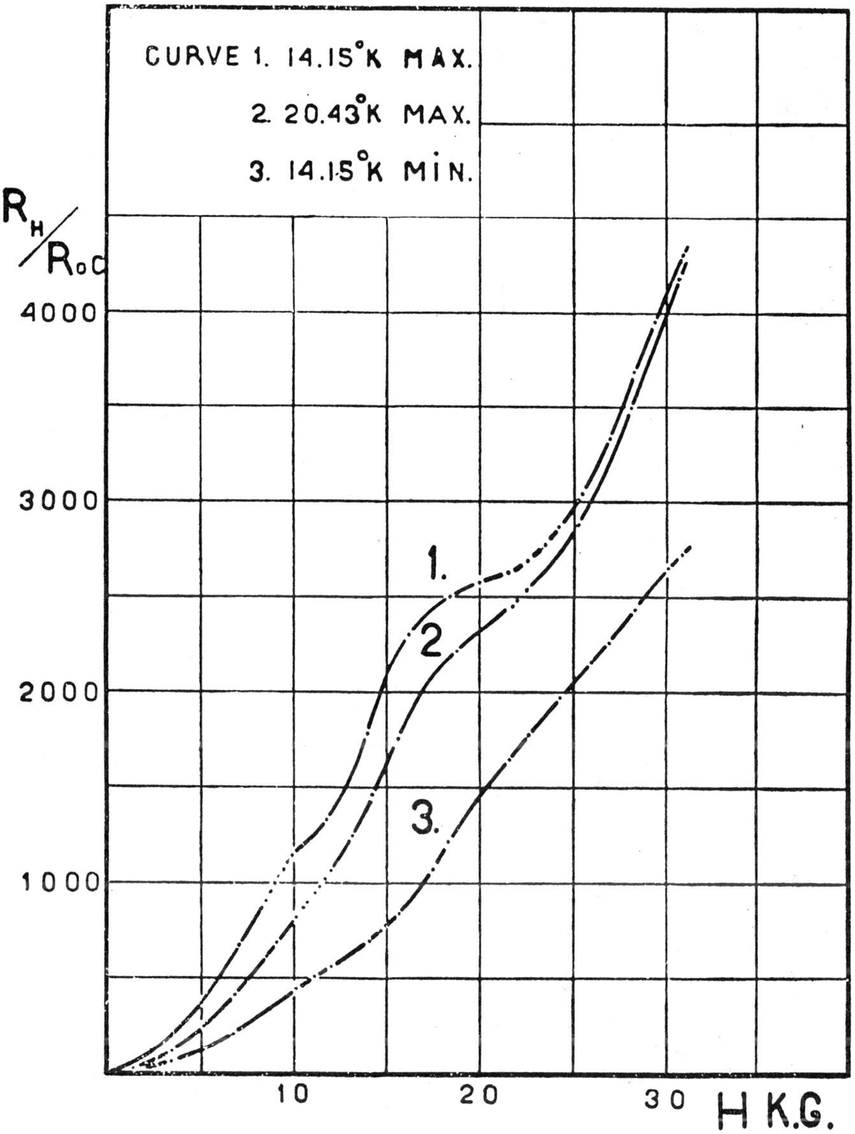

Fig. 2.

Den Widerstand von Kristall 674 haben wir im Gebiet bis $H = 22$ K.G. sowohl im grossen als im kleineren Elektromagneten des Laboratoriums durchgemessen. Der Verlauf blieb genau derselbe.

Der Widerstand von Kristall 622 wurde nur im kleineren Magneten bis $H = 22$ K.G. gemessen.

In diesem Gebiet, sind wie man sieht die Formen der Kurven für beide Kristalle sehr ähnlich.

Der Widerstand von Kristall 674 wurde für einige Felder (von 4 K.G. bis 31 K.G.) für alle mögliche Orientierungen gemessen bei denen die Drahtachse (ist gleich Hauptachse des Kristalls) senkrecht zum H-Feld bleibt.

Alle diese Kontrollversuche machen es vorläufig wahrscheinlich, dass dem komplizierten Verlauf der Kurven ein reeller Effekt zugrunde liegt.

Schlieslich möchen wie Herrn J. W. BLOM herzlich danken für seine Hilfe bei den Messungen.

MAGNETORESISTANCE

by

R. G. Chambers

H. H. Wills Physics Laboratory
University of Bristol, England

ABSTRACT

After a brief review of the properties of spherical, ellipsoidal and cylindrical energy surfaces, the general magnetoresistance problem is discussed for the low-field, intermediate-field and high-field regions in turn. The Jones-Zener solution for low fields can give some information about the Fermi surface, but only if some assumption is made about the anisotropy of relaxation time. At intermediate fields, where the transport equation cannot be solved by series expansion in either $\omega\tau$ or $(\omega\tau)^{-1}$, the variational method is particularly useful, and with this approach it may also be possible to separate out Fermi surface anisotropy from relaxation time anisotropy.

At high fields, as Lifshitz and Peschanskii have shown, the magnetoresistance behavior is largely determined by the presence or absence of open electron orbits, and a discussion is given of Fermi surface topologies and types of electron orbit. The theory of high-field behavior is then outlined, and the results of Alekseevskii and Gaidukov discussed.

The paper concludes with a brief survey of recent work on the theory of oscillatory magnetoresistance in the quantum region.

I. INTRODUCTION

Since Peierls (1930, 1931) first pointed out that magnetoresistance effects were essentially due to departures from the isotropic free-electron model, it has been recognized that these effects might tell us a great deal about the anisotropy of the Fermi surface and of the scattering process, if the wealth of information they contain could be interpreted, and Blochinzev and Nordheim (1933) and Jones and Zener (1934b) were already attempting such an analysis almost thirty years ago. Thus magnetoresistance is certainly the oldest weapon in the Fermi surface armory, but until recently it has not proved a particularly powerful one. In the past two years, however, the situation has changed radically. The theoretical work of Lifshitz and his school, and the experimental results of Alekseevskii and Gaidukov, have already given us much new information about the Fermi surface and promise to yield a great deal more. It is unfortunate that they cannot be here to describe this work for themselves.

In this review, we follow the usual convention of taking the direction of the field $\underline{H}$ as the z axis. The quantities usually observed experimentally are the resistivities $\rho(\underline{H})$ with current flow $\underline{J}$ parallel to $\underline{H}$ (longitudinal case) or perpendicular to $\underline{H}$ (transverse case). For brevity we shall denote the ratios $[\rho(\underline{H}) - \rho(0)]/\rho(0) = \Delta\rho(\underline{H})/\rho(0)$ by $D_L(H)$ and $D_T(H)$ for the longitudinal and transverse cases respectively.

If $\omega = eH/m^*c$ is the cyclotron frequency of a representative electron in field H (where m^* is the cyclotron mass, defined below), and τ is its relaxation time, we can distinguish several different regimes; in order of increasing H these are:

(i) $\omega\tau \ll 1$: the low-field region.

(ii) $\omega\tau \sim 1$: the intermediate-field region.

(iii) $\omega\tau \gg 1$: the (classical) high-field region.

(iv) $\hbar\omega \gtrsim kT$: the quantum oscillation region.

(v) $\hbar\omega > \epsilon_0$: the quantum limit (ϵ_0 is the Fermi energy).

Region (v) is only reached, for most metals, in fields $H \gtrsim 10^8$ G, and is therefore (at present) only of academic interest, for metals. In theoretical work on the quantum effects in regions (iv) and (v), it is often assumed that condition (iii) holds still in the 'low-field' region $\hbar\omega < kT$; the classical high-field limit then corresponds to the quantum low-field limit.

We shall assume the Lorentz force equation

$$\hbar\dot{\underline{k}} = e(\underline{E} + \underline{v} \times \underline{H}/c) \tag{1}$$

to hold throughout regions (i) - (iii), so that (neglecting the small perturbation due to $\underline{E}$) the electron moves in an orbit in $\underline{k}$-space given by k_H $(=k_z)$ = constant, ϵ = constant. If the orbit is closed, it is easily shown that the cyclotron frequency is given by $\omega = eH/m^*c$ with $m^* = (\hbar^2/2\pi)\, dA/d\epsilon$, where A is the area enclosed by the orbit in $\underline{k}$-space.

Early theoretical work was largely confined to the low-field region $\omega\tau \ll 1$, because of Jones and Zener's (1934a) conclusion that (1) broke down at higher fields. It is now generally accepted that this conclusion was invalid (cf. Chambers 1956a for a discussion and references), and in fact the Onsager-Lifshitz treatment of orbital quantisation in region (iv) involves the use of (1) up to the highest fields.

Before discussing the theory of magnetoresistance for arbitrary Fermi surface $\epsilon(\underline{k})$ and arbitrary relaxation time $\tau(\underline{k})$, it is convenient to review briefly the results for quadratic energy surfaces with isotropic relaxation, i.e. $\tau = \tau(\epsilon)$ only. In this case the transport equation is readily solved, and the resulting equations are applicable throughout regions (i) - (iii).

II. QUADRATIC ENERGY SURFACES

For a spherical Fermi surface with isotropic relaxation, of course, the magnetoresistance effects are vanishingly small: $D_L = 0$, $D_T \simeq 0$ for all H. The resistivity tensor in a field H_z has components

$$\rho_{xx} = \rho_{yy} = \rho_{zz} = \rho_0, \quad -\rho_{xy} = \rho_{yx} = \omega\tau\rho_0 = A_H H_z, \tag{2}$$

where $\rho_0 = m^*/ne^2\tau$ and $A_H = 1/nec$ (m^* : effective mass; n : number of conduction electrons per cm^3). The other components of ρ_{ij} vanish. The quantity usually calculated theoretically is the conductivity tensor σ_{ij}, reciprocal to ρ_{ij} : this has the non-vanishing components

$$\sigma_{xx} = \sigma_{yy} = \sigma_0/(1+\omega^2\tau^2), \quad \sigma_{xy} = -\sigma_{yx} = \sigma_0\omega\tau/(1+\omega^2\tau^2), \quad \sigma_{zz} = \sigma_0 \tag{3}$$

where $\sigma_0 = ne^2\tau/m^*$. Thus σ_{xx}, σ_{yy} vanish as H^{-2} for $\omega\tau \gg 1$, though ρ_{xx}, ρ_{yy} are independent of H. This is illustrated in Fig. 1 : at high fields, $|E|/|J|$ grows as H, and the angle Φ between $\underline{J}$ and $\underline{E}$ approaches $\pi/2$ ($\tan\Phi = \omega\tau$), so that $J_{//}/E \equiv \sigma_{xx}$ approaches zero, though $E_{//}/J \equiv \rho_{yy}$ remains unchanged.

For an ellipsoidal Fermi surface, the magnetoresistance effects also vanish. It can be shown from Bronstein's (1932) solution that if we take as axes 1, 2, 3 the principal axes of the ellipsoid, we have

$$\rho_{ii} = m^*_i/ne^2\tau; \quad -\rho_{ij} = \rho_{ji} = H_k/nec, \tag{4}$$

from which it follows that the resistivity $\rho_{\alpha\alpha}$ measured in any direction is independent of $\underline{H}$. This result is most simply obtained by using a drift-velocity argument (I am indebted to Dr. MacDonald for suggesting this approach). From the Boltzmann equation we have

$$e\,[\underline{E}\,df_0/d\epsilon - (\hbar c)^{-1}(df_1/d\underline{k}) \times \underline{H}]\cdot\underline{v} = -f_1/\tau \tag{5}$$

where $f_1 = f - f_0$ is the perturbation to the Fermi function. If we write $\underline{E} = \underline{E}_1 + \underline{E}_2$, where

$$\underline{E}_2\,df_0/d\epsilon - (\hbar c)^{-1}(df_1/d\underline{k}) \times \underline{H} = 0, \tag{6}$$

we are left with

$$e\,(df_0/d\epsilon)\,\underline{E}_1\cdot\underline{v} = -f_1/\tau, \tag{7}$$

so that $df_1/d\underline{k} = -e\,\underline{E}_1\cdot(\underline{m}^*)^{-1}_{ij}\,\tau\hbar\,(df_0/d\epsilon)$, where $(\underline{m}^*)^{-1}_{ij} = \hbar^{-2}d^2\epsilon/dk_i dk_j$ is the reciprocal mass tensor. Now $\underline{E}_1$, acting alone, would produce a mean drift velocity $\underline{v}_D = e\,\underline{E}_1\cdot(\underline{m}^*)^{-1}\tau$, so that (6) becomes

$$\underline{E}_2 + \underline{v}_D \times \underline{H}/c = 0 \tag{8}$$

Thus if all electrons have the same reciprocal mass tensor (as for quadratic energy surfaces) and the same relaxation time, all will acquire the same drift velocity in a

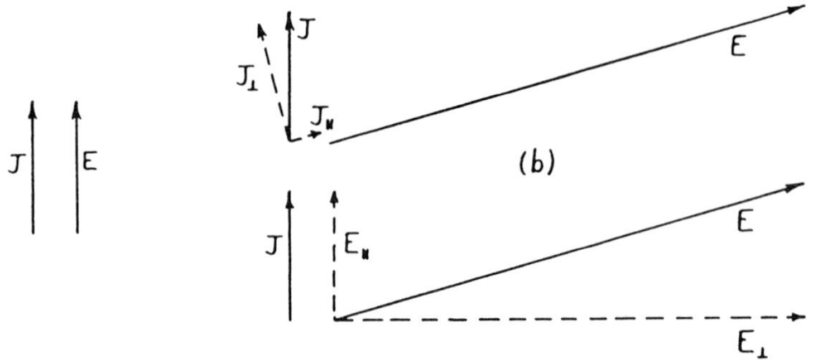

Fig. 1 Relation between $\underline{J}$ and $\underline{E}$ in a free-electron metal (a) without, (b) with magnetic field applied. Field direction normal to paper.

given applied field, so that a common resolution of E into E_1 and E_2 applies to the whole assembly. Also $\underline{J} = e \int \underline{v} f_1 \, d\underline{k}/4\pi^3 = e \int \underline{v}_D f_o \, d\underline{k}/4\pi^3$ (assuming τ constant, and integrating by parts), so that

$$\underline{J} = ne^2\tau\,(\underline{m}^*)^{-1}\cdot\underline{E}_1 \ , \tag{9}$$

and from (8),

$$\underline{E}_2 = -\underline{J} \times \underline{H}/nec \ . \tag{10}$$

The general solution (9), (10) reduces at once to (4) if we choose axes in which $(\underline{m}^*)^{-1}$ is diagonal. From this point of view, magnetoresistance is due to variation of $(\underline{m}^*)^{-1}$ or τ over the Fermi surface, leading to different drift velocities for different groups of electrons, so that (8) cannot be satisfied for the whole assembly simultaneously by a single choice of $\underline{E}_2$.

For later reference, we note that for a cylindrical Fermi surface, with $m^*_2 = m^*_3 = m^*$, $m^*_1 = \infty$, equation (4) yields for the conductivity tensor

$$\sigma_{22} = \sigma_{33} = \sigma_o/(1 + \omega_1^2\tau^2) \ ; \quad \sigma_{23} = -\sigma_{32} = \sigma_o\omega_1\tau/(1 + \omega_1^2\tau^2) \ , \tag{11}$$

where $\sigma_o = ne^2\tau/m^*$ and $\omega_1 = eH_1/m^*c$; the other tensor components vanish. Thus such a surface gives a two-dimensional conductor, incapable of carrying current in the direction of the cylinder axis, and if the applied field is perpendicular to the axis, so that $H_1 = 0$, the conduction properties are completely unaffected by the field: we have simply $\sigma_{22} = \sigma_{33} = \sigma_o$. Finally, if we transform from the principal axes 1, 2, 3 of the cylinder to axes x, y, z, such that $H = H_z$ and the cylinder axis 1 lies in the x-z plane at an angle θ to x, we find for the conductivity tensor

$$\sigma/\sigma_o = \begin{matrix} \alpha \sin^2\theta & \beta\sin\theta & -\alpha\sin\theta\cos\theta \\ -\beta\sin\theta & \alpha & \beta\cos\theta \\ -\alpha\sin\theta\cos\theta & -\beta\cos\theta & \alpha\cos^2\theta \end{matrix} \tag{12}$$

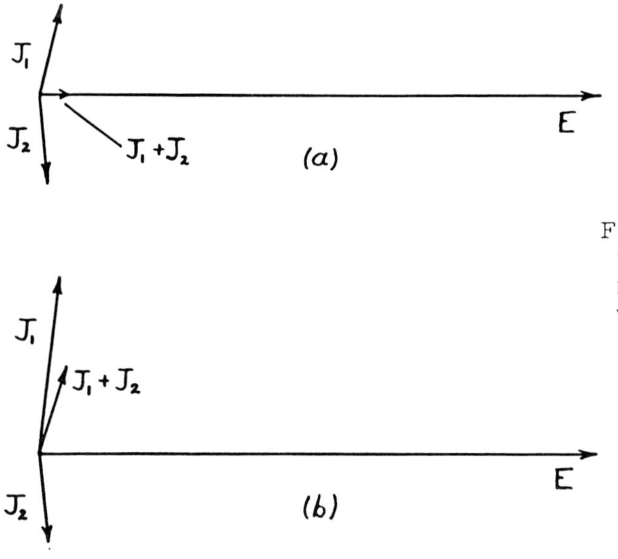

Fig. 2 Relation between $\underline{J}_1$, $\underline{J}_2$ and $\underline{E}$ in a two-band metal, (a) with equal numbers of electrons and holes, (b) with unequal numbers.

where $\alpha = (1 + \omega_0{}^2\tau^2\sin^2\theta)^{-1}$, $\beta = \alpha\,\omega_0\tau\sin\theta$, and $\omega_0 = eH_z/m^*c$; $\omega_0\sin\theta$ is the actual cyclotron frequency of the electrons in their elliptical orbits in $\underline{k}$-space. Thus if θ is small, i.e. if H_z is almost perpendicular to the cylinder axis, we have σ_{yy}, $\sigma_{zz} \simeq \sigma_0$ until $\omega_0\tau$ becomes comparable with $1/\theta$; at fields so high that $\omega_0\tau \gg 1/\theta$, we have σ_{yy}, $\sigma_{zz} \sim H^{-2}$. The resistivity components ρ_{ii} remain unaffected by H, of course, as for the general ellipsoid.

The simplest model to show a non-vanishing magnetoresistance effect is the 'two-band' model, with two sets of quasi-free carriers of different mobilities (Ariyama 1938, Sondheimer and Wilson 1947, Sondheimer 1948, Chambers 1952b). The Hall angles $\phi = \tan^{-1}\omega\tau$ between $\underline{E}$ and $\underline{J}$ will then differ for the two bands (if one band contains electrons and one holes, ϕ_1 and ϕ_2 will differ in sign), so that in a transverse magnetic field the currents $\underline{J}_1$ and $\underline{J}_2$ produced by the common electric field $\underline{E}$ will add vectorially to produce a smaller resultant than $|J_1| + |J_2|$. If the components $J_\perp$ (Fig. 1) become equal and opposite, we are left with the small components $J_{//}$ (Fig. 2a), which vanish as H^{-2}; correspondingly the resistivity grows as H^2 without limit. The condition for this is that $H(\sigma_{xy1} + \sigma_{xy2}) \rightarrow 0$ in high fields, so that from (3) we require $A_{H1} + A_{H2} = 0$, or $n_1 = n_2$ if one band contains electrons and one holes. In the resultant conductivity tensor, $\sigma_{xy} = \sigma_{xy1} + \sigma_{xy2}$ then vanishes as H^{-3}. If $A_{H1} + A_{H2} \neq 0$, the resultant σ_{xy} only vanishes as H^{-1}, as for a single band, and the qualitative behaviour at high fields reverts to that described by (3) : the resistivity tends to a constant saturation value, and the Hall angle between $\underline{J}_1 + \underline{J}_2$ and $\underline{E}$ tends to $\pi/2$ instead of to zero. We then have the curious result that the current in one band flows in almost the opposite direction to the net flow (Fig. 2b). The high-field saturation resistivity exceeds the zero-field value by a factor proportional to $(A_{H1} + A_{H2})^{-2}$, and is reached when $H \gg (\rho_{01} + \rho_{02}) / (A_{H1} + A_{H2})$.

By choosing the parameters of the two-band model to fit the experimental results on transverse magnetoresistance and Hall effect, it may be possible to arrive at useful order-of-magnitude estimates of the band structure, in appropriate cases (Ariyama 1938). It is possible to increase the number of parameters available by going to a four-band model (Borovik 1955, Borovik and Volotskaya 1959), but this is certainly pushing the idea too far. The model of two spherical surfaces is incapable of producing a longitudinal magnetoresistance effect, since there is then no Hall effect and $\underline{J}_1$, $\underline{J}_2$ remain parallel to $\underline{E}$ and to each other, unaffected by H.

A much more realistic model is obtained if we use a set of ellipsoids instead of spheres. Blochinzev and Nordheim (1933) first showed that such a model could account quite well in order of magnitude for the value of D_T observed experimentally. Frenkel and Kontorowa (1935) rederived their results much more simply by assuming the validity of (9) and (10) for the individual ellipsoids, and also showed that on this model $D_L \neq 0$: the longitudinal effect no longer vanishes, in general (cf. also Kohler 1938). This is because the current contributions $\underline{J}_n$ from the various ellipsoids will not be parallel to $\underline{E}$, even for H = 0, unless the direction of E happens to coincide with a principal axis of each ellipsoid. Thus for $\underline{H} \parallel \underline{E}$, $\underline{J}_n \times \underline{H}$ will not vanish, so that from (9), (10) the angle between J_n and E will be altered by the applied field, and a non-zero D_L arises from essentially the same 'two-band' mechanism that we have discussed above.

The many-ellipsoid model has been applied to Bi by Jones (1936), and more recently by Okada (1955, 1957), Abeles and Meiboom (1956) and Mase and Tanuma (1960); the effective mass ratios so deduced are in fair agreement with the dHvA values. It has also been applied with considerable success to Soule's (1958) data on graphite by McClure (1958), taking the Fermi surface to consist simply of a set of cylinders obeying (11). In this work McClure has also used an ingenious variant on the Kramers-Kronig relations, due to E.N. Adams, to separate out the numbers of electrons and holes. Finally, the many-valley model has of course been applied by many authors to semiconductors, where it usually gives good agreement with experiment (Lax 1958). A basic weakness of this method of Fermi surface determination is that the parameters involved are not the effective masses but the mobilities $e\tau/m^*$; the effective masses themselves cannot be determined directly, and even their ratios cannot be found unless it is assumed that $\tau = \tau(\epsilon)$ only. This difficulty persists in the more general theories, which we consider next.

III. GENERAL ENERGY SURFACES. LOW-FIELD REGION

If a relaxation time $\tau(\underline{k})$ is assumed to exist, the magnetoresistance problem for arbitrary $\epsilon(\underline{k})$ is most simply formulated by using the kinetic method (Chambers 1952a). It should no longer need emphasising that when $\tau(\underline{k})$ exists this method is no less exact than the Boltzmann equation method, and gives identical results. Briefly, we have

$$\underline{J} = \frac{e}{4\pi^3}\int \underline{v}\, f_1\, d\underline{k} = -\frac{e}{4\pi^3}\int \underline{v}\,\frac{df_o}{d\epsilon}\,\Delta\epsilon\; d\underline{k}$$

$$= -\frac{e}{4\pi^3}\int \underline{v}\,\frac{df_o}{d\epsilon}\, d\underline{k}\int_{-\infty}^{o} dt\, \underline{E}\cdot\underline{v}(t)\exp\left[-\int_t^{o} ds/\tau(s)\right] \qquad (13)$$

where $\Delta\epsilon$ is the mean energy gained by the electron from the field $\underline{E}$ since its previous collision. Its velocity $\underline{v}(t)$ at time t differs from $\underline{v}$ (= $\underline{v}(0)$) because of its orbital motion in $\underline{H}$; similarly $\tau(s) \neq \tau(0)$ if τ varies with $\underline{k}$.

In low fields, we put $\int_t^0 ds/\tau(s) = u$, and expand $\underline{v}(t)\,\tau(t)$ as a Taylor series in u about $\underline{v}(0)\tau(0)$: we then obtain at once

$$\sigma_{ij} = -\frac{e^2}{4\pi^3}\int v_i \frac{df_0}{d\epsilon}\, d\underline{k} \int_0^\infty du \left[(v_j\tau) + u(\dot{v_j\tau}) + u^2(\ddot{v_j\tau})/2 \ldots\right] e^{-u}$$

$$= -\frac{e^2}{4\pi^3}\int v_i \frac{df_0}{d\epsilon}\, d\underline{k} \left[(v_j\tau) + (\dot{v_j\tau}) + (\ddot{v_j\tau}) \ldots\right] \qquad (14)$$

where $v_j\tau = v_j(0)\tau(0)$, and $(\dot{v_j\tau})$ etc. are its derivatives with respect to u around the electron orbit. Writing $du = -dt/\tau = d\theta/\omega\tau$, where θ is the 'orbit angle' and $\oint d\theta = 2\pi$ for a closed orbit, we see that (14) is equivalent to an expansion of σ_{ij} in powers of ω, i.e. in powers of H. This series was first derived by Jones and Zener (1934b), assuming $\tau = \tau(\epsilon)$ only, and evaluated by them up to terms in H^2 for the Fermi surface of Li, which they had calculated to be a slightly distorted sphere. The theoretical values of D_L and D_T so obtained were only about an order of magnitude smaller than those found experimentally: quite creditable agreement. This pioneer work remained for many years the only serious attempt at detailed comparison of theory and experiment for a real metal.

Davis (1939) evaluated (14) up to terms in H^2 for a metal of cubic symmetry, expressing $\epsilon(\underline{k})$ and $\tau(\underline{k})$ in terms of the cubic harmonic Y_4 as $k = k_0(\epsilon) + k_4(\epsilon)Y_4(\theta,\phi)$, $\tau = \tau_0(\epsilon) + \tau_4(\epsilon)Y_4(\theta,\phi)$. This should give a reasonably good description if the distortion from cubic symmetry is not too great; in particular, if the Fermi surface does not touch the zone boundary. He gave explicit expressions for B_L and B_T in the low-field region where $D_L = B_L H^2$, $D_T = B_T H^2$, in terms of the ratios k_4/k_0, τ_4/τ_0 and their energy-dependence, and showed that a reasonable choice of these parameters could account for the observed values of B_L and B_T in group I metals. As with quadratic energy surfaces, the effect of anisotropy of $\tau(\underline{k})$ can be simulated by appropriate anisotropy of $\epsilon(\underline{k})$ (though the converse is not true), so that the two cannot be determined separately by experiment in the low-field region, unless two scattering mechanisms, assumed to be of different anisotropy, are present in variable proportions (Cooper and Raimes 1959). Davis was disturbed to find that the maximum value of B_L/B_T expected theoretically was 0.24, whereas experimental values usually exceeded this. This is not surprising, however: the experimental work has generally been done on polycrystalline samples, whereas Davis's calculation was confined to a single crystal with H parallel to a cube axis. For this field direction, B_L is particularly small, for much the same reason that it vanishes in the many-ellipsoid model when each ellipsoid has a principal axis in the direction of H.

García-Moliner (1958b) has used Seitz's (1950) phenomenological expression for the magnetoresistance of cubic metals in low fields, derived from symmetry arguments, to deduce average values of B_T and B_L for polycrystalline material from Davis's results, and shown that this difficulty then vanishes. He has then used these average values to interpret the experimental data on the alkali metals. Assuming an isotropic relaxation time, and making various other assumptions to eke out the rather meagre experimental data, he concludes that in Na the Fermi surface is indeed spherical to within 1 - 2 per cent, whereas in Li the surface must be far less spherical than some calculations suggest, and closer to the form originally estimated by Jones and Zener (1934b).

Olson and Rodriguez (1957) have used essentially Davis's method, extended to terms in Y_6, to evaluate their low-field experimental results on Cu, Ag and Au. They conclude that their results cannot be fitted by assuming anisotropy in $\tau(\underline{k})$ only, with $\epsilon(k)$ spherical; on the other hand, if $\tau(\underline{k})$ is assumed to be isotropic, the results suggest that the Fermi surface may touch the zone boundary in the [111] directions - a suggestion which has been amply confirmed by other work. To justify their assumption that $\tau = \tau(\epsilon)$ only, they show that if a unique relaxation time (independent of the form of f_1) exists at all for elastic collisions, it must depend only on energy and not directly on $\underline{k}$. This, though true, could equally well be used to argue that the whole relaxation-time assumption is inadequate, and that collisions cannot be properly represented by a parameter $\tau(\underline{k})$ independent of f_1. It certainly does not justify the abandonment of $\tau(\underline{k})$ as a better approximation to the collision integral than $\tau(\epsilon)$. They also argue qualitatively that point defect scattering will be fairly isotropic, and that for this reason $\tau(\underline{k})$ will in the residual resistance region indeed approximate to $\tau(\epsilon)$ simply, but this is certainly untrue of dislocation scattering (van Buren 1957), and is probably untrue of point defect scattering also (Ziman 1959). (It is probably untrue also, as one would expect, of phonon scattering: cf. Cooper and Raimes 1959).

IV. THE VARIATIONAL METHOD

In the intermediate-field region, where $\omega\tau \sim 1$, it is difficult to draw any general conclusions about σ_{ij} from (13). It is convenient to discuss here an alternative approach, in principle far more powerful, which does not depend on the relaxation-time assumption and which covers the whole range of fields from $\omega\tau \ll 1$ to $\omega\tau \gg 1$ with equal facility. Garcia-Moliner and Simons (1957) and Tsuji (1958) have shown how the variational method of Kohler (1949b), Sondheimer (1950) and Ziman (1956) can be extended to include the presence of a magnetic field (see also Ziman 1960). In this method, a plausible trial function is chosen for f_1 in (5), containing a certain number of adjustable parameters, and the best value of these parameters found by minimizing the appropriate variational function. If the form of the trial function has been well chosen, the value of f_1 and more particularly the resultant value of σ_{ij} will approximate well to the exact solution. The right side of (5) can be replaced by the general collision integral without making the problem completely intractable. García-Moliner (1958a) has discussed metals of cubic symmetry by taking

$$f_1 = \Sigma_i \left[a_i + b_i Y_4 (\theta, \phi) \right] v_i \, df_0/d\epsilon \tag{15}$$

as his trial function, which is appropriate if $\epsilon(\underline{k})$ and the scattering are not too anisotropic. The variational result for σ_{ij} then involves eight field-independent integrals, all of which involve the form of $\epsilon(\underline{k})$ and three, in addition, the form of the collision operator. Treating these as parameters, Garcia-Moliner finds that Seitz's (1950) phenomenological expression for σ_{ij} is formally applicable at all fields, with coefficients which are now functions of $|H|^2$, and that at high fields $\rho(H)$ should saturate and become isotropic. At intermediate fields, the observed angular dependence of D_T in Na and W can be simulated quite closely by appropriate choice of the eight parameters available. At low fields his results approximate to those of Davis, for H // [100].

It is not clear from this work how far it may be possible to separate out the effects of scattering anisotropy and $\epsilon(\underline{k})$ anisotropy by measurements over a wide range of fields, but it seems that if and when (15) is a good approximation to the form of f_1, a useful amount of information may be derivable from experiment, in terms of the eight parameters involved. For more realistic forms of f_1, applicable for instance to open Fermi surfaces, the computational labour involved looks rather forbidding, and the detailed interpretation of the fitting parameters in terms of the collision integral and the form of $\epsilon(\underline{k})$ will not be easy.

V. FERMI SURFACE TOPOLOGY; TYPES OF ORBIT

At low fields, an electron in its orbit in $\underline{k}$-space traverses only a small part of the Fermi surface before colliding; the magnetoresistance effects are essentially determined by integrals over the local properties, and the large-scale structure of the surface is not important. At high fields, where $\omega\tau \gg 1$, the electron explores its whole orbit or a large part of it before colliding, and the topology of the surface becomes of paramount importance.

In discussing the topology, it is essential to think of the energy surfaces $\epsilon(\underline{k})$ in each band as extending periodically throughout the whole of $\underline{k}$-space, and not simply confined within the Brillouin zone about $\underline{k} = 0$. Thus the Brillouin zone itself must be thought of as repeated periodically to fill $\underline{k}$-space completely, so that when an electron approaches the 'boundary' of a given zone, it simply passes through into the adjacent repetition of the same zone. This was familiar enough thirty years ago (cf. Slater 1934) but later became overlooked, so that a number of zone-structures were proposed which lacked this basic space-filling property (cf. Chambers 1956b, Barron and Fisher 1959). There are indeed only fourteen types of Brillouin zone for electrons, as for phonons, corresponding to the fourteen Bravais lattices, though if the lattice has a basis, symmetry may lead to the zones 'sticking together' in pairs, as in the c.p.h. structure.

If a given sheet of the Fermi surface lies entirely within the zone boundaries, it must clearly form a closed surface. If it consists of sections localized on separate zone faces or around separate corners, these sections will again combine to form simple closed surfaces, containing electrons or holes, in the periodic zone scheme. But if the sections extend from one face of the zone to another, or from one corner to another, the resultant structure in the periodic zone scheme will be a multiply-connected or 'open' surface which extends throughout $\underline{k}$-space. This concept again was familiar thirty years ago: Bethe, in the source-book, gave a picture of such an open surface for a simple cubic metal (Sommerfeld and Bethe 1933, Fig. 24), and pointed out that the noble metals might well have such surfaces (ibid., Fig. 25c). But this too became overlooked, because of the attractive simplicity of the electron-hole concept, until Lifshitz, Azbel' and Kaganov (1955, 1956; hereafter called LAK) drew attention to the important properties of such surfaces in high magnetic fields.

We have seen that the orbit of an electron between collisions is given by ϵ = constant, k_z = constant (where $H = H_z$). We can visualize such an orbit by thinking of the periodically repeated Fermi surface, sliced through by a plane normal to $\underline{H}$. If the Fermi surface consists simply of closed regions, this will clearly give rise only to closed orbits, but if the surface is open, the possibilities are much more complex. Fig. 3 shows one possible type of open Fermi surface for a cubic metal (s.c., b.c.c. or f.c.c.: in the last case, there would be two separate, interwoven surfaces of this type). This is not the open surface which occurs for instance in the noble metals, but it is the simplest to draw and to visualize. The figure shows, for H_z along the [100] axis, two sections for different values of k_z. The shaded regions show the filled states, and the boundaries of the shaded regions show the electron orbits on the Fermi surface for these values of k_z. If H_z is out of the paper, the electron moves around the orbit keeping the filled states on its right (cf. (1)), so that the orbits on the left move clockwise, and those on the right - 'hole' orbits - anticlockwise. Correspondingly, the cyclotron mass $m^* = (\hbar^2/2\pi)\, dA/d\epsilon$ is positive for the electron orbits and negative for the hole orbits. For one particular value of k_z, intermediate between those shown, there will be a transition between electron and hole orbits, and open orbits will arise which run diagonally through the whole of k-space. But at the points of intersection of these orbits, the energy $\epsilon(k_x, k_y)$ has a saddle-point, so that the velocity in the x-y plane vanishes; correspondingly, for values of k_z close to the critical value, $dA/d\epsilon$ and m^* become very large, and $\omega = eH/m^*c$ very small. Lifshitz and Kaganov (1959), in a useful review of conduction electron dynamics, have discussed the properties of such saddle points, and have also given examples of some really complicated Fermi surfaces.

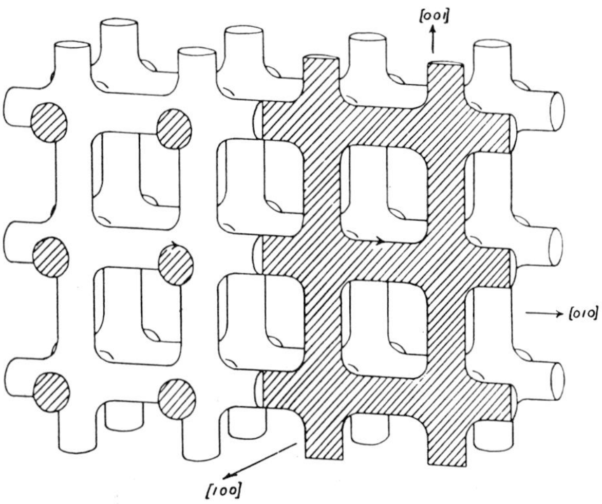

Fig. 3 One possible type of open Fermi surface for a cubic metal, showing two sections of constant k_z, for H_z along [100]. Left: electron orbits; right: 'hole orbits'.

Thus for H_z normal to a principal plane of the reciprocal lattice (i.e. along a principal axis of the crystal lattice), only closed electron and hole orbits exist, except for one particular value of k_z, which gives intersecting open orbits with $\omega = 0$. If we now tilt H_z about the [010] axis, a typical slice of the Fermi surface looks like Fig. 4: we now have closed electron orbits at the top and bottom of the figure, and between them open orbits, unbounded in $\underline{k}$-space and again having $\omega = 0$. In this case such orbits exist not just for one unique value of k_z, but for a range of k_z in front of and behind that shown. Alternatively, if the Miller indices of the slice plane are incommensurable, we need consider only one slice, which will eventually 'sample' every point on the Fermi surface

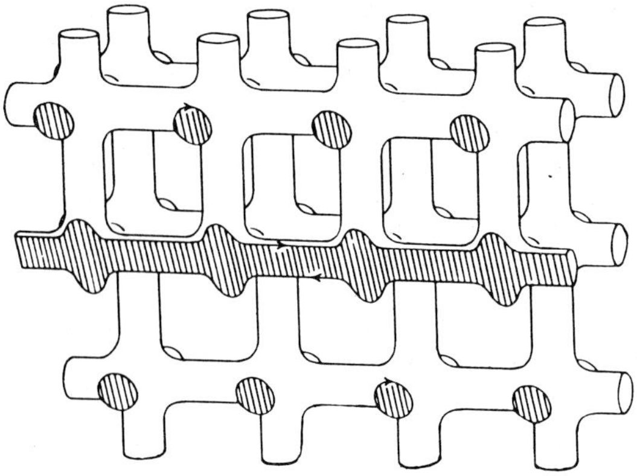

Fig. 4 Section of Fermi surface for H_z in (010) plane, showing the open orbits bounding the central shaded strip. Electron orbits above and below.

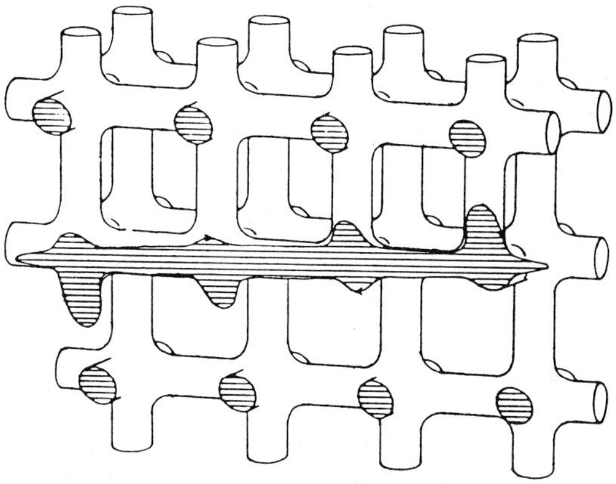

Fig. 5 As Fig. 4, with H_z tilted slightly out of the (010) plane. The two open orbits have coalesced into an extended orbit - basically an elliptical orbit along [010].

as it passes through the periodic zone scheme.

We shall call open orbits of the type shown in Fig. 4 'type A' open orbits. If we next tilt H_z at a slight angle θ to the (010) plane, we arrive at the skewed slice shown in Fig. 5: the open orbits of Fig. 4 have coalesced into an extended electron orbit which runs through several cycles of the zone scheme before closing, and is basically a section of the [010] cylindrical surface (as, too, are the open orbits of Fig. 4). The cyclotron frequency will thus be approximately $\omega_0 \sin\theta$, where ω_0 is the frequency for the simple electron orbits of Fig. 3 (cf. (11) and (12)).

Finally and most important, what happens if we tilt H_z slightly away from a principal crystal axis in an arbitrary direction? Clearly some of the resultant orbits may be extremely tortuous (Shockley 1950, Chambers 1956b), but Lifshitz and Peschanskii (1958, hereafter called LP 1), with remarkable insight, have shown that (i) these orbits will be open, (ii) however tortuous in detail, their overall path follows a straight line in k-space, and (iii) the direction of this line is the intersection of the x-y plane normal to H_z with the adjacent reciprocal lattice plane. Fig. 6 attempts to illustrate this. H_z has been tilted away from [100] slightly, in roughly the [110] direction, and the resultant slice passes from a region of electron orbits into a region of hole orbits; separating these two is the 'type B' open orbit 00', running in the general direction of the dashed line, which we choose as the x axis. If θ is the angle between H_z and [100], it is clear that for small θ the regions of electron and hole orbits will be large, and the transitions from one to the other infrequent, so that the open orbits will be few and far between; the proportion of the Fermi surface covered by open orbits will grow linearly with sin θ. As θ increases, the region of hole orbits between two open orbits will shrink until eventually the hole orbits vanish, at some angle θ', and the open orbits with them. According to LP 1, the proportion of Fermi surface covered by open orbits falls linearly to zero as θ approaches θ', but it seems more probable to me that the proportion will continue to increase as sin θ up to θ', and then vanish abruptly. Possibly two adjacent open orbits occasionally coalesce into an archipelago of extended orbits at some slightly smaller angle θ'', which then vanish at θ', but this seems unlikely.

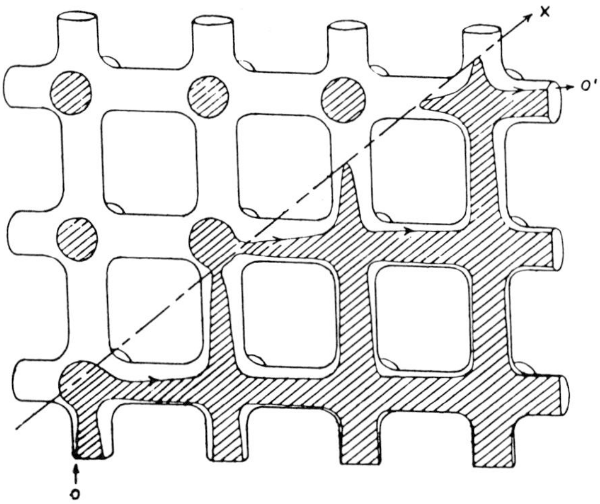

Fig. 6 As Fig. 3, with H_z tilted slightly away from [100] in an arbitrary direction. Regions of electron orbits (top left) and hole orbits (bottom right), separated by an open orbit 00'. Direction of open orbit taken as x axis.

It becomes clear that open Fermi surfaces are relatively complex compared with closed ones: they may give rise to open orbits of type A when H is normal to a principal axis (as in Fig. 4); in this case extended orbits arise for adjacent directions of H (Fig. 5), or of type B when H is almost normal to a principal plane (Fig. 6); in this case the open orbits separate regions of electron and hole orbits and vanish abruptly at the angle θ' where the electron or hole orbits vanish. As H approaches the normal to the principal plane, the type B open orbits become rarer, until in the limit they form a set of measure zero, occurring only for isolated values of k_z. In the general case, 'principal axis' here means any axis which can be drawn to lie entirely within the periodically repeated Fermi surface (Priestley 1960), and 'principal plane' means any plane which gives only electron orbits, but which for different k_z gives only hole orbits.

If the Fermi surface consists of a 'corrugated plane', as may occur in uniaxial metals, open orbits will exist for all directions of H, except the direction normal to the plane (LAK; Lifshitz and Peschanskii 1960, hereafter called LP 2). Conversely, there exist Fermi surfaces which, though themselves open, never give rise to open orbits for any field orientation: one such surface, proposed by LAK, arises in the third zone of Al according to Harrison (1960). An open surface of this type will not be revealed by magnetoresistance measurements.

VI. THE HIGH-FIELD REGION: THEORY

We can now discuss the high-field behaviour of closed and open Fermi surfaces. According to (1), we have $\hbar \underline{k} = e(\underline{r} - \underline{r}_0) \times \underline{H}/c$, so that the x-y projections of the real-space orbits are similar to the $\underline{k}$-space orbits, but rotated through $\pi/2$. Thus for closed or extended orbits, the average velocities $\overline{v_x} = \overline{v_y} = 0$, though $\overline{v_z} \neq 0$ generally. But for an open orbit running through $\underline{k}$-space in the x direction, the real-space motion is unbounded in the y direction though still bounded in x: $\overline{v_x} = 0$; $\overline{v_y}$, $\overline{v_z} \neq 0$. This is the essential difference between open and closed orbits.

At high fields, an electron will traverse a closed orbit many times, or 'sample' a large part of an open orbit, between collisions, so that v(t) and $1/\tau(s)$ in (13) are rapidly varying functions of time, and in the limit we can replace $1/\tau(s)$ by its mean value $\overline{\tau^{-1}}$ around or along the orbit. We then have

$$\sigma_{ij} = -\frac{e^2}{4\pi^3} \int v_i \frac{df_o}{d\epsilon} \, d\underline{k} \int_{-\infty}^{o} dt \, v_j(t) \exp(\overline{\tau^{-1}} t) \tag{16}$$

If $\overline{v_i}$, $\overline{v_j} \neq 0$, we can also replace $v_j(t)$ by $\overline{v_j}$, since $\exp(\overline{\tau^{-1}} t)$ is only falling slowly, and thus we at once obtain for σ_{zz}

$$\sigma_{zz}(\infty) = -\frac{e^2}{4\pi^3} \int v_z \frac{df_o}{d\epsilon} \, d\underline{k} \; \overline{v_z} / (\overline{\tau^{-1}}), \tag{17}$$

independent of H, and smaller than the zero-field value

$$\sigma_{zz}(o) = -\frac{e^2}{4\pi^3} \int v_z^2 \, \tau \frac{df_o}{d\epsilon} \, d\underline{k} \tag{18}$$

by an amount depending on the anisotropy of v_z and τ^{-1} around the orbits. If open orbits are present, $\sigma_{yy}(\infty)$ and $\sigma_{yz}(\infty)$ will contain terms independent of H similar to (17), though generally smaller since the open orbits cover only part of the Fermi surface.

If $\overline{v_j} = 0$, no terms survive in H^0, and we have to proceed a little more carefully to find the leading terms: it is not difficult to show that in the high-field limit σ_{xx}, $\sigma_{yy} \sim H^{-2}$, and $\sigma_{ij} \sim H^{-1}$ for all other components, except for σ_{zz}, which is independent of H and given by (17) (LAK 1955, 1956; McClure 1956, Chambers 1956a; LAK also show that these conclusions are unaffected if $\tau(\underline{k})$ is replaced by a general collision operator). In particular, if the Fermi surface is closed, we have $\sigma_{xy} = -\sigma_{yx} = (n_1 - n_2)\,ec/H$, where n_1 and n_2 are the numbers of electrons and holes per cm^3. This result was first obtained by Kohler (1949a).

Thus if only closed orbits are present, the field-dependence of the various terms of σ_{ij} is qualitatively the same as for the free-electron model (3), and so the resistivity tensor has precisely the same qualitative behaviour (2): the diagonal components ρ_{ii} saturate, $-\rho_{xy} = \rho_{yx}$ tends to $H/(n_1 - n_2)ec$, and the other non-diagonal components saturate at fairly small values, depending on the anisotropy of the Fermi surface and the direction of H. If $n_1 = n_2$, the highest term in σ_{xy} becomes of order H^{-2}, and we then have typical two-band behaviour: on inverting σ_{ij} we find that ρ_{xx}, $\rho_{yy} \sim H^2$. The only change is that ρ_{xy}, ρ_{yx} also contain a term in H^2, in general, in addition to the Hall term linear in H.

For a single closed Fermi surface it is not difficult to estimate a plausible upper limit for the quantities $D_L(\infty) = \rho_{zz}(\infty)/\rho_{zz}(0) - 1$ and $D_T(\infty) = \rho_{xx}/\rho_{xx}(0) - 1$. Since $\rho_{zz}(\infty) = 1/\sigma_{zz}(\infty)$, we have from (17), (18) $\rho_{zz}(\infty)/\rho_{zz}(0) = \langle \overline{v_z}^2 \tau \rangle / \langle \overline{v_z^2/\tau^{-1}} \rangle$ where the two averages are first around a single orbit, and then over all orbits. For a spherical Fermi surface with $\tau = \tau(\underline{k})$, Kohler had previously found $\rho_{zz}(\infty)/\rho_{zz}(0) = \overline{\tau} \cdot \overline{\tau^{-1}}$: cf. Justi and Kohler (1939). Kohler (1949a) also studied $\rho_{xx}(\infty)$, and showed that it saturated; his conclusion that the saturation value was independent of the direction of H is not quite correct, and arises from the use of Peierls' (1931) expression for f_1 in the limit $\tau \to \infty$: $f_1 = (\hbar c/eH^2)(df_0/d\epsilon)\, \underline{H} \times \underline{E} \cdot \underline{k}$. This solution is not unique, and the actual form of f_1 in the limit includes an extra term. This term however will be small for cubic metals if $\epsilon(\underline{k})$ is not too anisotropic; if we neglect it, we find $\rho_{xx}(\infty)/\rho_{xx}(0) \simeq \int (k^2/\ell)\, dS. \int \ell\, dS/9V^2$, where $\ell = \tau v$ and the integrations are over the Fermi surface, of enclosed volume V (Chambers 1956a). For reasonable anistropies of $\epsilon(\underline{k})$ and $\tau(\underline{k})$, these expressions lead to estimates of $\sim 2 - 3$ at most for both $D_L(\infty)$ and $D_T(\infty)$. For a two-band or many-band model of closed surfaces, $D_L(\infty)$ will be unaffected, and we may still expect $D_T(\infty) \lesssim 10$, unless the numbers of electrons and holes are closely equal.

If open orbits are present, the situation is completely altered by the zero-order term in σ_{yy} (LP 1, LP 2). Because of this, we now find that $\rho_{xx} \sim H^2$ at high fields, though ρ_{yy} and ρ_{zz} still saturate, and $\rho_{xy} \sim H$. (If $\sigma_{yz} \sim H^0$, we also find that $\rho_{xz} \sim H$). It follows that

$$\rho_{\alpha\alpha} = \rho_{xx} \sin^2 \phi + \rho_{yy} \cos^2_\phi \sim H^2, \text{ for } \phi \neq 0; \qquad (19)$$

i.e. $\rho_{\alpha\alpha}$ rises as H^2, for large enough H, for all directions of current flow in the x-y plane, except along the y axis itself. The extraordinary anisotropy of D_T often found experimentally in single crystals, as $\underline{H}$ is rotated about a fixed direction of $\underline{J}$, is at once accounted for: in some directions, open orbits exist and $D_T \sim H^2$; in other directions, only closed orbits occur and D_T then saturates at a small value (unless the metal then simulates a two-band metal, with $\sigma_{xy} \sim H^{-2}$ instead of H^{-1}: cf. discussion of experimental results in next section). The coefficient of the term in H^2 will clearly depend on the proportion of open orbits present, and will fall roughly as $\sin\theta$, for instance, as H becomes normal to a principal plane.

Physically, we can interpret the effect of open orbits as follows. Since $\overline{v_x} = 0$ for the open orbits, they act as two-dimensional conductors, capable of carrying a current $\underline{J}_o$ only in the y-z plane. They act in fact qualitatively like a cylindrical Fermi surface with H perpendicular to the cylinder axis. The remaining closed orbits give a contribution to σ_{ij} qualitatively like that of a free-electron gas, (3), as we have seen. Thus we have something like a two-band model, and if the resultant current $\underline{J}_o + \underline{J}_c$ is constrained to flow in any but the y direction, the total field $\underline{E}$ adjusts itself to lie almost along x. It then produces currents J_o and J_c which tend to cancel, leaving only a small resultant in the required direction (Fig. 7), rather as for the ordinary two-band model with $n_1 = n_2$ (Fig. 2a).

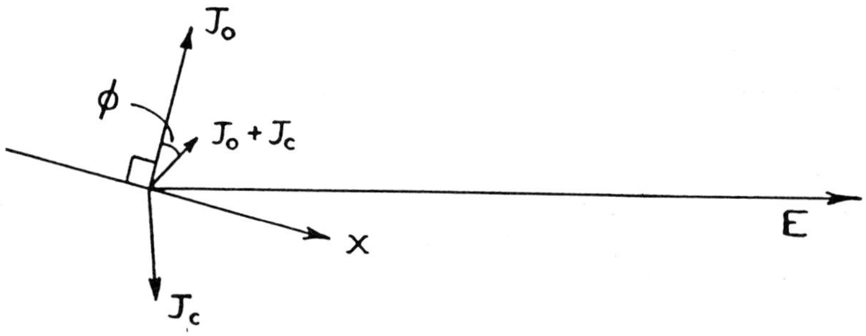

Fig. 7 Relation between $\underline{J}_c$, $\underline{J}_o$ and $\underline{E}$ in a metal containing both closed orbits and open orbits running along the x axis.

There is one case in which saturation occurs, even in the presence of open orbits: this is when two groups of type A orbits exist, running in different directions in the x-y plane. Then the two groups between them can carry a current in any direction in the x-y plane; the 'two-dimensional conductor' effect is lost, and both σ_{xx} and $\sigma_{yy} \sim H^0$. In this case both ρ_{xx} and ρ_{yy} will saturate at relatively modest values. Such a situation cannot arise for the simple Fermi surface shown in Figs. 3 - 7, but it can arise with e.g. the Fermi surfaces of the group I metals, for a few singular field directions.

It remains to discuss what happens if we have not open but highly extended orbits present (LP 1). This we can do at once, since such orbits behave qualitatively like a cylindrical Fermi surface with the cylinder axis at a small angle θ to the x-y plane. The contribution of the extended orbits is thus of the form (12); again the closed orbits will give a term of the form (3). Thus we now have $\sigma_{xx} \sim \alpha \sin^2\theta$ and $\sigma_{yy} \sim \alpha$, where $\alpha = (1 + \omega_0^2 \tau^2 \sin^2\theta)^{-1}$; σ_{xx} will be very small for small θ, but σ_{yy} remains little affected by H until we reach fields such that $\omega_0\tau \sim 1/\theta$; only for $\omega_0\tau >> 1/\theta$ does σ_{yy} finally fall as H^{-2}. In other words, a metal containing closed orbits of frequency $\sim \omega_0$ and extended orbits of frequency $\sim \omega_0 \sin\theta$ will look like a metal containing open orbits, in the field region $\omega_0\tau >> 1 >> \omega_0\tau\sin\theta$; ρ_{xx} will continue to grow as H^2 in this region, and only saturate finally at such high fields that $\omega_0\tau >> 1/\theta$. Physically, a highly-extended orbit 'looks like' an open orbit to an electron which only travels some way along one side of the orbit before colliding; for such an electron $\overline{v_y}$ averaged over a mean free path remains finite until the field is high enough for it to complete the orbit before colliding.

Lifshitz and Peschanskii have given stereograms showing the field directions for which open orbits exist for various types of Fermi surface having cubic symmetry. Figure 8 shows one example, for a Fermi surface consisting essentially of interpenetrating cylinders running in the [111] directions. Then type A open orbits exist for all field directions around the [111] zones, as shown by the full lines, and type B orbits around the [110] axes, as shown by the shaded regions. Thus when $\underline{H}$ lies in any of these directions, we expect a quadratic increase of D_T at high fields, except along the

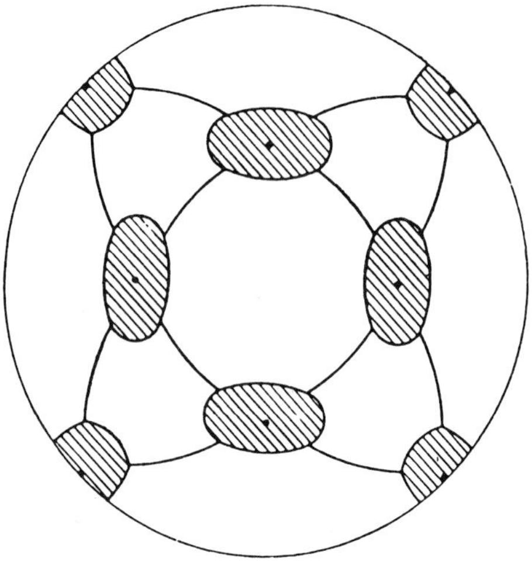

Fig. 8 Stereogram of magnetic field directions giving open orbits, for a Fermi surface consisting of interpenetrating cylinders along the [111] directions (from LP 2).

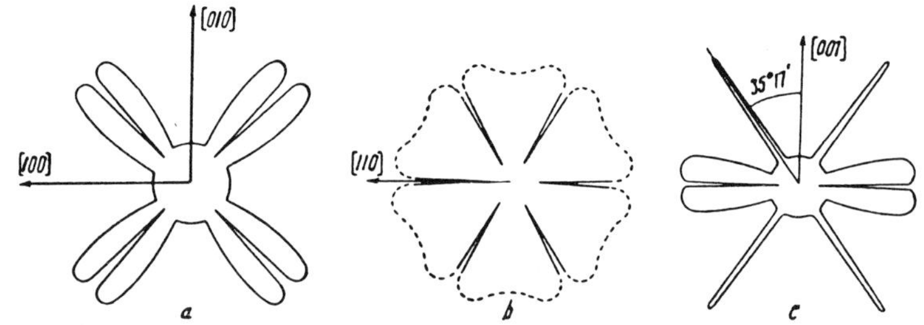

Fig. 9 Polar diagram of the expected variation of ρ ($\underline{H}$) with field direction, for constant field strength, corresponding to Fig. 8. $\underline{H} \perp \underline{J}$; (a) : $\underline{J}$ // [001] ; (b) : $\underline{J}$ // [111] ; (c) $\underline{J}$ // [110]. (from LP 2).

[110] directions themselves, shown by dots; there and in the rest of the diagram only closed orbits exist and D_T should saturate. The expected behaviour is illustrated in Fig. 9. In Fig. 9c, the sharp peaks occur where $\underline{H}$ crosses the [111] zone line; the peaks are not indefinitely sharp, because near this line the orbits become very extended, but they get progressively sharper in higher fields.

VII. HIGH-FIELD REGION: EXPERIMENTAL RESULTS

It is clear from the work of Lifshitz and Peschanskii that high-field studies of $D_T(\theta)$ on single crystals can provide exceedingly useful information on the presence of open orbits and thus on the connectivity of the Fermi surface. A great deal of experimental work on magnetoresistance has been published in the past 25 years, including

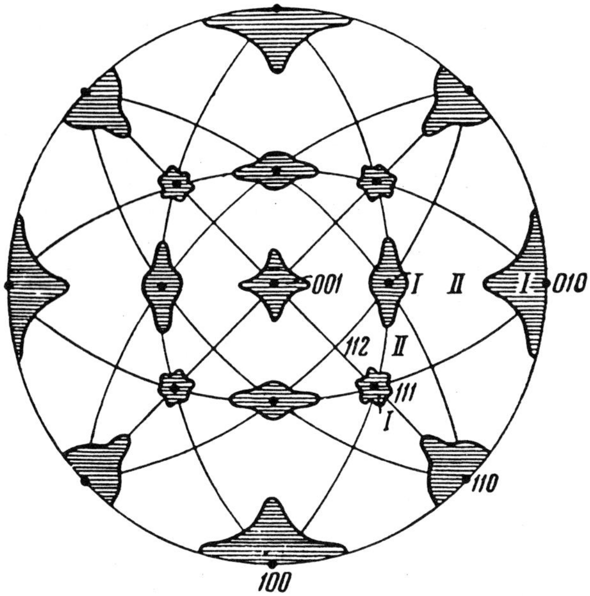

Fig. 10 Stereogram of magnetic field directions giving quadratic increase of D_T with H, in Au (from Gaidukov 1959).

a fair amount on single crystals, but much of this work can be interpreted only tentatively in terms of the LP theory. Alekseevskii and Gaidukov (1958, 1959a, b, 1960 : hereafter called AG 1, 2, 3, 4) have in the past two years begun a detailed series of single-crystal measurements in the light of the LP theory, and have already accumulated a remarkable amount of information, and in particular Gaidukov (1959) has made a full survey of Au, which we consider first. He finds that D_T fails to saturate for the field directions shown in Fig. 10. For example, Fig. 11 shows his results for a specimen in which $\underline{J}$ // [110], so that as θ varies from 0 to 90^0, the field direction varies along the straight line from [$\bar{1}$10] to [001] in Fig. 10. This line itself corresponds to a zone of open orbits of type A, according to Gaidukov, but these orbits must presumably be few in number, since their contribution to ρ(H) seems to be small, in the field used (23.5 kG). The peaks in Fig. 11 correspond to the boundaries of the shaded regions in Fig. 10; the deep minima between pairs of peaks correspond to the heavy dots in Fig. 10, where H is normal to a principal plane and only closed orbits exist. At θ = 55^0, type A orbits exist along both [110] and [111] ; as we have seen, the simultaneous presence of two open orbits with different axes gives saturation instead of a quadratic rise, and indeed there is a slight dip in Fig. 11 at this angle, corresponding to the cancellation of the small contribution from the [110] open orbits.

As Gaidukov has remarked, a Fermi surface of the type found by Pippard (1957) in Cu would give rise to precisely the open-orbit scheme shown in Fig. 10. Priestley (1960) has in fact shown that there is very satisfactory quantitative agreement between Gaidukov's results and those expected from Shoenberg's (1960) dHvA measurements on Au; indeed the agreement is rather too good: Gaidukov has plotted the positions of the

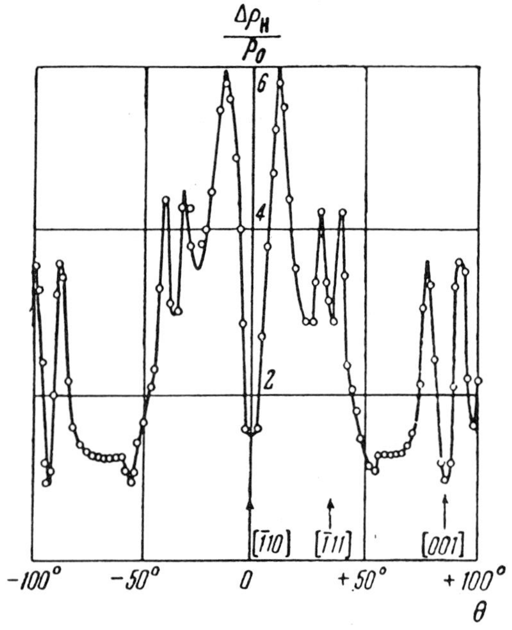

Fig. 11 Variation of D_T with field direction, in a field of 23.5 kG, for a single crystal Au specimen with J // [110] (from Gaidukov 1959).

maxima in D_T, and if LP 1 are correct in expecting a gradual fall in D_T towards the boundaries of the open-orbit regions, those regions must extend considerably further than is indicated in Gaidukov's figure. If we are right in thinking, on the other hand, that the open orbits should disappear abruptly, and D_T fall abruptly, at the edge of each region, the agreement between Fig. 10 and Priestley's calculations is explained. Against this, Gaidukov's Fig. 9 indicates that D_T falls smoothly to its low saturation value as we approach the edge of the region, just as it does when H becomes normal to a principal plane, and indeed his Fig. 9 shows a remarkable symmetry between the two cases. Just outside the open-orbit region, some of the orbits are still rather extended, and possibly these extended orbits account for this behaviour. In general, however, we can say that the agreement between the dHvA results and the D_T results in highly satisfactory.

Alekseevskii and Gaidukov have also studied Cu, Ag, Sn, Pb, Tl and Ga in some detail (AG 3), and Na (AG 2), Mg and Pt (AG 4) more briefly. We can also re-examine in the light of the LP theory the earlier single-crystal measurements of Justi, Grüneisen, Borovik and others, to which Jan (1957) gives extensive references, and the more recent work of Yahia and Marcus (1959: Ga), Connell and Marcus (1957: Bi), Babiskin (1957: Bi) and Borovik and Volotskaya (1960: In). Work on polycrystals is also relevant, however; as Ziman (1958) has pointed out, and as LP and AG have confirmed theoretically and experimentally, a polycrystalline sample of a metal with an open Fermi surface may give rise to precisely the linear $D_T(H)$ variation so frequently found experimentally over a wide range of fields. But this result must be handled with care: Lüthi, (1960), in an admirable survey of his high-field experiments and their interpretation, reports that in polycrystalline Zn and Sn, $D_T \sim H^{1.7}$ and $H^{1.8}$ respectively, and that in the highest fields used D_T reached values of 10^4, 10^3, respectively, from which one might conclude that in these metals we had closed Fermi surfaces with equal numbers of holes and electrons; $n_1 = n_2$. But the single-crystal work of AG on Sn and Borovik (1956) on Zn makes it clear that these metals have open Fermi surfaces; the rapid rise of D_T occurs because for many orientations where there are no open orbits, the metals 'look like' simple two-band metals with $n_1 = n_2$. The presence of open orbits in such a metal can be detected by making measurements in various current directions: for a two-band metal, $D_T \sim H^2$ for all J directions, but if $D_T \sim H^2$ because of open orbits, saturation should be found (possibly only in very high fields) when $\underline{J}$ is normal to the orbit axis (cf. (19)). Thus in interpreting work on polycrystals, we can conclude that open orbits exist if $D_T \sim H$ for large values of D_T, as found for instance in Cu by de Launay et al. (1959), but if $D_T \sim H^2$ in metals of even valency, we cannot tell whether the Fermi surface is open or closed.

On examining the literature with this in mind, and rejecting work in which the highest value achieved for D_T is too low to draw reliable conclusions, we find that little is left apart from the work of AG themselves, confirmed in some cases by Lüthi and others; in the following list I have indicated the sources in other cases.

(i) Closed Fermi surfaces: unequal numbers of electrons and holes.

In, Al, Na; (Li: Lüthi).

(ii) Closed surfaces: equal numbers of electrons and holes.

Bi (many authors), Sb (Yntema 1953), Be ? (Grüneisen and Adenstedt 1938), W (Grüneisen and Adenstedt 1938, Justi and Scheffers 1936, 1937, de Nobel 1957), Cd ? (Justi, Kramer and Schulze 1940).

(iii) Open surfaces.

Cu, Ag, Au, Mg, Zn, Cd?, Ga, Tl, Sn, Pb, Pt.

AG 3 conclude from their results that Cu and Ag have Fermi surface topologies similar to that of Au, and suggest tentatively that in Pb and Sn the surfaces resemble sets of broad interpenetrating cylinders running along the [111] axes in Pb, and along the [010] and [110] axes in Sn. They have not yet given a detailed interpretation for other metals. They class Cd under (iii); Justi et al. found $D_T \sim H^2$ in all orientations, but the anisotropy was very large, and it is not yet clear to which group it belongs. AG 3 make the interesting suggestion that Ga and Tl may have Fermi surfaces of the 'corrugated plane' type, and it may be that Cd also has a surface of this type, and possibly Be too.

In Li, Lüthi found a remarkably large difference between the properties at 80^0K and 4^0K: when plotted on a Kohler diagram, the value of D_T at 4^0K falls below the extrapolation of the 80^0K results by a factor of ten or more. From this he concludes that the close-packed low-temperature modification approximates much better to the free-electron model than the high-temperature phase, in agreement with the conclusions of Dugdale and Gugan (unpublished). Dr. Gugan points out to me that this result is puzzling, even if the low-temperature phase has $D_T = 0$, since only about 50 per cent of the material normally transforms. The discrepancy may be due to a breakdown of Kohler's rule, but Lüthi has shown that this is extremely well obeyed for Cu, Ag and Au between 80^0K and 4^0K.

Luthi also finds $D_T(\infty) \sim 1 - 3$ for In and Al, and $D_L(\infty) \lesssim 3$ for all the metals he has studied, in good agreement with theoretical expectation. In Na, MacDonald and others have sometimes found D_T to rise linearly with H, and Babiskin and Siebenmann

(1957) find that this behaviour persists up to $\omega\tau \sim 200$, where $D_T \sim 1.7$: this certainly cannot be taken to indicate an open Fermi surface, since apart from the very small value of D_T other workers (Justi and Kohler 1939, MacDonald 1950, AG 2) find saturation with $D_T \sim 1$ at much lower values of $\omega\tau$; it is probably due to spurious effects of the type discussed by Alekseevskii, Brandt and Kostina (1958). In this connection we should also mention the disturbing conclusion of Berlincourt (1958), that 'size effects' may arise in D_T and A_H (Hall coefficient) measurements at much greater specimen thicknesses than one would expect theoretically: this result remains puzzling.

We have not so far discussed Hall effect measurements, of which relatively few have been made. Borovik (1955) has found that in single-crystal Sn and Zn, A_H shows anisotropy as marked as that of D_T itself, as would be expected from the LP theory, and Borovik and Volotskaya (1960) find that $A_H(\infty)$ in In and Al corresponds to one free carrier per atom, within 10 per cent. In the case of Al, the evidence from A_H is decisive: Lüthi (1959) finds that the closed Fermi surface models of Heine (1956) and Harrison (1959, 1960) both lead to calculated values of D_T in rough agreement with his single-crystal measurements, but the positive A_H is compatible only with Harrison's model.

In summary, the remarkable anisotropy of D_T in single crystals, for many years an outstanding puzzle, is now understood, and found to be a fruitful source of information on the Fermi surface topology. From the results so far available, it is clear that the traditional picture of electrons and holes is totally inadequate for metals generally: in a remarkable number of them, open Fermi surfaces exist, and the 'hole' concept ceases to be well-defined.

VIII QUANTUM OSCILLATION REGION

The oscillatory behaviour which occurs for $\hbar\omega \gtrsim kT$, though of great importance, need not detain us long: from the physics of the problem, it is clear that oscillations in the transport properties have the same origin as the dHvA susceptibility oscillations, and they will clearly have essentially the same period and temperature-dependence; the only remaining question concerns their amplitude. Here there has been some theoretical disagreement, but this now seems to be resolved, and it appears that Ţiţeiça's (1935) remarkable pioneer work was in all essentials correct.

Ţiţeiça considered a free-electron model with phonon scattering, and showed that whereas D_L could be evaluated by a Boltzmann-like approach, treating E_z as a perturbation, D_T had to be evaluated by setting up initial wave-functions which included E_x, and corresponded to a steady-state current J_y, and then introducing the collisions as a perturbation which displaced the electron orbits in the x direction, so producing a current $J_x \sim 1/\tau$ (cf. (3)). Lacking the Poisson summation formula, he was unable to evaluate the oscillatory terms, but he showed that in the 'low-field' limit $\hbar\omega << kT$, $\omega\tau >> 1$, the resistivity was given exactly by the Grüneisen-Bloch equation (a conclusion also reached by Kohler (1940), using a more dubious semi-classical analysis), and he showed that the non-oscillatory part of D_L and D_T (corresponding to the Landau diamagnetism) was extremely small, except in the quantum limit $\hbar\omega > \epsilon_0$, where D_L, $D_T \sim |H|$. Akhieser (1939) derived the oscillatory terms by applying Poisson's formula to Ţiţeiça's formulation, still considering phonon scattering; Davydov and

Pomeranchuk (1940), more realistically, assumed point defect scattering and ellipsoidal energy surfaces, but otherwise followed essentially Ţiţeica's method, as did Zil'berman (1955), who also evaluated A_H and the thermoelectric effect. Davydov and Pomeranchuk found a logarithmic divergence in σ_{xx}, periodic in 1/H, unless collision broadening of the energy levels was taken into account; this has also been encountered by Adams and Holstein (1959), who give a full treatment of D_L and D_T for various scattering mechanisms, and by Kubo et al. (1959), who treat the problem by Kubo's fluctuation correlation technique. Because of this divergence, it is difficult to calculate accurately the amplitude of the oscillatory terms, but it is certainly much greater than that found by Lifshitz (1957), Lifshitz and Kosevich (1957) and Argyres (1958); Adams and Holstein point out that these authors neglected the effect of $\underline{E}$ itself on τ, and also (Lifshitz) of the quantization on τ (cf. also Argyres and Roth, 1959, Kosevich and Andreiev 1960, Skobov 1960, Pippard 1960).

Apart from Lifshitz and Kosevich, these authors have so far considered only quadratic $\epsilon(\underline{k})$, but there is no reason to doubt that for general $\epsilon(\underline{k})$ the oscillations will have comparable amplitude, and that their period and temperature-dependence will be essentially identical with those of the dHvA oscillations. One important result, justifying the semi-classical treatment used in the preceding sections, is that the non-oscillatory component of σ is practically unaffected by quantization, except in the quantum limit, where all the electrons occupy the lowest quantum state; we have already stressed that for metals (though not for semiconductors) this limit is beyond the reach of attainable fields.

As expected, the experimentally observed oscillations in the transport properties agree closely in period with the dHvA oscillations (for references see Soule 1958 or Kahn and Frederikse 1959). The one outstanding discrepancy, in Bi along the trigonal axis, where the dHvA period of 1.17×10^{-5} G^{-1} agreed poorly with the transport period of 1.58×10^{-5} G^{-1} (Babiskin 1957), appears to have been resolved by Brandt's (1960) discovery of dHvA oscillations from the holes, which for this orientation have a period of 1.57×10^{-5} G^{-1}. It remains to be explained why the electrons show up so much more clearly in the dHvA effect, and the holes in the transport oscillations. If this complementary behaviour is characteristic, studies of transport oscillations will prove a particularly useful adjunct to dHvA studies.

In this review, we have not considered the information theoretically obtainable by studying single-crystal thin films in a magnetic field (Kaner 1958, Gurevich 1958), since the practical preparation of such films is a forbidding problem; nor have we considered the quantum oscillations of the external photo-electric effect in a magnetic field: according to Zil'berman and Kulik (1960) these offer the enticing prospect of determining $\epsilon(\underline{k})$ not only on the Fermi surface but also below it. But it is clear that straight-forward measurements of the transverse magnetoresistance of single crystals in high fields can now give, and are already giving, a very useful contribution to our knowledge of the Fermi surface.

REFERENCES

Abeles, B. and Meiboom, S., 1956, Phys. Rev. 101, 544.

Adams, E.N. and Holstein, T., 1959, J. Phys. Chem. Solids 10, 254.

Akhieser, A., 1939, Zh. eksper. teor. Fiz. 9, 426; 1939, C.R. Acad. Sci. U.S.S.R. 23, 874.

Alekseevskii, N.E., Brandt, N.B. and Kostina, T.I., 1958, Zh. eksper. teor. Fiz. 34, 1339 (translation: 1958, Soviet Physics - J.E.T.P. 7, 924).

Alekseevskii, N.E. and Gaidukov, Yu.P., 1958, Zh. eksper. teor. Fiz. 35, 554 (translation: 1959, Soviet Physics - J.E.T.P. 8, 383).

Alekseevskii, N.E. and Gaidukov, Yu.P., 1959a, Zh. eksper. teor. Fiz. 36, 447 (translation: 1959, Soviet Physics - J.E.T.P. 9, 311).

Alekseevskii, N.E. and Gaidukov, Yu.P., 1959b, Zh. eksper. teor. Fiz. 37, 672 (translation: 1960, Soviet Physics - J.E.T.P. 10, 481).

Alekseevskii, N.E. and Gaidukov, Yu.P., 1960, Zh. eksper. teor. Fiz. 38, 1720 .

Argyres, P., 1958, J. Phys. Chem. Solids 4, 19: 1958, Phys. Rev. 109, 1115.

Argyres, P. and Roth, L.M., 1960, J. Phys. Chem. Solids 12, 89.

Ariyama, K., 1938, Sci. Pap. Inst. Phys. Chem. Res (Tokyo) 34, 344.

Babiskin, J., 1957, Phys. Rev. 107, 981.

Babiskin, J. and Siebenmann, P.G., 1957, Phys. Rev. 107, 1249.

Barron, T.H.K. and Fischer, G., 1959, Phil. Mag. 4, 826.

Berlincourt, T.G., 1958, Phys. Rev. 112, 381.

Blochinzev, D. and Nordheim, L., 1933, Z. Phys. 84, 168.

Borovik, E.S., 1955, Izv. Akad. Nauk. S.S.S.R. Ser fiz 19, 429; 1955, Columbia Tech. Trans. 19, 383.

Borovik, E.S., 1956, Zh. eksper. teor. Fiz. 30, 262 (translation: 1956, Soviet Physics-J.E.T.P. 3, 243).

Borovik, E.S. and Volotskaya, V.G., 1959, Zh. eksper. teor. Fiz. 36, 1650 (translation: 1959, Soviet Physics - J.E.T.P. 9, 1175).

Borovik, E.S. and Volotskaya, V.G., 1960, Zh. eksper. teor. Fiz. 38, 261 (translation: 1960, Soviet Physics - J.E.T.P. 11, 189).

Brandt, N.B., 1960, Zh. Eksper. teor. Fiz. 38, 1355.

Bronstein, M., 1932, Phys. Z. Sowjet. 2, 28.

Chambers, R.G., 1952a, Proc. Phys. Soc. A 65, 458.

Chambers, R.G., 1952b, Proc. Phys. Soc. A 65, 903.

Chambers, R.G., 1956a, Proc. Roy. Soc. A 238, 344.

Chambers, R.G., 1956b, Can. J. Phys. 34, 1395.

Connell, R.A. and Marcus, J.A., 1957, Phys. Rev. 107, 940.

Cooper, J.R.A. and Raimes, S., 1959, Phil. Mag. 4, 145, 1149.

Davis, L., 1939, Phys. Rev. 56, 93.

Davydov, B. and Pomeranchuk, I., 1940, J. Phys. U.S.S.R. 2, 147.

de Launay, J., Dolecek, R.L. and Webber, R.T., 1959, J. Phys. Chem. Solids 11, 37.

de Nobel, J., 1957, Physica 23, 261, 349.

Frenkel, J. and Kontorowa, T., 1935, Phys. Z. Sowjet. 7, 452.

Gaidukov, Yu.P., 1959, Zh. eksper. teor. Fiz. 37, 1281 (translation: 1960, Soviet Physics - J.E.T.P. 10, 913).

García-Moliner, F., 1958a, Proc. Roy. Soc. A 249, 73.

García-Moliner, F., 1958b, Proc. Phys. Soc. 72, 996.

García-Moliner, F. and Simons, S., 1957, Proc. Cambridge Phil. Soc. 53, 848.

Grüneisen, E. and Adenstedt, H., 1938, Ann. Phys. (Leipzig) 31, 714.

Gurevich, L., 1958, Zh. eksper. teor. Fiz. 35, 668; 1959, Soviet Physics - J.E.T.P. 8, 464.

Harrison, W., 1959, Phys. Rev. 116, 555.

Harrison, W., 1960, Phys. Rev. 118, 1182.

Heine, V., 1956, Proc. Roy. Soc. A 240, 340.

Jones, H., 1936, Proc. Roy. Soc. A 155, 653.

Jones, H. and Zener, C., 1934a, Proc. Roy. Soc. A 144, 101.

Jones, H. and Zener, C., 1934b, Proc. Roy. Soc. A 145, 268.

Justi, E. and Kohler, M., 1939, Ann. Phys. (Leipzig) 36, 349.

Justi, E., Kramer, J. and Schulze, R., 1940, Phys. Z. 41, 308.

Justi, E. and Scheffers, H., 1936, Phys. Z. 37, 700.

Justi, E. and Scheffers, H., 1937, Phys. Z. 38, 891.

Kahn, A.H. and Frederikse, H.P.R., 1959, Solid State Physics 9, 257. Edited by F. Seitz and D. Turnbull (N.Y., Academic Press).

Kaner, E.A., 1958, Zh. eksper. teor. Fiz. 34, 658 (translation: 1958, Soviet Physics - J.E.T.P. 7, 454.

Kohler, M., 1938, Phys. Z. 39, 9.

Kohler, M., 1940, Ann. Phys. (Leipzig) 38, 283.

Kohler, M., 1949a, Ann. Phys. (Leipzig) 5, 99.

Kohler, M., 1949b, Ann. Phys. (Leipzig) 6, 18.

Kosevich, A.M. and Andreiev, V.V., 1960, Zh. eksper. teor. Fiz. 38, 882.

Kubo, R., Hasegawa, H. and Hashitsume, N., 1959, J. Phys. Soc. Japan 14, 56.

Lax, B., 1958, Rev. Mod. Phys. 30, 122.

Lifshitz, I.M., 1957, Zh. eksper, teor. Fiz. 32, 1509 (translation: 1957, Soviet Physics - J.E.T.P. 5, 1227 or 1958, J. Phys. Chem. Solids 4, 11.)

Lifshitz, I.M., Azbel', M.Ya. and Kaganov, M.I., 1956, Zh. eksper. teor. Fiz. 31, 63 (translation: 1957, Soviet Physics - J.E.T.P. 4, 41.)

Lifshitz, I.M., Azbel', M.Ya. and Kaganov, M.I., 1955, Zh. eksper. teor. Fiz. 30, 220 (translation: 1956, Soviet Physics - J.E.T.P. 3, 143.)

Lifshitz, I.M. and Kaganov, M.I., 1959, Uspekhi Fiz. Nauk. 69, 419 (translation: 1960, Soviet Physics - Uspekhi 2, 831.)

Lifshitz, I.M. and Kosevich, A.M., 1957, Zh. eksper. teor. Fiz. 33, 88 (translation: 1958, Soviet Physics - J.E.T.P. 6, 67 or 1958, J. Phys. Chem. Solids 4, 1.)

Lifshitz, I.M. and Peschanskii, V.G., 1958, Zh. eksper. teor. Fiz. 35, 1251 (translation: 1959, Soviet Physics - J.E.T.P. 8, 875).

Lifshitz, I.M. and Peschanskii, V.G., 1960, Zh. eksper. teor. Fiz. 38, 188 (translation: 1960, Soviet Physics - J.E.T.P. 11, 137).

Lüthi, B., 1959, Helv. Phys. Acta 32, 470.

Lüthi, B., 1960, Helv. Phys. Acta 33, 161.

MacDonald, D.K.C., 1950, Proc. Phys. Soc. A 63, 290.

Mase, S. and Tanuma, S., 1960, Sci. Rep. Res. Inst. Tohoku A 12, 35.

McClure, J.W., 1956, Phys. Rev. 101, 1642.

McClure, J.W., 1958, Phys. Rev. 112, 715.

Olson, R. and Rodriguez, S., 1957, Phys. Rev. 108, 1212.

Okada, T., 1955, Mem. Fac. Sci. Kyusyu B 1, 168.

Okada, T., 1957, J. Phys. Soc. Japan 12, 1327.

Peierls, R., 1930, Leipzig Vortrager, p. 75. Leipzig: S. Hirzel: or see: Interference of Electrons, p. 71. London: Blackie (1931).

Peierls, R., 1931, Ann. Phys. (Leipzig) 10, 97.

Pippard, A.B., 1957, Phil. Trans. Roy. Soc. A 250, 325.

Pippard, A.B., 1960, Rep. Prog. Phys. 23, 176.

Priestley, M.G., 1960, Phil. Mag. 5, 111.

Seitz, F., 1950, Phys. Rev. 79, 372.

Shockley, W., 1950, Phys. Rev. 79, 191.

Shoenberg, D., 1960, Phil. Mag. 5, 105.

Skobov, V.G., 1960, Zh. eksper. teor. Fiz. 38, 1304.

Slater, J.C., 1934, Rev. Mod. Phys. 6, 209.

Sommerfeld, A. and Bethe, H., 1933, Hdbuch d. Phys. 24/2, 400. Springer.

Sondheimer, E.H., 1948, Proc. Roy. Soc. A 193, 484.

Sondheimer, E.H., 1950, Proc. Roy. Soc. A 203, 75.

Sondheimer, E.H. and Wilson, A.H., 1947, Proc. Roy. Soc. A 190, 435.

Soule, D.E., 1958, Phys. Rev. 112, 698, 708.

Ţiţeiça, S., 1935, Ann. Phys. (Leipzig) 22, 129.

Tsuji, M., 1958, J. Phys. Soc. Japan 13, 979.

van Buren, H.G., 1957, Philips Res. Rep. 12, 1, 190.

Yahia, J. and Marcus, J.A., 1959, Phys. Rev. 113, 137.

Yntema, G.B., 1953, Phys. Rev. 91, 1388.

Zil'berman, G.E., 1955, Zh. eksper. teor. Fiz. 29, 762 (translation: 1956, Soviet Physics - J.E.T.P. 2, 650).

Zil'berman, G.E. and Kulik, I.O., 1960, Zh. eksper. teor. Fiz. 38, 1188.

Ziman, J.M., 1956, Can. J. Phys. 34, 1256.

Ziman, J.M., 1958, Phil. Mag. 3, 1117.

Ziman, J.M., 1959, Proc. Roy. Soc. A 252, 63.

Ziman, J.M., 1960, Electrons and Phonons. Clarendon Press, Oxford.

"Magnetoresistance," W.G. Chambers in "*The Fermi Surface*," edited by W.A. Harrison and M.B. Webb (Wiley, N.Y., 1960), pp. 100–124.

MAGNETO-OSCILLATORY CONDUCTANCE IN SILICON SURFACES

A. B. Fowler, F. F. Fang, W. E. Howard, and P. J. Stiles
IBM Watson Research Center, Yorktown Heights, New York
(Received 23 March 1966)

We have observed Shubnikov–de Haas[1] oscillations in a two-dimensional electron gas in the (100) surfaces of p-type silicon inverted by an electric field perpendicular to the surface.

The experiments were made on the field-effect structures shown in the inserts in Fig. 1. An electric field was applied between the 100-Ω-cm p-type silicon substrate and the aluminum gate electrode, so that degenerate electrons were induced in the silicon surface. The conductance between the diffused n^+ source and drain electrodes was measured. The distance between these coaxial contacts was 10 μ and the circumference of the gap between them was 500 μ. We have shown by Hall measurements[2] on similar surfaces that surface trapping was insignificant so that above the threshold for conduction the surface carrier density could be calculated from $n = \kappa(V_g - V_0)/4\pi\delta e$ in cgs units, where κ is the silicon-dioxide dielectric constant, V_g and V_0 are the voltage applied to the gate electrode and the threshold gate voltage, respectively, e is the electronic charge, and δ is the thickness of the oxide. Measurements were made on samples for which the oxide thicknesses were 5330 Å and 1150 Å, and the results were consistent.

The conductance and transconductance were measured as functions of magnetic field perpendicular to the surface up to 93 kOe. The conductance decreased to one-half of its zero-field value at 35 kOe at 1.4°K, at a gate voltage of 15 V for the thin oxide sample. The over-all variation agreed well with the usual expression $[1+(\mu^2H^2/C^2)]^{-1}$ for magnetoconductance,[3] yielding a mobility of 3000 cm²/V sec. This equaled the mobility calculated from the conductance at zero magnetic field. It is

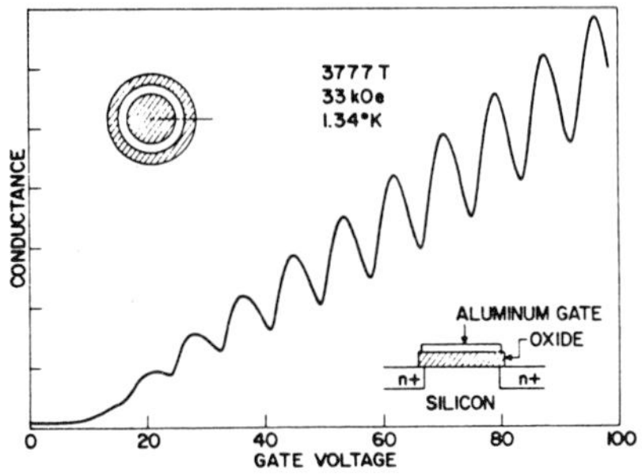

FIG. 1. The conductance as a function of gate voltage or surface field at 33 kOe. In the upper left-hand corner, a projection on the surface of the source-drain electrode configuration is shown. In the lower right-hand corner, a section through the structure is shown.

not surprising that Shubnikov–de Haas oscillations could be observed since $\omega_c\tau > 1$ above this field. The period of these oscillations, observed as low as 15 kOe at 1.3°K, varied linearly with the carrier density and inversely as the magnetic field.

Figures 1 and 2 show the conductance when the magnetic field was held constant while the gate voltage (and thus the surface field, surface carrier density, and Fermi energy) was varied. In contrast to the usual Shubnikov–de Haas experiments, which measure changes in the conductance as the Landau levels move through the Fermi surface, here we have moved the Fermi surface through the Landau levels.

The measurements indicate that because the period of the conductance maxima is constant, each Landau level contains the same number of states. This corresponds to a two-dimensional electron gas, for which the unperturbed density of states is independent of energy, rather than to a three-dimensional gas. In the magnetic field the distribution should consist of smeared δ functions.

It has been generally agreed[4-8] that the well created by high electric fields perpendicular to the surface of a semiconductor should result in quantization perpendicular to that surface. Howard and Fang[8] have argued that the four ellipsoids in silicon with their light mass perpendicular to the (100) surface should have zero-point energies much higher than for those with the heavy mass perpendicular. Thus, it was expected that the carriers would occupy two degenerate circular energy surfaces with an effective mass of approximately $0.2m_0$ parallel to the surface.

The degeneracy of the bands corresponding to the valley degeneracy, g_0, can be calculated

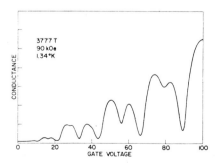

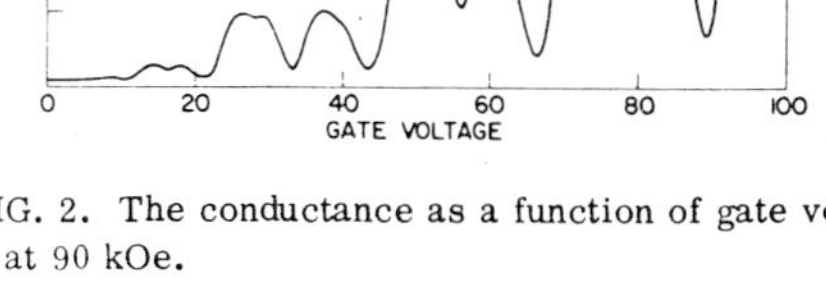

FIG. 2. The conductance as a function of gate voltage at 90 kOe.

from the period. This depends on the number of states between successive Landau levels being constant and equal to $n(E)E_L$, where $n(E)$ is the density of states at zero magnetic field, $2\pi g_0 g_s m^*/\hbar^2$; E_L is the Landau splitting, $eH\hbar/m^*c$; and g_s is the spin degeneracy. Thus, the number of electrons necessary to fill the surface bands between Landau levels or conductivity extrema is eHg_sg_0/hc. Since the carrier density is known as a function of gate voltage, the total degeneracy can be determined from the voltage period. It was 3.9 for the thin oxide samples and 4.32 for the thick oxide sample. This is satisfactory agreement with the expected value of 4.

There is good reason, however, to expect a removal of the valley degeneracy in the effective-mass approach. Lasher and Schultz[9] have pointed out that the two minima at k and $-k$ in the 100 direction, which in the infinite crystal correspond to states varying as $e^{\pm ikx}$, in a local potential well will provide degenerate bound states with $\sin kx$ and $\cos kx$ character. However, an ideal steep potential wall at a surface would allow only the function which behaves like $\sin kx$ as a solution. A real potential wall of finite extent might thus be expected to split the two types of states, thereby lifting the valley degeneracy. Otherwise, the effective-mass approximation is good.

The oscillation peaks could be traced as a function of magnetic field and electric field for each order. When this was done, the maxima were observed to split at the highest values of magnetic field and mobility or scattering time. Thus, in Fig. 2 the maxima at 60 and 50 V and at 83 and 74 V are split levels of the single peaks observed at 33 kOe at 20 and 28 V, respectively. A further splitting can be observed in the lowest levels. The apparent degeneracy of four is reduced to one at very high fields. Because the gate voltage is proportional to the surface carrier density rather than the Fermi energy, the split peaks occur halfway between the initial peaks once they are well resolved. The four levels must correspond to the two spins and two valleys. The valley splitting due to the wall cannot be larger than the Landau splitting at the lower fields. Thus, it cannot be larger than 1 meV.

The position of the conductance maxima may be plotted as a function of gate voltage and magnetic field. The extrapolation of the peaks to zero magnetic field intersect at the gate voltage for which the Fermi energy corresponds to the conduction-band minima. Thus, the threshold voltage is 4 V for the 5330-Å oxide sample. This is below the apparent threshold. The electrons have low mobility near the bottom of the bands where impurity scattering should be most important so that some oscillations are lost —below 10 V in Figs. 1 and 2. This result is consistent with our other observations of conductance in these samples at zero magnetic field. The Fermi energy and the order of the maxima can also be determined from these plots. If a mass of $0.19m_0$ is assumed, at a gate voltage of 100 V for the 5330-Å samples the Fermi energy is about 23 meV above the conduction-band minima. The phase of the oscillations indicates that the conductance maxima coincide with the maxima in the density of states.

The amplitude of the oscillations was measured as a function of temperature for gate voltages from 60 to 90 V. When the density of states is periodic in energy with the period of the Landau levels and is independent of temperature (conductance in zero field is observed to be independent of temperature) up to 4.3°K, then the amplitude of the conductance oscillations should be proportional to $T/\sinh \times(2\pi^2kT/\hbar\omega_c)$, which is the same as the corresponding expression for de Haas-van Alphen oscillations.[10] An accurate fit to this expression was found for a mass of $0.2m_0$.

Of many subsidiary observations, a damping of the oscillations when the source-drain voltage exceeded 0.5 mV was most interesting. At a source-drain voltage of 4 mV, corresponding to a field of 4 V/cm, the oscillations were 19% of those seen at 0.1 mV. Dumke has suggested that this is a result of electron heating since the energy relaxation time might be expected to be much longer than the momentum relaxation time.[11] The observed temperature dependence of the oscillation depth between 1.37 and 4.2°K may be used to estimate the electron temperature. When the lattice temperature was 1.37°K, the source-drain voltage was 2 mV, and the gate voltage was 90 V; the electron temperature was 3.2°K in a thick oxide sample.

Oscillatory effects could not be measured in (111) surfaces, presumably because our samples did not have the very high mobilities observed on (100) surfaces at 4.2°K. They could barely be seen on (110) surfaces but could not be interpreted.

In summary, we have demonstrated that a two-dimensional band model describes these surfaces and that there is surface quantization, that the valley degeneracy of the energy surfaces is two, with some small splitting, and that the mass parallel to the surface is $0.2m_0$. The reduction of the valley degeneracy predicted by Howard and Fang was substantiated.

We should like to thank J. Cummings for technical assistance.

---

[1]For review of this subject, see, for instance, D. Shoenberg, in Progress in Low Temperature Physics, edited by C. J. Gorter (North-Holland Publishing Company, Amsterdam, 1957), Vol. 2, p. 226.

[2]A. B. Fowler, F. F. Fang, and F. Hochberg, IBM J. Res. Develop. 8, 427 (1964).

[3]For a review of this subject, see, for instance, H. Brooks, Advances in Electronics and Electron Physics (Interscience Publishers, Inc., New York, 1955).

[4]J. R. Schreiffer, in Semiconductor Surface Physics, edited by R. H. Kingston (University of Pennsylvania Press, Philadelphia, Pennsylvania, 1957), p. 68.

[5]N. St. J. Murphy, in Solid Surfaces, edited by H. C. Gatos (North-Holland Publishing Company, Amsterdam, 1964), pp. 86 f.

[6]R. F. Greene, ibid., pp. 101 f.

[7]P. Handler and S. Eisenhour, ibid., pp. 64 f.

[8]W. E. Howard and F. F. Fang, Bull. Amer. Phys. Soc. 11, 240 (1966).

[9]G. J. Lasher and T. D. Schultz, private communication.

[10]R. E. Peierls, Quantum Theory of Solids, (Oxford University, Press, New York, 1954), p. 150.

[11]A. Zylbersztejn, thesis, Université de Paris (unpublished).

Journal of the Physical Society of Japan, Vol. 39, No. 2, August, 1975

Theory of Hall Effect in a Two-Dimensional Electron System

Tsuneya Ando,* Yukio Matsumoto and Yasutada Uemura
Department of Physics, University of Tokyo, Bunkyo-ku, Tokyo 113

(Received February 14, 1975)

Hall conductivity σ_{XY} is studied in various approximations. Characteristics of σ_{XY} are obtained for the case of both short- and long-ranged scatterers in the self-consistent Born approximation, which is the simplest one free from the difficulty of divergence. In case of short-ranged scatterers, a relation is shown to hold between σ_{XY} and σ_{XX} within this approximation, if one uses a relaxation time under magnetic fields. Under strong magnetic fields, effects of higher Born scattering become important in low-lying Landau levels. They depend on the sign of scatterers and strongly on their concentrations. Effects of simultaneous scattering from many scatterers are calculated to the lowest order.

§ 1. Introduction

When a strong magnetic field is applied perpendicularly to a two-dimensional electron system such as inversion layers on semi-conductor surfaces, the energy spectrum becomes discrete because of the complete quantization of the orbital motion. Such a singular system provides an ideal tool for studying the quantum transport phenomena. In a series of previous papers,[1-4] which are referred to as I, II, III and IV in what follows, the transverse conductivity σ_{XX} was studied systematically. The present paper is concerned with another important quantity—the Hall conductivity σ_{XY}.

In order to see characteristics of σ_{XY}, we first employ the self-consistent Born approximation (SCBA) which is the simplest one free from the difficulty of divergence caused by the singular nature of our system.[1] One takes into account effects of scattering from an impurity in the lowest Born approximation, while the collision broadening is included in a self-consistent manner.

According to a simple phenomenological argument, the Hall conductivity σ_{XY} becomes

$$\sigma_{XY}=-\frac{nec}{H}\frac{\omega_c^2\tau^2}{1+\omega_c^2\tau^2}$$
$$=-\frac{nec}{H}+\frac{1}{\omega_c\tau}\frac{ne^2\tau}{m}\frac{1}{1+\omega_c^2\tau^2}, \tag{1.1}$$

where n is the total number of electrons in a unit area, ω_c is the cyclotron frequency given by $\omega_c=eH/mc$, and τ is the relaxation time. The above equation can be written as

$$\sigma_{XY}=-\frac{nec}{H}+\Delta\sigma_{XY}, \tag{1.2}$$

with

$$\Delta\sigma_{XY}=\frac{1}{\omega_c\tau}\sigma_{XX}. \tag{1.3}$$

In the SCBA such kind of relationship between $\Delta\sigma_{XY}$ and σ_{XX} holds in case of short-ranged scatterers if a relaxation time under the magnetic field is used in eq. (1.3). In contrast to the case of σ_{XX}, however, contributions from higher order approximations are relatively important and can be crucial especially under extremely strong magnetic fields. Those higher order corrections are also investigated.

In § 2, starting from the center migration theory of Kubo *et al.*[5,6] and using the technique of Green's function, we obtain necessary equations of $\Delta\sigma_{XY}$. In § 3, σ_{XY} is explicitly calculated in the SCBA under strong magnetic fields and characteristics in case of short- and long-ranged scatterers are obtained. Assuming scatterers with a short-ranged potential, we calculate the oscillatory σ_{XY} under magnetic fields of arbitrary strength and show that eq. (1.3) holds. In the first part of § 4, we investigate effects of higher Born scattering and show that they are important in low-lying Landau levels especially when the concentration of scatterers is not so large. Effects of simultaneous scattering from more than two scatterers are also investigated and are shown to be relatively important in low-lying Landau levels. Some discussions on the obtained results are given in § 5.

* Present address: Physik-Department der Technischen Universität München, 8046 Garching b. München, Fed. Rep. Germany.

§ 2. Hall Conductivity σ_{XY}

We consider a two-dimensional system described by the Hamiltonian

$$H=\frac{1}{2m}\left(\boldsymbol{p}+\frac{e}{c}\boldsymbol{A}\right)^2+\sum_{\mu}\sum_{i} v^{(\mu)}(\boldsymbol{r}-\boldsymbol{R}_i, Z_i)\ , \tag{2.1}$$

where bold-faced letters represent two-dimensional vectors in the xy-plane, $\boldsymbol{A}=(-Hy/2, Hx/2)$, and $v^{(\mu)}(\boldsymbol{r}-\boldsymbol{R}_i, Z_i)$ is effective two-dimensional potential of a scatterer located at $(\boldsymbol{R}_i, Z_i)$. The potential is assumed to be cylindrically symmetric for simplicity. It can be written in a mixed representation[1,5)]

$$H=\sum_{NX} E_N a^{+}_{NX} a_{NX}+\sum_{\mu}\sum_{i}\sum_{NX}\sum_{N'X'}\sum_{m} 2\pi l^2 \varphi^{*}_{N+mX}(\boldsymbol{R}_i) \times \varphi_{N'+mX'}(\boldsymbol{R}_i)(Nm|v^{(\mu)}(Z)|N'm) a^{+}_{NX} a_{N'X'}\ , \tag{2.2}$$

where $E_N=(N+1/2)\hbar\omega_c$, $l^2=c\hbar/eH$, a^{+}_{NX} is the creation operator of an electron in the N-th Landau level having a center coordinate X, $\phi_{NX}(\boldsymbol{r})$ is its wave function, and $(Nm|v^{(\mu)}(Z)|N'm)$ is a matrix element of a potential between states with an angular momentum m around the position of a scatterer.

According to the center migration theory of Kubo *et al.*[5),6)] one has

$$\Delta\sigma_{XY}=\frac{e^2\hbar}{i\pi L^2}\int f(E)\,\mathrm{d}E\left\langle \mathrm{Tr}\,\dot{X}\left(\frac{\partial}{\partial E}\mathrm{Re}\frac{1}{E-H+i0}\right)\dot{Y}\ \mathrm{Im}\frac{1}{E-H+i0}-(\dot{X}\leftrightarrow\dot{Y})\right\rangle\ , \tag{2.3}$$

where L^2 is the area of the system, $f(E)$ is the Fermi distribution function, a trace should be taken over one-electron states, $\langle\cdots\rangle$ means an average over all configurations of scatterers, and

$$\begin{pmatrix}\dot{X}\\ \dot{Y}\end{pmatrix}=\frac{l}{\hbar}\sum_{\mu}\sum_{i}\begin{pmatrix} l\dfrac{\partial}{\partial y}\\ -l\dfrac{\partial}{\partial x}\end{pmatrix} v^{(\mu)}(\boldsymbol{r}-\boldsymbol{R}_i, Z_i)$$

$$=-\frac{l}{\hbar}\sum_{\mu}\sum_{i}\sum_{NX}\sum_{N'X'}\sum_{m}\sum_{\pm}\begin{pmatrix}\left(Nm\left| l\dfrac{\partial v^{(\mu)}(Z)}{\partial y}\right| N'm\pm 1\right)\\ -\left(Nm\left| l\dfrac{\partial v^{(\mu)}(Z)}{\partial x}\right| N'm\pm 1\right)\end{pmatrix} \times 2\pi l^2 \varphi^{*}_{N+mX}(\boldsymbol{R}_i)\varphi_{N'+m\pm 1X'}(\boldsymbol{R}_i) a^{+}_{NX} a_{N'X'}\ . \tag{2.4}$$

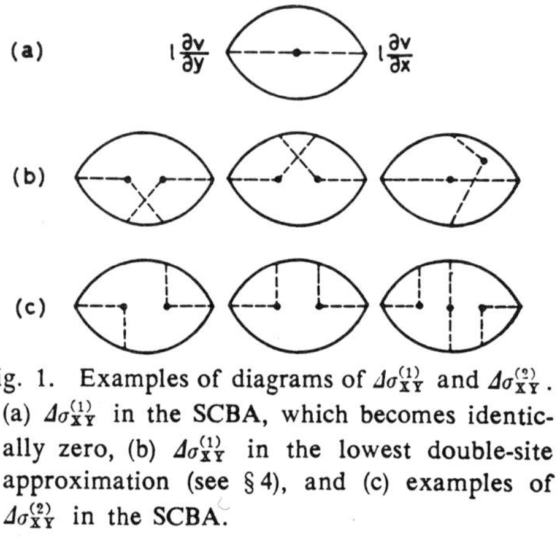

Fig. 1. Examples of diagrams of $\Delta\sigma^{(1)}_{XY}$ and $\Delta\sigma^{(2)}_{XY}$. (a) $\Delta\sigma^{(1)}_{XY}$ in the SCBA, which becomes identically zero, (b) $\Delta\sigma^{(1)}_{XY}$ in the lowest double-site approximation (see § 4), and (c) examples of $\Delta\sigma^{(2)}_{XY}$ in the SCBA.

We divide $\Delta\sigma_{XY}$ into two parts: $\Delta\sigma^{(1)}_{XY}$ and $\Delta\sigma^{(2)}_{XY}$. The former $\Delta\sigma^{(1)}_{XY}$ represents contributions from those diagrams which can not be cut into two parts by cutting two internal electron lines, and the latter $\Delta\sigma^{(2)}_{XY}$ contributions from other diagrams. Examples are shown in Fig. 1. As will be shown in the following, $\Delta\sigma^{(1)}_{XY}$ and $\Delta\sigma^{(2)}_{XY}$ are different in nature.

Introduce a quantity $\xi^{x}_{N+1,N}(E', E)$ or $\xi^{y}_{N+1,N}(E', E)$, which we call ξ-part. Each ξ-part is

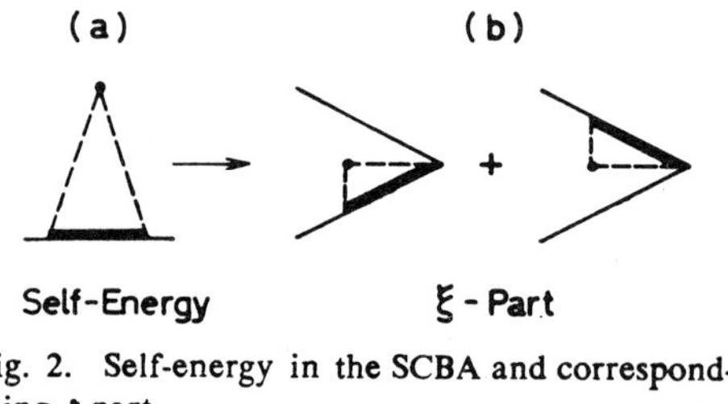

Fig. 2. Self-energy in the SCBA and corresponding ξ-part.

obtained graphically from self-energy diagrams by replacing one of matrix elements of the potential $v^{(\mu)}(\boldsymbol{r}, Z)$ by the corresponding $\partial v^{(\mu)}(\boldsymbol{r}, Z)/\partial x$ or $\partial v^{(\mu)}(\boldsymbol{r}, Z)\partial y$ as is shown in Fig. 2. From graphical consideration one gets

$$\Delta\sigma_{XY}^{(2)} = \frac{e^2}{8\pi^2\hbar}\int f(E)\,\mathrm{d}E \sum_{N_1}\sum_{\pm}\lim_{E'\to E}\mathrm{Im}_{(E)}\,\mathrm{Re}_{(E')}\frac{\partial}{\partial E'}[\xi^y_{N_1\pm 1N_1}(E, E') \times G_{N_1}(E)G_{N_1\pm 1}(E')\Xi^x_{N_1\pm 1N_1}(E', E) - (x\leftrightarrow y)]\,, \tag{2.5}$$

with

$$\Xi^x_{N_1\pm 1N_1}(E', E) = \xi^x_{N_1\pm 1N_1}(E', E) + \sum_{N_2}\gamma_{N_1\pm 1N_1, N_2\pm 1N_2}(E', E)G_{N_2\pm 1}(E')G_{N_2}(E)\Xi^x_{N_2\pm 1N_2}(E', E)\,, \tag{2.6}$$

where $\gamma_{N_1\pm 1N_1, N_2\pm 1N_2}$ is the proper vertex part defined in I and $G_N(E)$ is average Green's function

$$G_N(E) = \left\langle\left(0\left|a_{NX}\frac{1}{E-H}a^+_{NX}\right|0\right)\right\rangle. \tag{2.7}$$

Making use of

$$\left(Nm\pm 1\left|l\frac{\partial v^{(\mu)}(Z)}{\partial y}\right|N'm\right) = \pm i\left(Nm\pm 1\left|l\frac{\partial v^{(\mu)}(Z)}{\partial x}\right|N'm\right), \tag{2.8}$$

one can show that

$$\xi^y_{N\pm 1N}(E', E) = \mp i\xi^x_{N\pm 1N}(E', E)\,. \tag{2.9}$$

Further, as is easily seen, one has

$$\begin{aligned}&\xi^x_{N\pm 1N}(E', E) = \bar{\xi}^x_{NN\pm 1}(E, E')\,,\\ &\xi^x_{N\pm 1N}(E, E) = \xi^y_{N\pm 1N}(E, E) = 0\,,\end{aligned} \tag{2.10}$$

where $\bar{\xi}$ can be obtained from ξ by taking the complex conjugate of matrix elements of potentials appearing in ξ. By the use of above relations, eq. (2.5) becomes

$$\Delta\sigma_{XY}^{(2)} = \frac{e^2}{4\pi^2\hbar}\int\left(-\frac{\partial f}{\partial E}\right)\mathrm{d}E\sum_{N_1}\sum_{\pm}(\pm i)\bar{\xi}^x_{N_1N_1\pm 1}(E+i0, E-i0) \times G_{N_1}(E+i0)G_{N_1\pm 1}(E-i0)\Xi^x_{N_1\pm 1N_1}(E-i0, E+i0)\,. \tag{2.11}$$

Therefore, $\Delta\sigma_{XY}^{(2)}$ is determined only from quantities at the Fermi energy at zero temperature. Next let us consider the case under strong magnetic fields and assume that the Fermi level lies in the N-th Landau level. In this case, one has to retain terms with $N_1 = N$ and $N-1$ in eq. (2.11), and further one has $\Xi^x_{N\pm 1N} = \xi^x_{N\pm 1N}$ and $G_{N\pm 1} = \mp(\hbar\omega_c)^{-1}$. Therefore, eq. (2.11) is reduced to

$$\Delta\sigma_{XY}^{(2)} = \frac{e^2}{2\pi^2\hbar}\int\left(-\frac{\partial f}{\partial E}\right)\mathrm{d}E\frac{1}{\hbar\omega_c}\sum_{\pm}\xi^x_{N\pm 1N}(E-i0, E+i0)\bar{\xi}^x_{N\pm 1N}(E-i0, E+i0)\,\mathrm{Im}\,G_N(E-i0)\,. \tag{2.12}$$

§ 3. Self-Consistent Born Approximation (SCBA)

3.1 Case of strong magnetic fields

In the SCBA, $\Delta\sigma_{XY}^{(1)}$ is given by the diagram shown in Fig. 1(a) and becomes

$$\Delta\sigma_{XY}^{(1)} = \frac{e^2}{\pi^2\hbar}\int f(E)\,\mathrm{d}E\,2\pi l^2\sum_{\mu}\int\mathrm{d}Z\,N_i^{(\mu)}(Z)\sum_{N_1}\sum_{N_2}\sum_{m}\sum_{\pm}\frac{1}{2i}\left[\left(N_1m\left|l\frac{\partial v^{(\mu)}(Z)}{\partial y}\right|N_2m\pm 1\right)\times\left(N_2m\pm 1\left|l\frac{\partial v^{(\mu)}(Z)}{\partial x}\right|N_1m\right) - (x\leftrightarrow y)\right]\mathrm{Im}\,G_{N_1}(E+i0)\frac{\partial}{\partial E}\mathrm{Re}\,G_{N_2}(E+i0)\,, \tag{3.1}$$

where $N_i^{(\mu)}(Z)$ is the concentration of scatterers in a unit volume. With the aid of eqs. (2.9) of I and (2.7) of III, one gets

$$\sum_m \sum_\pm \left(N_1 m \left| l \frac{\partial v^{(\mu)}(Z)}{\partial y} \right| N_2 m \pm 1 \right) \left(N_2 m \pm 1 \left| l \frac{\partial v^{(\mu)}(Z)}{\partial x} \right| N_1 m \right)$$
$$= \int \frac{d^2 r}{2\pi l^2} \int \frac{d^2 r'}{2\pi l^2} \, l \frac{\partial v^{(\mu)}(\boldsymbol{r}, Z)}{\partial y} \, l \frac{\partial v^{(\mu)}(\boldsymbol{r}', Z)}{\partial x'} J_{N_2 N_2}(\boldsymbol{r}-\boldsymbol{r}') J_{N_1 N_1}(\boldsymbol{r}-\boldsymbol{r}') , \tag{3.2}$$

which becomes identically zero because of the cylindrical symmetry of the potential. Therefore, $\Delta\sigma_{\mathrm{XY}}^{(1)}$ vanishes in the SCBA.

Next let us consider $\Delta\sigma_{\mathrm{XY}}^{(2)}$. As the ξ-part one should include those diagrams shown in Fig. 2(b).

$$\xi^{x}_{N\pm1 N}(E', E) = 2\pi l^2 \sum_\mu \int dZ \, N_i^{(\mu)}(Z) \sum_{N'} \sum_m \left\{ \left(N\pm1\, m\mp1 \left| l \frac{\partial v^{(\mu)}(Z)}{\partial x} \right| N'm \right) (N'm | v^{(\mu)}(Z) | Nm) \right.$$
$$\left. \times G_{N'}(E) - (N\pm1\, m | v^{(\mu)}(Z) | N'm) \left(N'm \left| l \frac{\partial v^{(\mu)}(Z)}{\partial x} \right| Nm \pm 1 \right) G_{N'}(E') \right\} . \tag{3.3}$$

Again with the aid of (2.9) of I and (2.7) of III, one has

$$\sum_m (N\pm1\, m | v^{(\mu)}(Z) | N'm) \left(N'm \left| l \frac{\partial v^{(\mu)}(Z)}{\partial x} \right| Nm\pm1 \right)$$
$$= -\sum_m \left(N\pm1\, m\mp1 \left| l \frac{\partial v^{(\mu)}(Z)}{\partial x} \right| N'm \right) (N'm | v^{(\mu)}(Z) | Nm)$$
$$= -\int \frac{d^2 r}{2\pi l^2} \int \frac{d^2 r'}{2\pi l^2} \, l \frac{\partial v^{(\mu)}(\boldsymbol{r}, Z)}{\partial x} v^{(\mu)}(\boldsymbol{r}', Z) J_{NN\pm1}(\boldsymbol{r}-\boldsymbol{r}') J_{N'N'}(\boldsymbol{r}-\boldsymbol{r}') \exp[\mp i\theta(\boldsymbol{r}, \boldsymbol{r}')] . \tag{3.4}$$

Therefore, one sees that the ξ-part (3.3) satisfies eq. (2.10). Under strong magnetic fields, it becomes

$$\xi^{x}_{N\pm1 N}(E-i0, E+i0) = \frac{1}{4} (\Gamma_N{}^\pm)^2 [-2i \,\mathrm{Im}\, G_N(E+i0)] , \tag{3.5}$$

where

$$(\Gamma_N{}^\pm)^2 = 4 \cdot 2\pi l^2 \sum_\mu \int dZ \, N_i^{(\mu)}(Z) \sum_m (N\mp1\, m | v^{(\mu)}(Z) | Nm) \left(Nm \left| l \frac{\partial v^{(\mu)}(Z)}{\partial x} \right| Nm\pm1 \right) . \tag{3.6}$$

Substitution of eq. (3.5) into eq. (2.12) yields

$$\Delta\sigma_{\mathrm{XY}}^{(2)} = \frac{e^2}{\pi^2 \hbar} \int \left(-\frac{\partial f}{\partial E} \right) dE \frac{\Gamma_N}{\hbar\omega_c} \left(\frac{\Gamma_N{}^{xy}}{\Gamma_N} \right)^4 \left[1 - \left(\frac{E-E_N}{\Gamma_N} \right)^2 \right]^{3/2} , \tag{3.7}$$

with

$$(\Gamma_N{}^{xy})^4 = \sum_\pm |\Gamma_N{}^\pm|^4 , \tag{3.8}$$

$$\Gamma_N{}^2 = 4 \cdot 2\pi l^2 \sum_\mu \int dZ \, N_i^{(\mu)}(Z) \sum_m |(Nm | v^{(\mu)}(Z) | Nm)|^2 , \tag{3.9}$$

where use has been made of eq. (3.5) of I.

When scatterers are of sufficiently short range, one replaces the potential by a δ-potential,

$$v^{(\mu)}(\boldsymbol{r}, Z) = V^{(\mu)}(Z) \delta^{(2)}(\boldsymbol{r}) . \tag{3.10}$$

Then

$$\left(N0 \left| l \frac{\partial v^{(\mu)}(Z)}{\partial x} \right| N'\pm1 \right) = \mp \frac{V^{(\mu)}(Z)}{2\pi l^2} \sqrt{\frac{N'+1/2\pm1/2}{2}} , \tag{3.11}$$

$$(N0 | v^{(\mu)}(Z) | N'0) = \frac{V^{(\mu)}(Z)}{2\pi l^2} , \tag{3.12}$$

and other matrix elements are zero. One gets

$$\Delta\sigma_{XY}^{(2)}=\int\left(-\frac{\partial f}{\partial E}\right)\mathrm{d}E\,\frac{e^2}{\pi^2\hbar}\frac{\Gamma}{\hbar\omega_c}\left(N+\frac{1}{2}\right)\left[1-\left(\frac{E-E_N}{\Gamma}\right)^2\right]^{3/2}, \tag{3.13}$$

and

$$(\Gamma_N^{xy})^4=\left(N+\frac{1}{2}\right)\Gamma^4\,, \tag{3.14}$$

where Γ^2 can be expressed in terms of the relaxation time τ_f obtained in the Born approximation by assuming the same scatterers in the absence of magnetic fields.

$$\Gamma^2=4\sum_\mu\int\mathrm{d}Z\,N_i^{(\mu)}(Z)\,\frac{|V^{(\mu)}(Z)|^2}{2\pi l^2}=\frac{2}{\pi}\hbar\omega_c\frac{\hbar}{\tau_f}\,. \tag{3.15}$$

When scatterers are of sufficiently slowly-varying type, eq. (3.6) becomes

$$\begin{aligned}(\Gamma_N^{\pm})^2=&\,4\cdot 2\pi l^2\sum_\mu\int\mathrm{d}Z\,N_i^{(\mu)}(Z)\int\frac{\mathrm{d}^2r}{2\pi l^2}\int\frac{\mathrm{d}^2r'}{2\pi l^2}\,l\,\frac{\partial v^{(\mu)}(\boldsymbol{r},Z)}{\partial x}v^{(\mu)}(\boldsymbol{r}+\boldsymbol{R})\\&\times J_{NN\pm1}(\boldsymbol{R})J_{NN}(\boldsymbol{R})\exp\left(\mp i\theta(\boldsymbol{R})\right)]\\ \simeq&-4\cdot 2\pi l^2\sum_\mu\int\mathrm{d}Z\,N_i^{(\mu)}(Z)\int\frac{\mathrm{d}^2r}{2\pi l^2}\left[\,l\,\frac{\partial v^{(\mu)}(\boldsymbol{r},Z)}{\partial x}\right]^2\sqrt{\frac{N+1/2\pm1/2}{2}}\,,\end{aligned} \tag{3.16}$$

where $\theta(\boldsymbol{R})=\tan^{-1}R_y/R_x$. It can be written as

$$(\Gamma_N^{\pm})^2=-2\sqrt{\frac{N+1/2\pm1/2}{2}}\langle(l\nabla V(\boldsymbol{r}))^2\rangle\,, \tag{3.17}$$

where $V(\boldsymbol{r})$ is the local potential energy. Therefore, one has

$$(\Gamma_N^{xy})^4=4\left(N+\frac{1}{2}\right)\langle(l\nabla V(\boldsymbol{r}))^2\rangle^2\,, \tag{3.18}$$

and $\Delta\sigma_{XY}$ decreases in proportion to $(l/d)^4$, where d is of the order of the range of scatterers.

In order to see such range dependence explicitly, we assume scatterers with a Gaussian potential

$$v^{(\mu)}(\boldsymbol{r},Z)=\frac{V^{(\mu)}(Z)}{\pi d^2}\exp\left(-\frac{r^2}{d^2}\right). \tag{3.19}$$

After a little manipulation, one gets

$$(\Gamma_N^{\pm})^2=\Gamma^2h_N^{\pm}(\alpha)\,,$$

with

$$h_N^{\pm}(\alpha)=\pm\frac{1}{\sqrt{2N+1\pm1}}\times\int_0^\infty\mathrm{d}x\,xL_N^0(\alpha^2x)L^1_{N-1/2\pm1/2}(\alpha^2x)\,\mathrm{e}^{-(1+\alpha^2)x}\,, \tag{3.20}$$

where $\alpha=d/l$ and $L_N^m(x)$ is associated Laguerre's polynomial. Corresponding expression for Γ_N^2 is given by eqs. (2.24) and (2.26) in ref. 7. For the ground Landau level, for example, one gets

$$\left(\frac{\Gamma_0^{xy}}{\Gamma_0}\right)^4=\frac{1}{2(1+\alpha^2)^2}\,. \tag{3.21}$$

In Fig. 3, $(\Gamma_N^{xy}/\Gamma_N)^4/(N+1/2)$ is plotted against the range α for several Landau levels. Therefore,

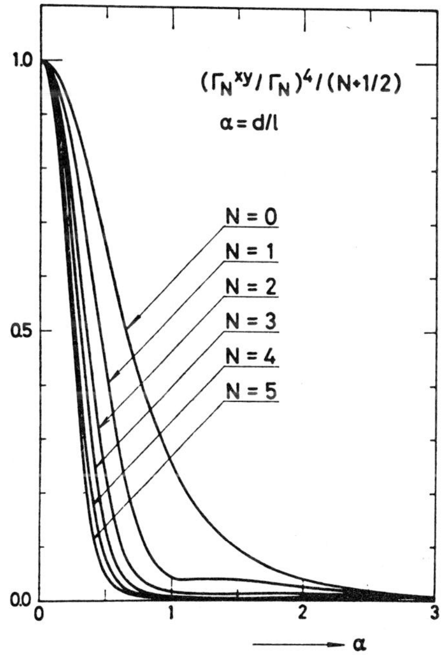

Fig. 3. $(\Gamma_N^{XY}/\Gamma_N)^4/(N+1/2)$ as a function of $\alpha=d/l$.

$\Delta\sigma_{XY}^{(2)}$ decreases very rapidly with increasing range if Γ_N is kept constant.

3.2 Short-ranged scatterers under magnetic fields of arbitrary strength

In this section we specialize ourselves to the case of short-ranged scatterers. By the use of eqs. (3.11) and (3.12), eq. (3.3) becomes

$$\xi^{\tau}_{N\pm1N}(E-i0,\,E+i0)=\pm 2i\sqrt{\frac{N+1/2\pm1/2}{2}}\,\mathrm{Im}\,\Sigma(E+i0)\ , \tag{3.22}$$

where

$$\Sigma(E)=\frac{1}{4}\Gamma^2\sum_N G_N(E)\ . \tag{3.23}$$

Since the vertex part $\gamma_{N_1\pm1N_1,\,N_2\pm1N_2}$ vanishes, the Hall conductivity becomes

$$\begin{aligned}\Delta\sigma_{XY}&=\frac{e^2}{2\pi^2\hbar}\int\left(-\frac{\partial f}{\partial E}\right)\mathrm{d}E\,2(\mathrm{Im}\,\Sigma(E+i0))^2\sum_N(N+1)\,\mathrm{Im}\,[G_N(E+i0)G_{N+1}(E-i0)]\\&=\frac{e^2}{2\pi^2\hbar}\int\left(-\frac{\partial f}{\partial E}\right)\mathrm{d}E\,\frac{-2\,\mathrm{Im}\,\Sigma(E+i0)}{\hbar\omega_c}\sum_N(N+1)(\hbar\omega_c)^2\,\mathrm{Im}\,G_N(E+i0)\,\mathrm{Im}\,G_{N+1}(E+i0)\ .\end{aligned} \tag{3.24}$$

On the other hand, the transverse conductivity σ_{XX} is written as

$$\sigma_{XX}=\frac{e^2}{2\pi^2\hbar}\int\left(-\frac{\partial f}{\partial E}\right)\mathrm{d}E\sum_N(N+1)(\hbar\omega_c)^2\,\mathrm{Im}\,G_N(E+i0)\,\mathrm{Im}\,G_{N+1}(E+i0)\ . \tag{3.25}$$

One notices that at zero temperature

$$\Delta\sigma_{XY}=\frac{-2\,\mathrm{Im}\,\Sigma(E_F+i0)}{\hbar\omega_c}\sigma_{XX}\ , \tag{3.26}$$

where E_F is the Fermi energy. If one puts $\mathrm{Im}\,\Sigma(E_F+i0)=-\hbar/2\tau$, eq. (3.26) is the same as the phenomenological formula (1.3).*

Let us consider the case that the Fermi level lies in high Landau levels at zero temperature. The discussion in IV applies directly to the present σ_{XY}. When the magnetic field is sufficiently strong ($\omega_c\tau_f\gg1$), one gets.

$$\Delta\sigma_{XY}=\frac{e^2}{\pi^2\hbar}\left(N+\frac{1}{2}\right)\frac{\Gamma}{\hbar\omega_c}\times\left[1+\left(\frac{\pi^2}{8}-1\right)\left(\frac{\Gamma}{\hbar\omega_c}\right)^2+\cdots\right], \tag{3.27}$$

or

$$\frac{\Delta\sigma_{XY}}{\sigma_{XX}}=\frac{\Gamma}{\hbar\omega_c}\left[1-\frac{\pi^2}{24}\left(\frac{\Gamma}{\hbar\omega_c}\right)^2+\cdots\right], \tag{3.28}$$

at each peak associated with the N-th Landau level. When the magnetic field is rather weak ($\omega_c\tau_f\lesssim1$), on the other hand, one has

$$\sigma_{XY}=-\frac{ne^2\tau_f}{m}\frac{\omega_c\tau_f}{1+\omega_c^2\tau_f^2}\times\left[1-\frac{2(1+3\omega_c^2\tau_f^2)}{\omega_c^2\tau_f^2(1+\omega_c^2\tau_f^2)}\cos\frac{2\pi X'}{\hbar\omega_c}\times\exp\left(-\frac{\pi}{\omega_c\tau_f}\right)\right], \tag{3.29}$$

where X' is defined in IV.

* As is easily seen, eq. (3.26) is applicable to three-dimensional systems. Further it is easy to show that eq. (3.26) still holds for $\Delta\sigma_{XY}^{(2)}$ even if effects of higher Born scattering are included within the single-site approximation (SSA).

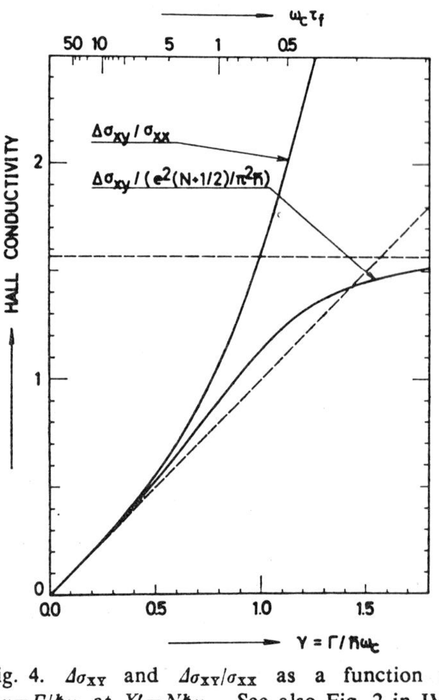

Fig. 4. $\Delta\sigma_{XY}$ and $\Delta\sigma_{XY}/\sigma_{XX}$ as a function of $\gamma=\Gamma/\hbar\omega_c$ at $X'=N\hbar\omega_c$. See also Fig. 2 in IV.

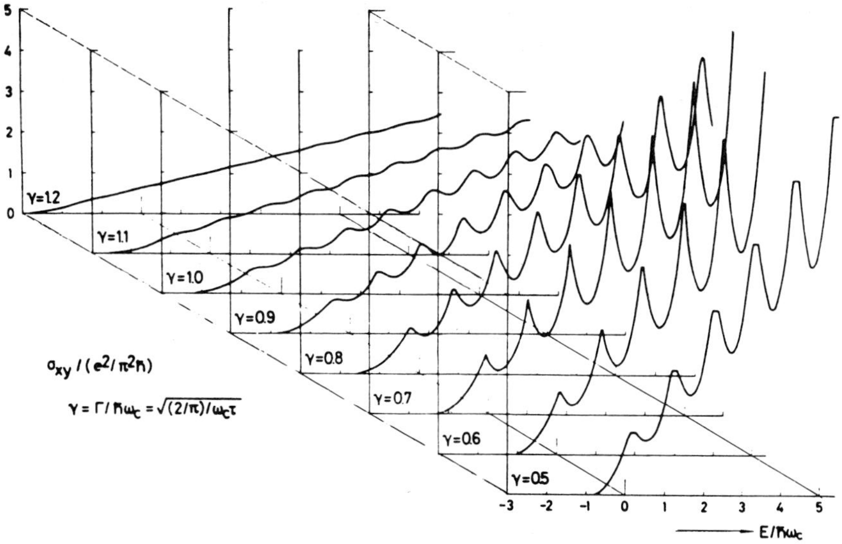

Fig. 5. Examples of oscillatory σ_{XY} for several $\gamma=\Gamma/\hbar\omega_c$.

In Fig. 4, $\Delta\sigma_{XY}$ and $\Delta\sigma_{XY}/\sigma_{XX}$ at $X'=N\hbar\omega_c$ for general $\gamma=\Gamma/\hbar\omega_c$, which are numerically calculated, are shown as a function of γ. The oscillatory conductivity σ_{XY} calculated numerically in a similar manner to IV is given in Fig. 5.

§ 4. Higher Approximations under Strong Magnetic Fields

First, we investigate effects of higher order scattering from a single scatterer employing the single-site approximation (SSA) as in II. When the magnetic field is sufficiently strong, one should include diagrams corresponding to those shown in Fig. 2 of II and one gets

$$\begin{aligned}\Delta\sigma_{XY}^{(1)}=&\frac{e^2}{\pi^2\hbar}\int f(E)\mathrm{d}E\,2\pi l^2\sum_{\mu}\int\mathrm{d}Z\,N_i^{(\mu)}(Z)\sum_m\sum_{\pm}\frac{1}{2i}\Big[\Big(Nm\pm1\Big|\,l\frac{\partial v^{(\mu)}(Z)}{\partial y}\Big|\,Nm\Big)\\&\times\Big(Nm\Big|\,l\frac{\partial v^{(\mu)}(Z)}{\partial x}\Big|\,Nm\pm1\Big)-(x\leftrightarrow y)\Big]\\&\times\mathrm{Im}\frac{1}{X_N-v_{Nm}^{(\mu)}(Z)}\frac{\partial}{\partial E}\mathrm{Re}\frac{1}{X_N-v_{Nm\pm1}^{(\mu)}(Z)}\,,\end{aligned}\tag{4.1}$$

where $X_N(E)=G_N(E)^{-1}$, $v_{Nm}^{(\mu)}(Z)=(Nm|v^{(\mu)}(Z)|Nm)$, and one should choose N in such a way that E_N is closest to E. Making use of eq. (2.8) and integrating by parts, one has

$$\begin{aligned}\Delta\sigma_{XY}^{(1)}=&\frac{e^2}{\pi^2\hbar}\int\Big(-\frac{\partial f}{\partial E}\Big)\mathrm{d}E\,2\pi l^2\sum_{\mu}\int\mathrm{d}Z\,N_i^{(\mu)}(Z)\sum_m\Big|\Big(Nm\Big|\,l\frac{\partial v^{(\mu)}(Z)}{\partial x}\Big|\,Nm-1\Big)\Big|^2\\&\times\Big\{[v_{Nm}^{(\mu)}(Z)-v_{Nm+1}^{(\mu)}(Z)]^{-2}\,\mathrm{Im}\Big[\ln\frac{X_N}{X_N-v_{Nm}^{(\mu)}(Z)}-\frac{v_{Nm}^{(\mu)}(Z)}{X_N-v_{Nm}^{(\mu)}(Z)}\\&-\ln\frac{X_N}{X_N-v_{Nm+1}^{(\mu)}(Z)}+\frac{v_{Nm+1}^{(\mu)}(Z)}{X_N-v_{Nm+1}^{(\mu)}(Z)}\Big]\\&+[v_{Nm}^{(\mu)}(Z)-v_{Nm+1}^{(\mu)}(Z)]^{-1}\Big[v_{Nm}^{(\mu)}(Z)\,\mathrm{Im}\frac{1}{X_N-v_{Nm}^{(\mu)}(Z)}\\&\times\mathrm{Re}\frac{1}{X_N-v_{Nm+1}^{(\mu)}(Z)}+v_{Nm+1}^{(\mu)}(Z)\,\mathrm{Re}\frac{1}{X_N-v_{Nm}^{(\mu)}(Z)}\,\mathrm{Im}\frac{1}{X_N-v_{Nm+1}^{(\mu)}(Z)}\Big]\Big\}\,,\end{aligned}\tag{4.2}$$

where such branch of logarithm should be chosen whose imaginary part lies between $-\pi$ and π. With the aid of eq. (2.9) of II and the relation

$$\left|\left(Nm \left| l\frac{\partial v^{(\mu)}(Z)}{\partial x} \right| Nm+1 \right)\right|^2 = \frac{1}{2}(N+m+1)[(Nm+1|v^{(\mu)}(Z)|Nm+1)-(Nm|v^{(\mu)}(Z)|Nm)]^2 \,, \tag{4.3}$$

which is given in Appendix of II, the Hall conductivity becomes

$$\begin{aligned}\sigma_{XY} &= -\frac{nec}{H}+\Delta\sigma_{XY}^{(1)}\,,\\ &= -\frac{e^2}{\pi^2\hbar}\int\left(-\frac{\partial f}{\partial E}\right)\mathrm{d}E\left\{\frac{\pi}{2}\left[N+1-\frac{1}{\pi}\,\mathrm{Im}\ln X_N\right]\right.\\ &\quad -2\pi l^2\sum_{\mu}\int \mathrm{d}Z\, N_i^{(\mu)}(Z)\sum_m\left|\left(Nm \left| l\frac{\partial v^{(\mu)}(Z)}{\partial x} \right| Nm+1 \right)\right|^2\\ &\quad \times[v_{Nm}^{(\mu)}(Z)-v_{Nm+1}^{(\mu)}(Z)]^{-1}\left[v_{Nm}^{(\mu)}(Z)\,\mathrm{Im}\frac{1}{X_N-v_{Nm}^{(\mu)}(Z)}\,\mathrm{Re}\frac{1}{X_N-v_{Nm+1}^{(\mu)}(Z)}\right.\\ &\quad \left.\left.+v_{Nm+1}^{(\mu)}(Z)\,\mathrm{Re}\frac{1}{X_N-v_{Nm}^{(\mu)}(Z)}\,\mathrm{Im}\frac{1}{X_N-v_{Nm+1}^{(\mu)}(Z)}\right]\right\}\,.\end{aligned} \tag{4.4}$$

The Green's function G_N or X_N is determined by the self-consistency equation (2.8) of II. From the above equation, one can conclude that $\Delta\sigma_{XY}^{(1)}$ vanishes and σ_{XY} becomes $-nec/H=-e^2(N+1)/2\pi\hbar$, when the Fermi level lies in energy gaps between adjacent N-th and $N+1$-th Landau levels at zero temperature.

Let us consider the case that concentrations of scatterers are sufficiently small and impurity bands are separated from the N-th Landau level. At the energy where the spectrum has a gap between two impurity bands or between an impurity band and the main Landau level, X_N becomes a negative real number in case of attractive scatterers and a positive one in case of repulsive scatterers. When the Fermi level lies in those spectral gaps at zero temperature, therefore, the Hall conductivity becomes $-e^2N/2\pi\hbar$ in case of attractive scatterers and $-e^2(N+1)/2\pi\hbar$ in case of repulsive scatterers.* This means that electrons which fully occupy impurity bands do not contribute to the Hall current, while those which occupy the main Landau level give rise to the same Hall current as that obtained when all i.e. $1/2\pi l^2$ electrons of the Landau level move freely.

In case of high concentrations of scatterers, the absolute value of X_N becomes large compared with $(Nm|v^{(\mu)}(Z)|Nm)$ and further X_N approaches a solution in the SCBA

$$X_N=\frac{1}{2}[E-E_N+i\sqrt{\Gamma_N^2-(E-E_N)^2}]\,, \tag{4.5}$$

Therefore, one has

$$\Delta\sigma_{XY}^{(1)}=-\frac{e^2}{3\pi^2\hbar}\int\left(-\frac{\partial f}{\partial E}\right)\mathrm{d}E\left(\frac{\Gamma_N'}{\Gamma_N}\right)^3\left[1-\left(\frac{E-E_N}{\Gamma_N}\right)^2\right]^{3/2}\,, \tag{4.6}$$

with

$$\begin{aligned}(\Gamma_N')^3 &= 2^3\cdot(2\pi l^2)\sum_{\mu}\int\mathrm{d}Z\,N_i^{(\mu)}(Z)\sum_m\left|\left(Nm \left| l\frac{\partial v^{(\mu)}(Z)}{\partial x} \right| Nm+1 \right)\right|^2\\ &\quad\times[(Nm|v^{(\mu)}(Z)|Nm)-(Nm+1|v^{(\mu)}(Z)|Nm+1)]\,.\end{aligned} \tag{4.7}$$

Especially in case of short-ranged scatterers, eq. (4.7) becomes

$$(\Gamma_N')^3=2^3\cdot 2\pi l^2\sum_{\mu}\int\mathrm{d}Z\,N_i^{(\mu)}(Z)\left[\frac{V^{(\mu)}(Z)}{2\pi l^2}\right]^3\,. \tag{4.8}$$

As can be seen from the above equations, the effects of higher Born scattering does not vanish in the limit of $\Gamma/\hbar\omega_c\to 0$, and further they depend on the sign of scatterers: $\Delta\sigma_{XY}^{(1)}$ becomes positive (negative) when scatterers are attractive (repulsive). It should be noticed also that $\Delta\sigma_{XY}^{(1)}$ does

* A repulsive scatterer can have impurity bands in our system.

not increase with the Landau level N in contrast to $\Delta\sigma_{XY}^{(2)}$ and that it is independent of N in case of short-ranged scatterers. With the increase of the concentration of scatterers, $\Delta\sigma_{XY}^{(1)}$ becomes small, which is consistent with the result obtained in II that the SSA approaches the SCBA in case of high concentrations of scatterers.

As an illustrative example, we consider a model system in which a single kind of scatterers with the Gaussian potential (3.19) is distributed randomly. In a similar manner to in § 3 of II, one can calculate σ_{XY} of the ground Landau level. Examples of the results are shown in Fig. 6.

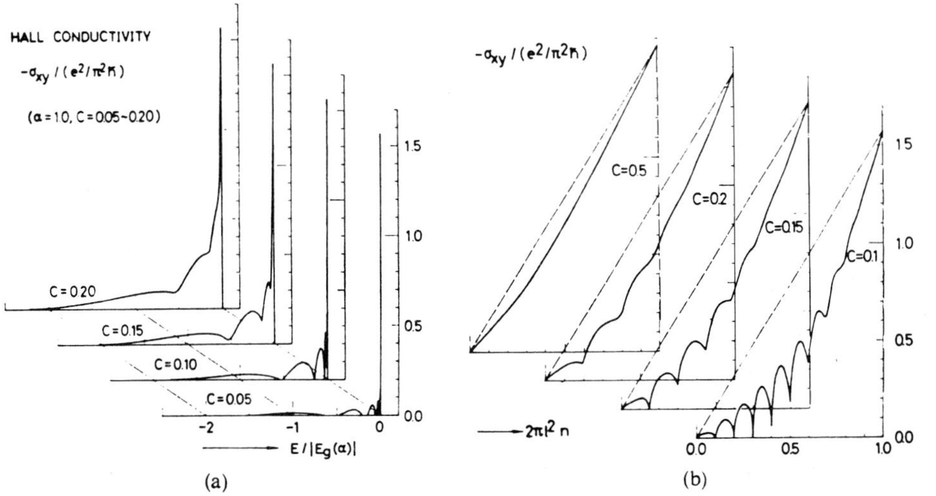

Fig. 6. Examples of σ_{XY} of the ground Landau level in case of scatterers with a Gaussian potential. $c=2\pi l^2 N_i$. See also Figs. 6~11. (a) σ_{XY} as a function of the energy. (b) σ_{XY} as a function of total number of electrons.

Next we investigate effects of simultaneous scattering from more than two scatterers to the lowest order, confining ourselves to the case of sufficiently high concentrations of short-ranged scatterers. One has to take into account those diagrams shown in Fig. 1(b) and gets

$$\Delta\sigma_{XY}^{(1)} = \frac{e^2}{\pi^2\hbar}\frac{\Gamma^4}{4^3}\int f(E)\,dE\,h_N^{(2)}\left[\mathrm{Im}\,G_N{}^3\frac{\partial}{\partial E}\mathrm{Re}\,G_N - \mathrm{Im}\,G_N\frac{\partial}{\partial E}\mathrm{Re}\,G_N{}^3\right], \tag{4.9}$$

with

$$h_N^{(2)} = 4\cdot\int\frac{d^2R}{2\pi l^2}\left\{\frac{N+1}{2}[J_{N+1N+1}(\boldsymbol{R})J_{NN}(\boldsymbol{R})+J_{N+1N}(\boldsymbol{R})^2] - \frac{N}{2}[J_{NN}(\boldsymbol{R})J_{N-1N-1}(\boldsymbol{R})+J_{NN-1}(\boldsymbol{R})^2]\right\}. \tag{4.10}$$

Partial integration of eq. (4.9) yields

$$\Delta\sigma_{XY}^{(1)} = \frac{e^2}{\pi^2\hbar}\frac{\Gamma^4}{4^3}\int\left(-\frac{\partial f}{\partial E}\right)dE\,h_N^{(2)}\left[\frac{1}{2}\mathrm{Im}\,G_N{}^2|G_N|^2 - \frac{1}{4}\mathrm{Im}\,G_N{}^4\right]. \tag{4.11}$$

If one uses G_N given by eq. (4.5) of the SCBA, it becomes

$$\Delta\sigma_{XY}^{(1)} = -\frac{e^2}{\pi^2\hbar}\frac{1}{2}h_N^{(2)}\int\left(-\frac{\partial f}{\partial E}\right)dE\frac{E-E_N}{\Gamma}\left[1-\left(\frac{E-E_N}{\Gamma}\right)^2\right]^{3/2}, \tag{4.12}$$

the integrand of which becomes zero at $E=E_N$ and becomes large at tails of the spectrum. One has, for example, $h_0^{(2)}=1$, $h_1^{(2)}=3/4, \cdots$, and $h_N^{(2)}$ decreases with the increase of N. Therefore, many-site corrections are not important for large Landau level index N.

§ 5. Discussion and Conclusion

The relation (3.26) between $\Delta\sigma_{XY}$ and σ_{XX} in the SCBA can physically be understood as follows. When an electric field E_y is applied in the y-direction, a center of the cyclotron motion of electrons moves in the x-direction with a drift velocity $v_x = cE_y/H$, which gives rise to the Hall conductivity $-nec/H$. Effects of scattering can be regarded as a frictional force acting on each electron, the strength of which is given by $F_x = -mv_x/\tau = -eE_y/\omega_c\tau$, where τ is a relaxation time. The current in the x-direction due to this force becomes

$$\Delta j_x = \sigma_{XX} \frac{F_x}{(-e)} = \frac{1}{\omega_c\tau} \sigma_{XX} E_y = \Delta\sigma_{XY} E_y \,. \qquad (5.1)$$

Therefore, one has eq. (3.26).

MOS inversion layers on Si (100) surface are a typical two-dimensional system. For usual electron concentrations ($N_{\rm inv} \gtrsim 10^{12}\,{\rm cm}^{-2}$), main scatterers of this system are considered to be of short range and their concentrations are considered to be relatively large. Therefore, the results in the SCBA together with eq. (4.6) are expected to apply to inversion layers. Measurements of both σ_{XX} and σ_{XY} will give useful informations on main scatterers and especially on the level broadening Γ. If one compares lineshapes obtained from eq. (3.6) of I and eqs. (3.13) and (4.6), one sees that the effective width of $\Delta\sigma_{XY}$ is smaller than that of σ_{XX}. Therefore, spin and valley splittings are expected to be seen in σ_{XY} at higher Landau levels than in σ_{XX}. Recently Igarashi, Wakabayashi and Kawaji measured these quantities.[8] The overall behavior of their results is satisfactorily in agreement with the present theory. Detailed comparison will be made in their paper.

Acknowledgements

The authors would like to thank Professor S. Kawaji, Mr. T. Igarashi and Mr. J. Wakabayashi for showing their experimental results prior to publication. They also thank Professor R. Kubo for valuable discussions. One of the authors (T.A.) is indebted to Sakkokai Foundation for financial support. Numerical computations were performed with the aid of HITAC 8800/8700 of the Computer Centre at the University of Tokyo.

References

1) T. Ando and Y. Uemura: J. Phys. Soc. Japan **36** (1974) 959.
2) T. Ando: J. Phys. Soc. Japan **36** (1974) 1521.
3) T. Ando: J. Phys. Soc. Japan **37** (1974) 622.
4) T. Ando: J. Phys. Soc. Japan **37** (1974) 1233.
5) R. Kubo, S. J. Miyake and N. Hashitsume: *Solid State Physics*, ed. F. Seitz and D. Turnbull (Academic Press, New York, 1965) Vol. 17, p. 269.
6) H. Shiba, K. Kanda, H. Hasegawa and H. Fukuyama: J. Phys. Soc. Japan **30** (1971) 972.
7) T. Ando: J. Phys. Soc. Japan **38** (1975) 989.
8) T. Igarashi, J. Wakabayashi and S. Kawaji: to be published in J. Phys. Soc. Japan; S. Kawaji, T. Igarashi and J. Wakabayashi: to be published in J. Phys. Soc. Japan.

Surface Science 58 (1976) 238–245

QUANTUM GALVANOMAGNETIC PROPERTIES OF n-TYPE INVERSION LAYERS ON Si(100) MOSFET

S. KAWAJI and J. WAKABAYASHI

Department of Physics, Gakushuin University, Mejiro, Toshima-ku, Tokyo, 171 Japan

Transverse magnetoconductivity σ_{xx} and Hall effect in n-type inversion layers of Si(100) MOSFET are measured for various source-drain fields between 0.08 and 40 V/cm under magnetic fields up to 150 kOe at 1.4 K. Conductivity peaks in low Landau levels are in good agreement with theory. Effect of the source-drain field in the magnetoconductivity is found to be very important in higher Landau levels as well as in the appearance of the lowest Landau level peak. Immobile electrons are clearly observed in conductivity bottoms. Electrode geometry effect for Hall effect measurement under strong magnetic fields is discussed.

1. Introduction

Since the first paper on magnetoconductivity experiments under strong magnetic field by Fowler et al. [1], many experimental and theoretical studies have been made on quantum galvanomagnetic properties of inversion layer electrons in silicon MOSFET [2–4]. Particularly, peak values of the transverse conductivity, σ_{xx}, under strong magnetic fields were studied theoretically and experimentally by Ando et al. [5] and showed fairly good agreement between theory and experiment. More detailed experiments on σ_{xx} were performed by Komatsubara et al. [6]. Recently, the Hall effect under strong magnetic field was also studied theoretically by Ando et al. [7] and experimentally by the authors' group [8,9].

In the present paper, results of precise measurements of non-sinusoidal oscillation of transverse magnetoconductivity σ_{xx} under magnetic fields up to 150 kOe at a temperature of 1.4 K for various source-drain fields will be described. Very clear spin and valley splittings were observed in σ_{xx} for Landau levels with index $N = 0$ and $N = 1$ at low source-drain field. Magnetic field dependence of peak values of σ_{xx} for $N = 0$ and 1 at low source-drain field is in very good agreement with theory. Immobile electrons were observed at the edges of each Landau level at low source-drain fields. Hall effect measurements on samples with large length to width ratios, when combined with the transverse conductivity σ_{xx}, were used unsuccessfully in evaluating the Hall conductivity σ_{xy}.

2. Experimental

Experiments were performed on n-channel Si MOSFET fabricated on (100) surfaces of p-type substrate with acceptor concentration of $1 \times 10^{14}/cm^3$. The thickness of the oxide layer was about 8000 Å (the number of induced electrons was 2.6×10^{10} electrons per 1 V of gate voltage). The peak value of the Hall mobility was about 12,000 $cm^2/V \cdot s$. The mobility dropped down to 3500 $cm^2/V \cdot s$ at a high gate voltage of 250 V. The interface state density, N_{ss}, evaluated from the shift of the conductance threshold voltage between 300 and 77 K was $5.2 \times 10^{10}/cm^2$.

Samples having circular electrodes with channel width of 50 μm and circumference-to-width ratio of 22 were used for transverse magnetoconductivity measurements. The Hall effect was measured on rectangular samples with length-to-width ratios of 20/3 and 1/3 fabricated on the same substrate. The magnetic field was applied perpendicular to the surface. The temperature was kept at 1.4 K through the experiments. The conductivity and Hall effect were measured at constant dc source-drain fields from 0.08 to 40 V/cm.

Non-sinusoidal oscillations in transverse magnetoconductivity σ_{xx} versus gate volt-

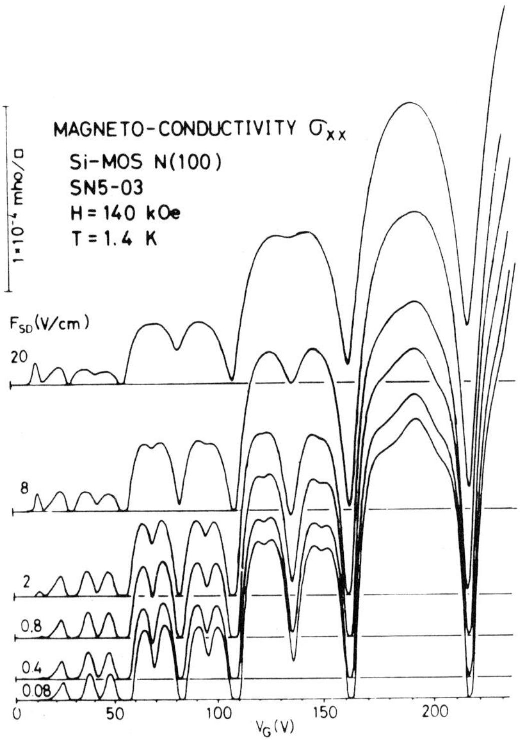

Fig. 1. Transverse magneto-conductivity σ_{xx} of an n-channel Si inversion layer under 140 kOe at 1.4 K for various source-drain fields F_{SD}.

age V_G at various source-drain fields F_{SD} under a magnetic field of 140 kOe are shown in fig. 1. In this figure, conductivity equal to zero (strictly speaking less than 10^{-8} mho/□) can be clearly observed, particularly in Landau levels with low level index N at low source-drain fields.

3. Peak values of σ_{xx}

The transverse magneto-conductivity σ_{xx} has been studied theoretically by Ando et al. [5], and in more detail by Ando [10]. When the interaction between different Landau levels can be neglected and the magnetic field is sufficiently strong ($\omega_c \tau_f \gg 1$), the peak conductivity at zero temperature is given by [10]

$$\sigma_{xx} = \frac{e^2}{\pi^2 \hbar}(N + 1/2)\left[1 - \left(1 - \frac{\pi^2}{12}\right)\left(\frac{\Gamma}{\hbar\omega_c}\right)^2 \dots\right]$$

$$= \frac{e^2}{\pi^2 \hbar}(N + 1/2)\left[1 - \frac{0.113}{\omega_c \tau_f} - \dots\right], \tag{1}$$

where Γ is the width of the Landau level, $\omega_c = eH/m^*c$, and τ_f the conduction electron relaxation time at zero magnetic field with a relation to Γ, $\Gamma/h\omega_c = (2/\pi\omega_c\tau_f)^{1/2}$.

The experimental results of Kobayashi and Komatsubara [5] and Komatsubara

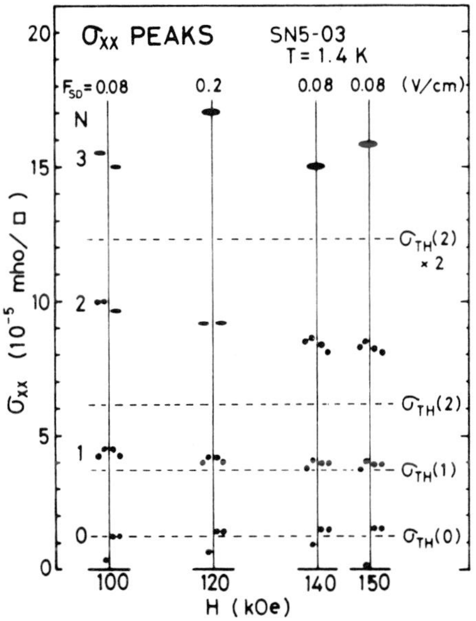

Fig. 2. Peak values of transverse magnetoconductivity σ_{xx} against magnetic field H; $T = 1.4$ K; $F_{SD} = 0.08$ V/cm for H = 100, 140 kOe and 150 kOe, and 0.2 V/cm for H = 120 kOe.

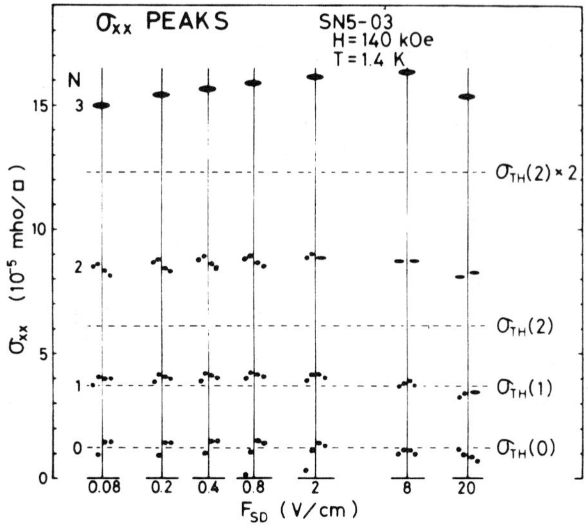

Fig. 3. Peak values of transverse magnetoconductivity σ_{xx} against source-drain field F_{SD}; H = 140 kOe, T = 1.4 K.

et al. [6] showed fairly good agreement with theory. However, separations between adjacent σ_{xx} peaks due to spin and valley splittings were incomplete in their results when compared precisely with theory. As is shown in fig. 1, the separation in oscillations in σ_{xx} is almost complete for N = 0 and is nearly complete for N = 1. Therefore, much better experimental examination of the theory can be made with the present results.

Peak values of σ_{xx} measured at the lowest source-drain fields in the present experiment are shown for magnetic fields of 100, 120, 140 and 150 kOe in fig. 2. Effects of the source-drain field, F_{SD}, on the conductivity peak values under a magnetic field of 140 kOe are shown in fig. 3. Higher order corrections in eq. (1) are unimportant because they are less than 1% and 3% even at 100 kOe for the highest electron mobility and the lowest electron mobility, respectively, in the present experiment. Therefore, the magnetic field dependence of the conductivity peaks comes from overlapping of adjacent Landau levels. Agreement between theory and experiment for N = 0 and N = 1 can be said to be very good. As is seen in fig. 2, valley-splitting in N = 0 Landau levels under 140 kOe looks complete. The situation under a magnetic field of 150 kOe is much better. However, experimental peak values of σ_{xx} are about 10% higher than the theoretical prediction. This fact shows clearly the existence of enhancement of valley-splitting, like the enhancement of the g-factor in spin splitting [11].

Fig. 3 shows that the effect of the source-drain field is important in the peak values of σ_{xx}. At zero magnetic field, the effect of the source-drain field is unimportant up to F_{SD} = 1 V/cm at gate voltages higher than V_G = 30 V. At low gate voltages, $V_G < 10$ V, source-drain fields of about 10 V/cm are necessary to overcome the so-called mobility edge [12]. Under strong magnetic fields, as shown particularly by the σ_{xx} peaks for $N = 3$ and $N = 2$ in fig. 3, the lowest source-drain field in the present experiment, 0.08 V/cm, is still not low enough to enable us to observe F_{SD}-independent conductivity. Conductivity peaks for $N = 0$ appear to be independent of F_{SD} for $F_{SD} < 0.4$ V/cm. For $F_{SD} > 0.8$ V/cm, conductivity peaks for down spin electrons (two peaks for lower gate voltages) start to increase concurrently as conductivity peaks for up spin electrons start to decrease. When the source-drain fields exceeds 2 V/cm, the peak conductivity in all Landau levels except the down spin levels in the Landau level with $N = 0$ starts to decrease.

All the behaviour described above suggests that (1) immobile electrons exist at each Landau level and (2) the source-drain field to mobilize immobile electrons at the edges of the Landau levels is lower for higher Landau levels; moreover (3) at higher source-drain field, $F_{SD} > 2$ V/cm, an effect of the source-drain field appaers to increase the width of the Landau levels and to decrease the peak values of the conductivity according to eq. (1) except for the lowest two peaks.

4. Immobile electrons under strong magnetic fields

In fig. 4, gate voltages for minima in the transverse conductivity σ_{xx} at low source-drain field are plotted for magnetic fields of 100, 120, 140 and 150 kOe. Particularly, the gate voltages for immobile electrons ($\sigma_{xx} = 0$) are shown by broad bars. The usual magnetoconductivity minima are shown by points. The lower edges of each Landau level are determined by inspecting the periodicity in the conductivity

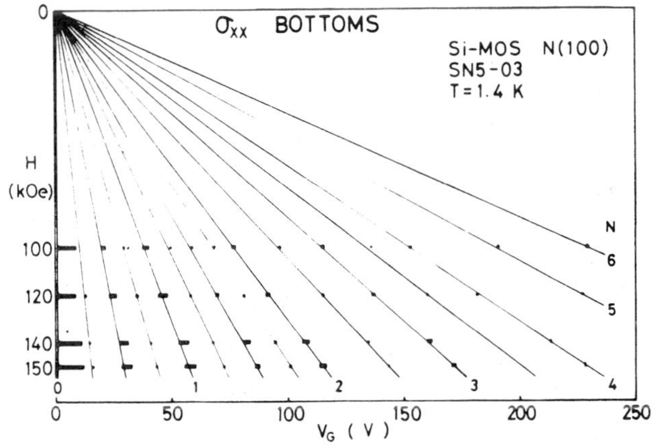

Fig. 4. Gate voltages for minima of transverse magnetoconductivity σ_{xx} under H = 100, 120, 140 and 150 kOe at 1.4 K. Broad bars indicate $\sigma_{xx} = 0$.

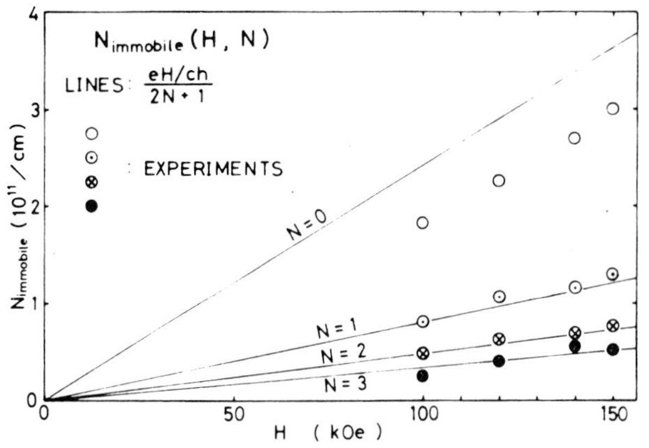

Fig. 5. Number of immobile electrons under strong magnetic field versus magnetic field H for different Landau level index N.

and connected by straight lines for each Landau level index N. Similarly, straight lines are shown for spin-splittings and valley-splittings.

Numbers of immobile electrons in σ_{xx} evaluated from the gate voltage width are shown in fig. 5. The number of immobile electrons ar the lower edge of a Landau level of index N is roughly expressed by

$$N_{\text{immobile}}(N) = \alpha \frac{eH/ch}{2N+1} , \qquad (2)$$

where eH/ch is the degeneracy of the Landau orbits. The straight lines of eq. (2) for $\alpha = 1$ are shown in fig. 5. The width of the gate voltage for immobile electrons is considered to be an upper bound. Therefore, the constant α is less than 1.

The density of interface states, N_{ss}, is approximately $5 \times 10^{10}/\text{cm}^2$. The number of immobile electrons $N_{\text{immobile}}(N)$ is magnetic field dependent and $N_{\text{immobile}}(N)$ for $N \leqslant 1$ is much larger than N_{ss}. This fact suggests that a model based on trapping of the electrons by conventional bound states associated with fixed interface charges is not applicable. Therefore a more generalized localization of the electrons by potential fluctuations is suggested.

The radius of a Landau orbit is given by $l_N = [c\hbar(2N+1)/eH]^{1/2}$, which leads to an area of a Landau orbit as $S_N = ch(2N+1)/2eH$. Therefore, eq. (2) can be rewritten $N_{\text{immobile}}(N) = \alpha/2S_N$ with $\alpha \leqslant 1$. This behaviour might be explained by a two-dimensional percolation model [13]. When the number of electrons is small, orbiting electrons under strong magnetic fields are well separated from each other by a shallow potential well. Therefore, they do not carry dc current. When the number of electrons increases, they tend to spread out homogeneously from each potential well. At some critical number of electrons, orbiting electrons flow out from a certain number of potential wells and an interconnection between potential wells which can support a dc current is established. According to the percolation theory, the critical electron numbers are determined from the ratio of the area occupied by total orbiting electrons to the sample area. The critical ratio depends on a model but it should be a number less than one. This is in accord with the experimental results in fig. 5 or eq. (2).

The F_{SD}-dependence of $N_{immobile}(N)$ is very weak up to 10 V/cm. Statistical distribution of potential fluctuations may account for this F_{SD}-dependence. In this regard the pinning of the electron crystal proposed by Wigner [14] may be a possible explanation of the experiment, particularly in the lowest Landau level.

5. Hall effect

The Hall conductivity σ_{xy}, under strong magnetic fields was evaluated from currents and Hall angles in samples with small length-to-width ratio, $L/W = 1/3$. Similar to previous results [8,9], the overall behaviour of σ_{xy} is in good agreement with theory [7], but quantitatively the experimental deviation from $-nec/H$ is still larger than the theory.

The evaluation of σ_{xy} from a combination of σ_{xx} measured in samples with circular electrodes and Hall angles measured in samples with large length-to-width ratio $L/W = 20/3$ was unsuccessful. Very high Hall fields generated in samples with large L/W ratio should make σ_{xx} in Hall samples different from that in circular electrode samples and prevent the meaningful evaluation of σ_{xy}.

6. Conclusions

Results of the experiments on the transverse magnetoconductivity σ_{xx} are presented. Behaviour of the conductivity peaks as well as the conductivity minima against the source-drain electric field and the magnetic field are qualitatively explained in terms of immobile electrons under a strong magnetic field. Immobile electrons exist at edges of each Landau level. Their number is dependent on the magnetic field, the Landau level index or the radius of the Landau orbit and the electric field in the plane parallel to the interface. Various models for immobilized electrons under a strong magnetic field may be possible. The weak but sensitive dependence of the conductivity peaks and minima on the source-drain field is probably a key to the solution of the problem.

One conclusion from the Hall effect experiments is that the recommended electrode geometry for Hall effect measurement under a strong magnetic field is one with a small length-to-width ratio, which is a poor geometry for the weak field case.

Further experimental studies are under progress on effects of temperatures and samples.

Acknowledgement

The authors are very grateful to Professor S. Tanuma and Dr. H. Suematsu, Institute for Solid State Physics, for their extending to us the use of high magnetic field facilities, to Dr. S. Sato and Mr. A. Yagi, Sony Semiconductor Development Division and Sony Research Center, for supplying the samples, and to Professor Y. Uemura, Drs. M. Tsukada, Y. Matsumoto and F. Ohkawa for their helpful discussions. A part of this work is supported by the grant-in-aids from Toray Science Foundation.

References

[1] A.B. Fowler, F.F. Fang, W.E. Howard and P.J. Stiles, in: Proc. Intern. Conf. on the Physics of Semiconductors, Kyoto; J. Phys. Soc. Japan Suppl. 21 (1966) p. 331.
[2] See Y. Uemura, in: Proc. 2nd Intern. Conf. on Solid Surfaces, Kyoto; Japan. J. Appl. Phys. Suppl. 2, Pt. 2 (1974) 17, and references cited there.
[3] See Y. Uemura, in: Proc. Intern. Conf. on Physics of Semiconductors, Stuttgart, 1974 (Teubner, Stuttgart, 1974) p. 665, and references cited there.
[4] P.J. Stiles, in: Proc. 2nd Intern. Conf. on Solid Surfaces, Kyoto; Japan. J. Appl. Phys., Suppl. 2, Pt. 2 (1974) 333.
[5] T. Ando, Y. Matsumoto, Y. Uemura, M. Kobayashi and K.F. Komatsubara, J. Phys. Soc. Japan 32 (1972) 859.
[6] K.F. Komatsubara, K. Narita, Y. Katayama, N. Kotera and M. Kobayashi, J. Phys. Chem. Solids 35 (1974) 723.
[7] T. Ando, Y. Matsumoto and Y. Uemura, J. Phys. Soc. Japan, to be published.
[8] T. Igarashi, J. Wakabayashi and S. Kawaji, J. Phys. Soc. Japan 38 (1975) 1549.
[9] S. Kawaji, T. Igarashi and J. Wakabayashi, Progr. Theoret. Phys. (Kyoto) Suppl. 57 (1975) 176.
[10] T. Ando, J. Phys. Soc. Japan 37 (1974) 1233.
[11] F. Ohkawa and Y. Uemura, Surface Sci. 58 (1976) 254.
[12] F. Stern, Phys. Rev. B9 (1974) 2762.
[13] V.K.S. Shante and S. Kirkpatrick, Advan. Phys. 20 (1971) 325.
[14] E.P. Wigner: Phys. Rev. 46 (1934) 1002.

Metrologia 22, 118–127 (1986)

metrologia

I. Fundamental Constants and Quantum Effects

The Discovery of the Quantum Hall Effect

G. Landwehr

Physikalisches Institut der Universität Würzburg, D-8700 Würzburg, Federal Republic of Germany

Abstract

After a brief explanation of the quantization of a two-dimensional electron gas in high magnetic fields the background of the discovery of the quantum Hall effect is given.

Introduction

The award of the 1985 Nobel Prize for physics to Klaus von Klitzing made me change the content of my talk at the Braunschweig symposium. I thought that, instead of quantum transport phenomena and their relevance for metrology, it was appropriate to speak about the background of the discovery of the quantum Hall effect. This seemed timely because my acquaintance with Klaus von Klitzing goes back to 1965 when he worked in the summer-time in my laboratory at the Physikalisch-Technische Bundesanstalt (PTB). But before I give a account of the history of the development of the quantum Hall effect, I shall briefly explain the properties of a two-dimensional (2D) electron gas in high magnetic fields.

Magneto-Transport Properties of a Two-Dimensional Electron Gas

The term "two-dimensional electron gas" has a precisely defined meaning in modern solid-state physics. It applies to strong inversion or accumulation layers at planar semiconductor-semiconductor or semiconductor-insulator interfaces. The high electric field present at the interface results in a narrow potential well with a typical thickness between 30 Å (3 nm) and 100 Å (10 nm), which is smaller than the de Broglie wavelength of the electrons. This gives rise to boundary quantization and standing-wave patterns. An example of the potential well of a *p*-type GaAs-(GaAl)As heterostructure is given in Fig. 1. It was obtained by solving self-consistently Schrödinger's and Poisson's equations [1]. Only a single bound state or electric subband E_0 is occupied. The charge distribution as a function of the distance from the interface is indicated by the dotted line; it was assumed beforehand that the wave function vanishes at the interface. The term "subband" was chosen because the electrons are free to move parallel to the interface and show band-like behaviour. As long as the electrons cannot make a transition to the first excited state, no motion perpendicular to the interface is possible and one talks about a two-dimensional electron gas.

The system can be fully quantized by applying a high magnetic field perpendicular to the interface.

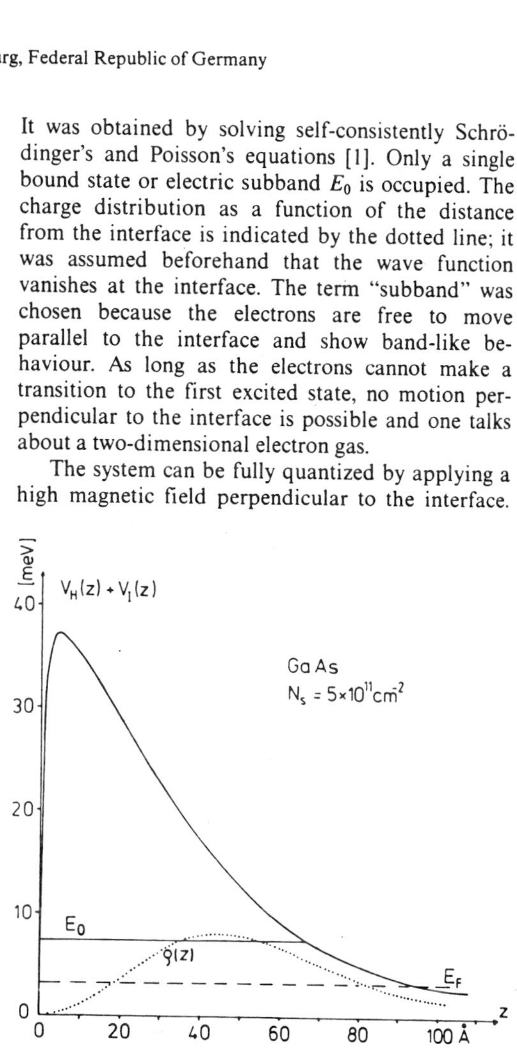

Fig. 1. Potential well of a *p*-type GaAs-(GaAl)As heterostructure. The sum of the potentials due to free electrons V_H and impurities V_I has been plotted as a function of the distance z from the interface. E_0 = energy of the only occupied electric subband; E_F = Fermi energy; $\varrho(z)$ = hole concentration

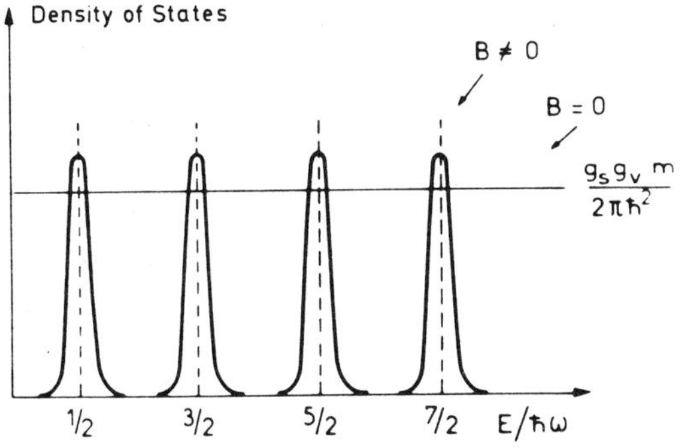

Fig. 2. Density of states of a two-dimensional electron gas as a function of energy, normalized to the cyclotron energy $\hbar\omega$, with and without a transverse magnetic field B

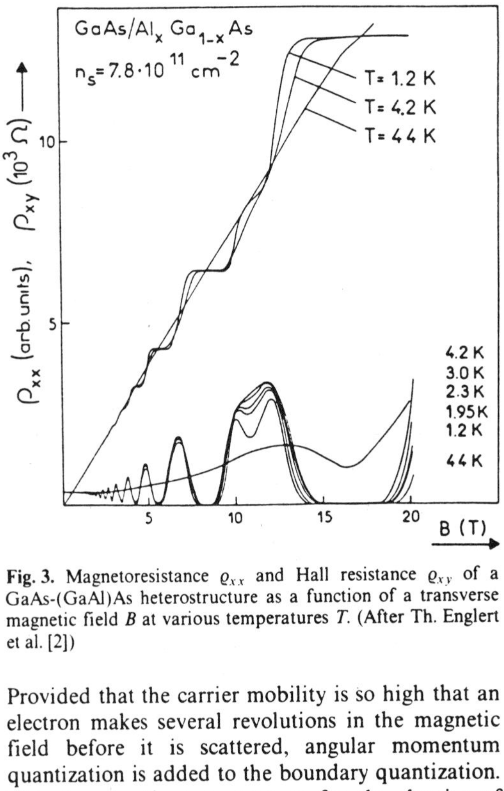

Fig. 3. Magnetoresistance ϱ_{xx} and Hall resistance ϱ_{xy} of a GaAs-(GaAl)As heterostructure as a function of a transverse magnetic field B at various temperatures T. (After Th. Englert et al. [2])

Provided that the carrier mobility is so high that an electron makes several revolutions in the magnetic field before it is scattered, angular momentum quantization is added to the boundary quantization. This has drastic consequences for the density of states. Whereas the density of states of a 2D electron gas is not dependent on the energy, this no longer holds in high transverse magnetic fields. In this case, the density of states has peaks at the Landau energies $E = (n + 1/2)/(\hbar\,\omega_c)$, where n is the Landau quantum number, $\hbar$ = Planck's constant h divided by 2π, and ω_c the cyclotron frequency eB/m_c (e = electron charge, B = magnetic field, m_c = cyclotron electron effective mass). If the product of cyclotron frequency ω_c and electron scattering time $\tau \gg 1$, the wings of the density-of-state peaks do not overlap and a gap in the density of states arises (Fig. 2). This has remarkable consequences for the transport properties. The electrical resistance oscillates as a function of the magnetic field (Shubnikov-de Haas effect) and vanishes in certain magnetic field ranges. This can be recognized at the bottom of Fig. 3, where the magnetoresistance ϱ_{xx} of an n-type GaAs-(GaAl)As heterostructure with an electron mobility of 60,000 cm^2/V·s at 4.2 K has been plotted. The data were obtained by Th. Englert and co-workers at the high-magnetic-field facility in Grenoble [2]. Results obtained at various temperatures have been drawn: one finds that the quantum oscillations have almost disappeared if the temperature is raised to 44 K. Under these circumstances, the condition $\omega_c\tau > 1$ is no longer obeyed. In addition, the Hall resistance ϱ_{xy}, the ratio of Hall voltage and the current (used in an arrangement useful for the measurement of the Hall effect), has been plotted. At 44 K the Hall resistance increases linearly with the magnetic field. At helium temperatures pronounced plateaux arise in the ranges of the magnetic field where the magnetoresistance ϱ_{xx} vanishes. The plateaux are quantized in inverse-integer fractions of h/e^2:

$$\varrho_{xy} = h/(e^2 \mathrm{i})\,, \quad \mathrm{i} = 1, 2, 3, \ldots \qquad (1)$$

This startling discovery was made by von Klitzing in the spring of 1980 [3]. The big surprise was that the plateau values could be expressed in terms of (1) with very high precision. Before, it was believed that the plateaux in ϱ_{xx} were caused by localization of electrons in bound states and that these electrons would not contribute to the Hall resistance.

I shall not discuss here the great importance of the discovery for metrology and for the realization of the ohm. This aspect is treated thoroughly in other contributions to this volume. I shall rather concentrate on the circumstances and the background of the important finding.

Early Work on Semiconductor Physics

When I became acquainted with Klaus von Klitzing in 1965 he was a physics student at the Technische Universität Braunschweig. At that time we were studying in the semiconductor laboratory of the PTB the Shubnikov-de Haas effect for Bi_2Te_3 [4] and Te [5] in order to determine the shape of the Fermi surface. Also, quantum phenomena in InSb were investigated [6]. Pulsed high magnetic fields up to 25 T were produced by discharging a high-voltage capacitor bank into a nitrogen-cooled copper coil.

The first task of von Klitzing was to calibrate magnet coils. In addition, he became acquainted for the first time with a 2D electronic system. In the mid-sixties we were also studying in Braunschweig the magneto-transport properties of *p*-type inversion layers adjacent to a medium-angle grain boundary in Ge bicrystals [7]. This work was started in 1960 at the semiconductor laboratory of the University of Illinois [8] where we looked already for electric subbands predicted by J. R. Schrieffer [9] in 1957. The search for boundary quantization in *p*-type Ge inversion layers was successful only many years later [10].

The Early Work in Würzburg

In 1968 I accepted a call to the Universität Würzburg, and Klaus von Klitzing followed me in 1969 after completion of his diploma thesis in Braunschweig. He began to work on a doctoral thesis immediately and was assigned to study the transport properties of Te single-crystals at helium temperatures in high magnetic fields in the quantum limit [11]. This condition can be achieved if only the lowest Landau level is occupied. In the course of this work, von Klitzing discovered a new effect which is now dubbed "magneto-impurity effect" [12]. If the electric field in Te samples of particular orientation is so high that the free carriers are slightly heated, weak resonance structures become visible in the magneto-resistance. The effect is so small that double electronic differentiation is necessary in order to reveal it properly. It arises if the energy difference of two impurity levels becomes equal to a characteristic energy, for instance the cyclotron energy. The effect – the first von Klitzing effect – can serve to characterize the impurity content of Te crystals. In Germany we still practice the old habit of presenting a new doctor immediately after his final examination with a fancy doctor's hat. On von Klitzing's hat one could read: "Recorded curve: Nothing; first derivative: nothing; second derivative: von Klitzing effect".

The Habilitationsschrift – a thesis which is required for appointment as a lecturer – was submitted in 1977 and was devoted to the elucidation of the magneto-impurity effect in tellurium. Included were data which were obtained during a one year's stay at the Clarendon Laboratory in Oxford 1975/76, where von Klitzing was associated with R. A. Stradling and R. Nicholas.

But there was another unexpected discovery during the PhD-thesis work. In 1971, Shubnikov-de Haas-like oscillations were observed by von Klitzing in the second derivative of the magneto-resistance of rather pure Te samples. The doping was so low that the usual Shubnikov-de Haas oscillations could be excluded. Because of my previous work on Ge bicrystals I suspected that surface layers were involved. Etching of the specimens immediately revealed that this conjecture was correct [13]. The surface quantum-oscillations were subsequently investigated in detail and were the beginning of extensive research on semiconductor interface properties in Würzburg.

Work on Silicon Inversion Layers

At the end of the sixties, Gerhard Dorda of the Siemens Forschungslaboratorien, München, had investigated the piezo-resistance effect in MOSFETs (Metal Oxide Semiconductor Field-Effect Transistors). His results differed substantially from data obtained from bulk material. He attributed his results to boundary quantization and was able to demonstrate that quantum effects were relevant even at room temperature [14].

In order to obtain further information he wanted to extend his work to helium temperatures and to apply high magnetic fields. Therefore he visited us in Würzburg in the spring of 1973. During our discussions it turned out that, in the Siemens Laboratories, *p*-channel MOSFETs were available which seemed suitable for measuring Shubnikov-de Haas oscillations from which the electronic band structure could possibly be deduced.

Since 1966, when the pioneering work on Shubnikov-de Haas oscillations in Si-MOSFETs of (100) orientation by Fowler, Fang, Howard and Stiles was published [15], it had become evident that in high-quality MOSFETs a 2D electron gas can be realized. The carrier concentration in the channel between source and drain can be varied at will by changing the gate voltage V_g, the electron concentration may be adjusted between 10^{11} cm^{-2} and 10^{13} cm^{-2}. Consequently, the Fermi energy E_F can be varied by two orders of magnitude because there is a linear relationship between E_F and V_g. This allows a new type of magnetoresistance experiment. Instead of keeping the carrier concentration constant and sweeping the magnetic field B, as in a usual Shubnikov-de Haas experiment, one can keep B constant and sweep the carrier concentration (an appropriate term for this would be: Magneto field-effect experiment). Spectacular data obtained on a (100) Si-MOSFET in a magnetic field of 20 T and at a temperature of 0.4 K by Th. Englert [16] at the high-magnetic-field facility of the Max-Planck-Institut für Festkörperforschung in Grenoble are shown in Fig. 4. The transverse component of the resistance ϱ_{xx} shows sharp spikes

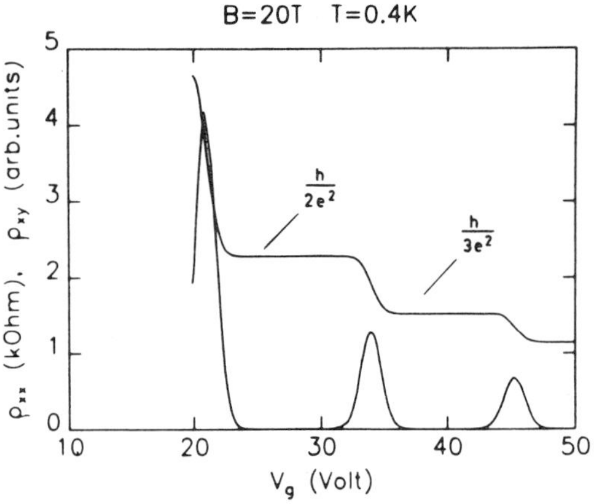

Fig. 4. Magnetoresistance ϱ_{xx} and Hall resistance ϱ_{xy} of a (100) silicon MOSFET as a function of the gate voltage V_g at a fixed magnetic field of 20 T and a temperature of 0.4 K. (After Th. Englert [16])

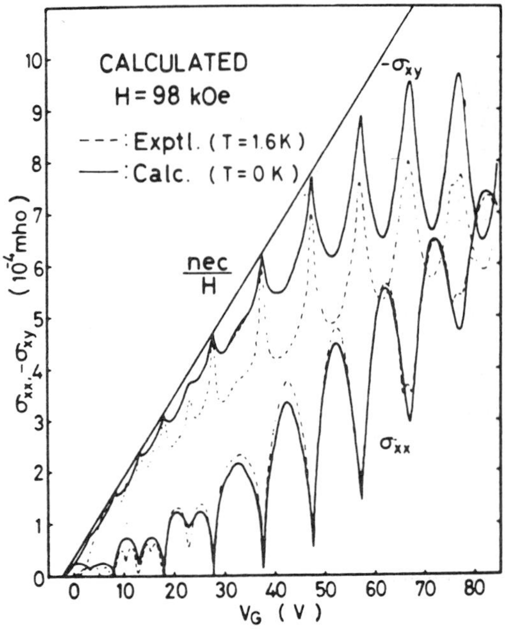

Fig. 5. Magnetoconductivity σ_{xx} and negative Hall conductivity $-\sigma_{xy}$ of a (100) Si-MOSFET as a function of the gate voltage V_g at a fixed magnetic field of $H = 98$ kOe ≘ 9.8 T. Dashed curve: experimental data obtained at $T = 1.6$ K. Full curves: theoretical calculations for $T = 0$ K. Straight line: classical Hall curve $\sigma_{xy} = nec/H$ (in c.g.s. units). (After S. Kawaji et al. [17])

when the Fermi level crosses a Landau level. As in GaAs, the Hall resistance is quantized in inverse-integer fractions of h/e^2 (see, for comparison, Fig. 3). One should note, however, that the plateaux in ϱ_{xy} and the wide ranges in which ϱ_{xx} vanishes exist in Si only at very high magnetic fields. At lower magnetic fields and $T = 1.6$ K the plateaux are absent. In Fig. 5, data obtained by S. Kawaji and co-workers [17] in Tokyo in a magnetic field of 9.8 T are reproduced. The magnetoconductivity σ_{xx} and the Hall conductivity $-\sigma_{xy}$ have been plotted as a function of the gate voltage V_g. The components of the magnetoconductivity tensor are obtained by inverting the magnetoresistance tensor. Both experimental (dashed lines) and theoretical curves (solid lines) have been plotted. No plateaux are visible; at gate voltages below 30 V, σ_{xx} vanishes only at particular V_g values. The Hall conductivity increases proportionally to V_g, with garland-like oscillations superimposed. One can note that the Hall conductivity reaches its classical value and amounts to $i\,e^2/h$, whenever σ_{xx} has vanished.

Returning now to G. Dorda's visit in Würzburg 1973, it was obvious that nobody had observed Shubnikov-de Haas oscillations in *p*-channel MOSFETs until then and that it seemed very rewarding to learn more about this system. Attempts to see the Shubnikov-de Haas effect in *p*-type devices in the United States had been unsuccessful because the Hall mobility was not large enough to fulfil the condition $\omega_c\tau > 1$ in the magnetic fields available. It turned out, however, that in the best specimens produced by the Siemens technologists the hole mobility of (110) devices at helium temperatures was between 2000 and 3000 cm²/V · s, sufficient to meet the condition for angular momentum quantization in magnetic fields of 10 T. K. von Klitzing was asked to do the experiments and was immediately successful. His differentiation techniques revealed well-developed Shubnikov-de Haas oscillations at rather modest magnetic fields. An analysis of the temperature dependence of the amplitude of the oscillations showed that the effective mass differed substantially from the bulk values and was strongly dependent on the electric field at the interface $Si-SiO_2$ [18, 19]. Self-consistent calculations of the subband structure performed independently by E. Bangert in Würzburg [20] and J. Ohkawa and Y. Uemura in Tokyo [21] confirmed the experimental data. At that time a public controversy developed. An American group of experimentalists published effective-mass data about a factor of 2 higher than those obtained by the Würzburg/München group. When, in a theoretical paper coming from the American West Coast, effective-mass values were calculated which agreed fairly well with the American experiments it was charged that we had obviously erred by a factor of 2. However,

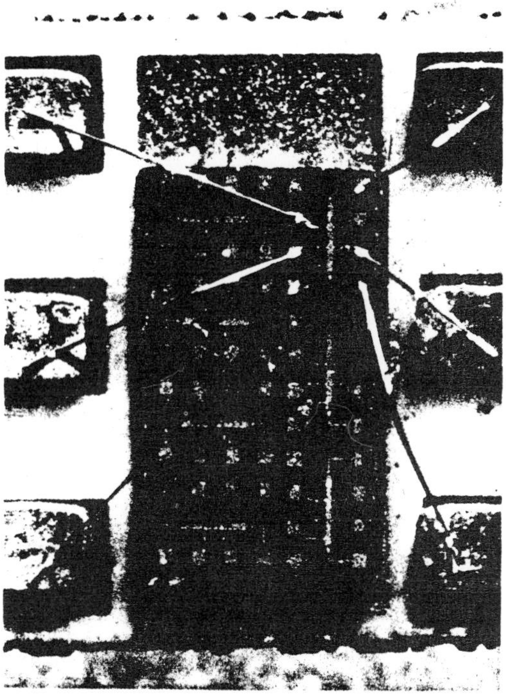

Fig. 6. A chip with 7 MOSFETs for magnetoresistance and Hall-effect measurements. Length and width of the devices: 400 μm and 40 μm, respectively. The separate current and potential leads can be recognized, also the gate electrode. Contact is made with gold wires. The samples were made by the Siemens Forschungslaboratorien, München

Fig. 7. Superconducting magnet with a maximum field of $B = 14.6$ T, in which the first precision measurements of the quantum Hall effect were made

the authors had not taken into account that the experiments in Würzburg were done by K. von Klitzing who usually is not wrong by a factor of 2! Later on, we were able to demonstrate that the theoretical approach chosen by our American colleagues was inadequate [22].

In the mid-seventies we found out that *n*-channel MOSFETs had not been investigated as thoroughly as we thought before. Measurements of Shubnikov-de Haas oscillations in *n*-type inversion layers of (110) and (111) orientation revealed an anomalous valley-degeneracy factor of 2 although values of 4 and 6 were expected for the two orientations, respectively [23]. The origin of the discrepancy is not clear at present in spite of subsequent work in several laboratories. Also, the valley-splitting problem was addressed in Würzburg [24].

We also started work on *n*-channel MOSFETs of (100) orientation. Through the collaboration with the Siemens Forschungslaboratorium München, especially with G. Dorda, we had access to very-high-quality devices. Special samples with separate current and potential leads were available which were very suitable for checking the just-developing magneto-transport theory for silicon inversion layers. Figure 6 shows a chip with several Hall samples, which have a length of 400 μm. The peak electron-mobility at 4.2 K was around 17,000 $cm^2/V \cdot s$, the highest value achieved in the world at that time. The high mobility allowed us to obtain unusually high $\omega_c \tau$-values in the superconducting magnets available in Würzburg. An especially powerful one had been added in 1976 when we participated in a special program on semiconductor electronics of the Deutsche Forschungsgemeinschaft. The magnet consisted of concentric Nb_3Sn and NbTi coils and allowed one to generate 14.6 T with an excitation time to the peak value of only 10 min. The magnet – in which the first precision measurements of the quantum Hall effect were made in 1980 – is shown in Fig. 7.

Thomas Englert – then a graduate student – investigated high-quality (100) *n*-channel MOSFETs in the new magnet in order to check the predictions of theory. In Tokyo, Y. Uemura and his co-workers did a great deal of work on 2D magneto transport. We had very close relations with our Japanese colleagues. Especially, three international conferences held in Würzburg in 1972, 1974 and 1976 with the title "The Application of High Magnetic Fields in Semiconductor Physics" helped to establish very good contacts with colleagues from overseas and from Europe. Only invited papers were given and ample time was provided for discussion. Our meetings had a good reputation because of the relevant subjects addressed and because of the very frank discussions which helped to clear many actual problems. Especially, the 1976 conference had an emphasis on MOS physics. At this meeting, T. Ando from the University of Tokyo gave a very nice review of his work on two-dimensional transport [25] and Y. Uemura talked about localization in a 2D system in high magnetic fields [26].

One of the predictions of Ando's theory was that the peak values of the magneto conductivity σ_{xx} should only depend on the Landau quantum number and on h/e^2 [27]. Qualitatively, the prediction turned out to be correct; however, there was a quantitative discrepancy when Englert made the first experiments with the new magnet. S. Kawaji and co-workers, who have studied magneto-transport in Si-MOSFETs extensively in the last decade, also had difficulties in verifying Ando's predictions [28]. From their results they drew the conclusion that their Hall samples did not seem suitable for deducing the correct magnetoconductivity tensor.

In 1977 Englert observed for the first time in Würzburg the vanishing of ϱ_{xx} in magnetoresistance measurements at 14.2 T in certain ranges of the gate voltage, accompanied by plateaux in the Hall resistance. The data are reproduced in Fig. 8 which shows not only ϱ_{xx} and ϱ_{xy} but also σ_{xx} and σ_{xy} obtained by inversion of the magnetoresistance tensor. A close collaboration with von Klitzing had been established, also aimed at the investigation of the electronic g-factor which should oscillate as a function of the occupation of the Landau levels according to theory [29]. Inspection of the plateau around 7 V in Fig. 8b shows that the Hall resistance has a value around 6.4 kΩ. Precise measurements would without doubt have revealed a value of 6453.2 Ω, corresponding to $h/(4e^2)$. The results were

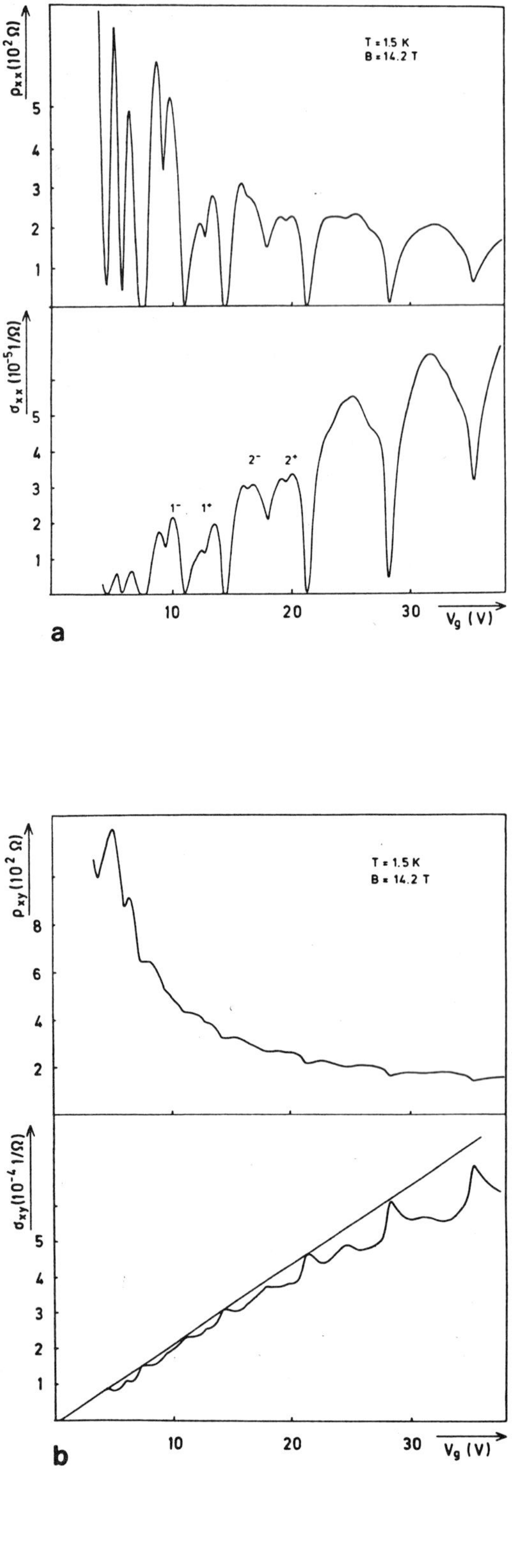

Fig. 8. a Upper curve: magnetoresistance ϱ_{xx} of a (100) Si-MOSFET as a function of the gate voltage V_g in a transverse magnetic field of 14.2 T. Lower curve: magnetoconductivity σ_{xx} as a function of V_g, with Landau quantum numbers; + (plus) and − (minus) characterize the spin splitting. Note that both ϱ_{xx} and σ_{xx} vanish at the same, particular gate voltages. **b** Upper curve: Hall resistance ϱ_{xy} as a function of the gate voltage V_g for the same specimen. Note the plateau at 6.4 kΩ corresponding to h/e^2. Lower curve: Hall conductivity σ_{xy}, calculated from the resistivity tensor. According to our present knowledge, the straight line representing the classical Hall conductivity should cross the plateau at $V_g \approx 7$ V in the middle. (After Th. Englert, Dissertation, Univ. Würzburg 1977, published in [30])

presented in September 1977 at the international conference on two-dimensional electronic systems in Berchtesgaden, Germany. They were published in the conference proceedings in the following year [30].

From the above it can be concluded that the quantum Hall effect could have been discovered already in 1977 and it is interesting to reflect why it was not. It has been explained already that, according to the widely accepted theory by Ando [25, 27], the magnetoconductivity and the magnetoresistance should vanish at particular gate voltages whenever a Landau level is fully occupied. At these voltages the Hall resistance was predicted to have a value of $h/(e^2 i)$. Under these conditions, the resistivity is zero because, due to the gap in the density of states, there are no final states for scattering. Consequently, the centres of the cyclotron motion move on a cycloidal path along a Hall sample under the combined action of the Hall field and the transverse magnetic field. If the gate voltage is slightly enhanced, the Fermi level enters the next Landau level with the consequence that there are final states for the scattering and that the resistance has a finite value. From the plateaux in ϱ_{xx} in Fig. 8 it was concluded that carriers were trapped in localized states, which could not contribute to the conductivity and that the Fermi level was pinned. In spite of the localization the resistivity remained unmeasurably small because there was a sufficiently large number of unscattered, free electrons. Because the Hall effect measures the density of free electrons, it was tacitly assumed that the localized electrons would not contribute to ϱ_{xy}. Therefore nobody compared the results with theory. It should be noted that well-developed plateaux in the Hall conductivity of (100) MOSFETs were also observed by Kawaji et al. [28] in magnetic fields of 15 T prior to the discovery of the quantum Hall effect. It took about 2½ years to find out what was really happening. Indeed, something unexpected was occurring: although electrons are localized in bound states, the Hall resistivity in the plateaux has the value predicted by the free-electron theory. The localized electrons do not contribute to the Hall resistivity but the remaining free electrons make up for the deficit!

The Discovery of the Quantum Hall Effect

The discovery that the Hall resistance of high-quality MOSFETs is quantized in inverse-integer fractions of h/e^2 was made in the night of February 4/5 in the high-magnetic-field facility of the Max-Planck-Institut für Festkörperforschung in Grenoble. The laboratory is jointly operated with the French research organization Centre Nationale de la Recherche Scientifique (CNRS). High magnetic fields up to 25 T can be generated by water-cooled copper coils with a power supply of 10 MW. The facility is not only available for staff members but also for guest scientists from all over the world. Th. Englert accompanied me when I became director of this laboratory in 1978. The work on the magneto-transport of 2D systems was continued. In the autumn of 1979 K. von Klitzing joined the group. In the meantime he had obtained a Heisenberg fellowship which allowed him to do research free of teaching duties. In collaboration with R. Nicholas from Oxford, unsolved problems of the valley splitting in (100) MOSFETs were investigated [31]. Also, experiments on the g-factor were performed because our previous results had been questioned [32].

In the course of this work the plateaux in the Hall resistance, which had been observed in Würzburg before, showed up again. However, because of the higher fields available in Grenoble they were better pronounced. This can be recognized in Fig. 9. For the experiments on the valley splitting, samples of different origin seemed useful; for this reason, M. Pepper (Cavendish Laboratory and Plessey Company, U.K.) was asked to provide high-quality MOSFETs of different geometry. During the experiments in January/February 1980 practically all specimens investigated showed well-defined Hall plateaux which surprised von Klitzing and Englert and let them make a simple calculation. It turned out that the plateau no. 4 always had a value of about 6450 Ω corresponding to $h/(4e^2)$. Subsequently, von Klitzing called V. Kose, the director of the

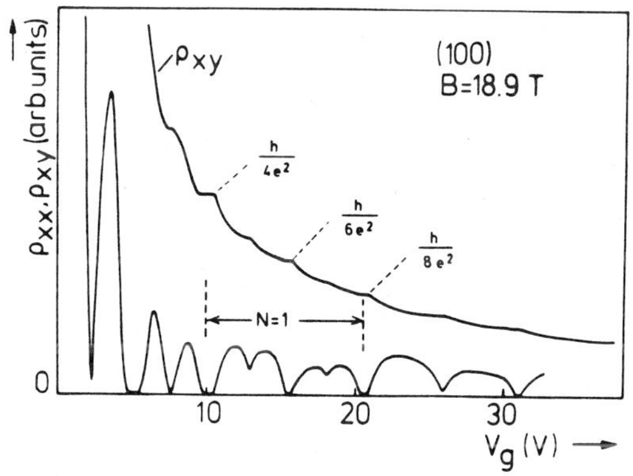

Fig. 9. Magnetoresistance ϱ_{xx} and Hall resistance ϱ_{xy} for a (100) Si-MOSFET as a function of the gate voltage V_g in a transverse magnetic field of $B = 18.9$ T at $T = 1.6$ K. Plateaux in the Hall resistance are developed at $h/(4e^2)$, $h/(6e^2)$ and $h/(8e^2)$ (von Klitzing effect). See also [3]

division of the PTB Braunschweig concerned with electrical standards, in order to find out whether the PTB was interested in a precise determination of h/e^2. The answer was that the PTB was very interested indeed, provided that a precision of better than 10^{-6} could be achieved! The first experiments in Grenoble had an accuracy of about 1% and an improvement there seemed difficult to achieve, because the enhancement of the precision required long-term measurements for which Bitter magnets are not especially suitable. One reason is their limited lifetime, another one the high costs for electricity. For precision experiments the 14.6 T superconducting magnet in Würzburg seemed more suitable, which only requires a generous supply of liquid helium. Therefore, von Klitzing returned to the Physikalisches Institut Würzburg at the end of February 1980 in order to try precision experiments on selected high-mobility samples. During the following weeks the high-field magnet was operated close to its maximum field for about 200 hours. He succeeded in reducing the uncertainty in the measurement of the Hall resistance to about 5×10^{-6} and to show that it is really quantized in (inverse-integer) fractions of h/e^2. The values given in the famous publication in Phys. Rev. Lett. [3] do not deviate substantially from the latest results [34].

In May 1980 the article was written and submitted to Physical Review Letters. The title was "Realization of a resistance standard based on fundamental constants". The paper was returned with the remark that the method for the time being did not seem suitable for a precision determination of the ohm, for which higher accuracy would be required. Subsequently, the manuscript was re-written and modified in such a way that emphasis was on the determination of the fine-structure constant, introduced into atomic physics by A. Sommerfeld a long time ago. After this switch from "applied" physics to "general" physics, the paper was accepted for publication in Phys. Rev. Lett. One can see what remarkable consequences a multiplication of e^2/h with the light velocity c and μ_0 can have!

At an international meeting on the precision determination of fundamental constants at the PTB in June 1980, von Klitzing gave a notable talk on the quantum Hall effect. The limits of the new method were not known because no theory of the quantum Hall effect existed at that time. In order to find out whether the QHE method could meet the tough requirements of precision metrology, it was necessary to perform further accurate experiments as soon as possible. For this reason, E. Braun and coworkers came to Würzburg in August in order to perform new experiments in collaboration with von Klitzing. It was possible to reduce the standard deviation of h/e^2 to about 1.3×10^{-6} [33]. The recently obtained precision with GaAs-(GaAl)As heterostructures [34] – the reproducibility is now 10^{-8} – has opened the possibility of maintaining the ohm by means of the quantum Hall effect.

I shall not say anything about the impact of the discovery on metrology, because this is well documented in this issue. It should, however, be mentioned that the resonance in the physics community was enormous. Obviously, a new macroscopic quantum effect had been discovered for which no theoretical prediction existed. The situation is properly characterized by a statement of D. Thouless, who began an invited paper on the theory of the quantum Hall effect at the fifth international conference on the electronic properties of two-dimensional systems, Oxford 1983, in the following way: "The discovery by von Klitzing, Dorda and Pepper, that the Hall-conductance of a two-dimensional electron system can be, with very high precision, an integer multiple of e^2/h was a triumph of experimental physics. In most comparable cases, such as the quantization of flux in superconductivity or the quantization of circulation for superfluid helium, there have been previous theoretical suggestions of the existence of the effect even if there were unexpected features in the experimental result. In this case, there was no more than approximate quantization suggested, and so there was no reason for the experimentalists to examine the transverse voltage in their device with the precision which they used. Once the discovery had been made, we theorists rushed in to show why the result had been obvious all the time. ... [35]". The approximate quantization mentioned concerns a paper of Ando, Matsumoto and Uemura from 1975 [36] which was only taken seriously many years after its publication.

Sometimes, the opinion is expressed that the discovery of the quantum Hall effect happened by chance. However, the preceding remarks on the circumstances should have made clear that this is not true at all. Only an intense study of magneto-transport problems in 2D systems could eventually result in the discovery of the quantum Hall effect. An important aspect was the thoroughness of K. von Klitzing and his drive for clarity. The existence of Hall plateaux was manifested in the literature since 1978, and everybody had the chance to arrive at the correct conclusions from the published experiments. According to the accepted theory it could not be expected that the Hall plateaux could be expressed precisely in terms of h/e^2. In this case, the motto which has been recommended by the famous Dutch low-temperature physicist Heike Kamerlingh Onnes to all experimentalists: *door meten tot weten* ("through measurements to knowledge") turned out to be significant. One should realize that only 2½ years passed after the plateaux were observed until their fundamental meaning was recognized. It should be recalled that after the first liquefaction of helium it took about 20 years before anomalously large heat conductivity was realized. The increase of the specific heat at the λ-point was measured by Kamerlingh Onnes shortly before he died, but he mistrusted the data and did not publish them. The recognition of the superfluidity of helium took even more time.

It is evident that for the discovery of the quantum Hall effect a favourable scientific environment was necessary. Only extensive studies of MOS problems in high magnetic fields – as practised in Würzburg – by a whole team of PhDs, graduate and undergraduate students could lead to the final success. Also, a close cooperation between industry and university was important. Only in MOSFETs with a very high mobility at helium temperatures could complete quantization leading to a vanishing resistance and Hall plateaux be achieved. It should be emphasized that our colleagues from industry not only provided us with samples; the exchange of ideas was equally essential. Whenever possible, our requirements for special samples were taken into account.

It is also evident that the PTB played a major role in the discovery of the quantum Hall effect. Through my long association with this institution I knew that the precision determination of fundamental constants is a never-ending challenge. In 1975 we considered determining e^2/h from the period of Shubnikov-de Haas oscillations observed on (100) MOSFETs, together with the charge on the surface condenser which is proportional to the difference between gate voltage and threshold voltage. However, we came to the conclusion that the threshold voltage is not stable enough to obtain PTB precision. This was actually the subject of a talk in the PTB colloquium which I gave in 1976. It should also be noted that Klaus von Klitzing had a special interest in precision measurements. As a student in Braunschweig he not only worked in the semiconductor laboratory but also in the laboratory for electrical standards. In the process of "Habilitation" one has not only to deliver a thesis but also to give a public lecture on a subject of general scientific interest. The title of von Klitzing's talk was: "Das internationale Einheitensystem". In his lecture he showed that precision experiments can be very exciting. He actually demonstrated the Josephson effect with equipment which he had borrowed from the PTB.

The discovery of the quantum Hall effect and its subsequent development for precision metrology bear a particular lesson. From my time at the PTB I recall that the line of work which we pursued in our laboratory was considered by many colleagues as not being especially relevant for the mission of the PTB; fortunately not so by the president and the directors. It was argued that quantum oscillations observed in semiconductors at helium temperatures in high magnetic fields had no potential for precision metrology and that this kind of research should better be performed at universities or Max-Planck institutes. These colleagues believed that it would have been better, for instance, to study the long-term annealing behaviour of manganin in order to obtain better standard resistors! The quantum Hall effect has again demonstrated that significant progress in the realization of the units is usually achieved only by entirely new approaches.

The conclusion one can draw from this is that not only applied work should be pursued in the institutions devoted to precision metrology and fundamental investigations should not be entirely neglected. The practical benefit of the previous activities of the PTB semiconductor laboratory is quite obvious: Without the existing equipment and expertise it would hardly have been possible to measure h/e^2 with the accuracy achieved in a relatively short period!

References

1. E. Bangert, G. Landwehr: Superlattices and Microstructures **1**, 363 (1985)
2. Th. Englert, D. C. Tsui, A. C. Gossard, Ch. Uihlein: Surf. Sci. **113**, 295 (1982)
3. K. von Klitzing, G. Dorda, M. Pepper: Phys. Rev. Lett. **45**, 494 (1980)
4. G. Landwehr, P. Drath: Z. Angew. Phys. **20**, 392 (1966)
5. E. Braun, G. Landwehr: Z. Naturforsch. **21 a**, 495 (1966)
6. L. M. Bliek, G. Landwehr, M. von Ortenberg: In: Proc. Int. Conf. on the Physics of Semiconductors, Moscow 1968 (Academy of Sciences USSR, Nauka, Leningrad 1968)
7. G. Landwehr: Phys. Status Solidi **3**, 440 (1963)
8. G. Landwehr, P. Handler: J. Phys. Chem. Solids **23**, 891 (1962)
9. J. R. Schrieffer: In: *Semiconductor Surface Physics*, R. H. Kingston (ed.) (Univ. of Pennsylvania Press, Philadelphia 1957), p. 55
10. G. Landwehr, E. Bangert, S. Uchida: Solid-State Electron. **28**, 171 (1985)
11. K. von Klitzing, G. Landwehr: Phys. Status Solidi (b) **45**, K119 (1971)
12. K. von Klitzing: Solid-State Electron. **21**, 223 (1978)
13. K. von Klitzing, G. Landwehr: Solid State Commun. **9**, 2201 (1971)
14. G. Dorda: Appl. Phys. Lett. **17**, 406 (1970); also in: *Festkörperprobleme XIII*, H. J. Queisser (ed.) (Vieweg Verlag, Braunschweig 1973), p. 215; J. Appl. Phys. **42**, 2053 (1971)
15. A. B. Fowler, F. F. Fang, W. E. Howard, P. J. Stiles: Phys. Rev. Lett. **16**, 901 (1966)
16. Th. Englert: In: *Application of High Magnetic Fields in Semiconductor Physics, Grenoble 1982*, G. Landwehr (ed.), Lecture Notes in Physics **177**, Springer, Berlin, Heidelberg, New York, Tokyo 1983, p. 87

17. S. Kawaji, T. Igarashi, J. Wakabayashi: Prog. Theor. Phys., Suppl. **57,** 176 (1975)
18. K. von Klitzing, G. Landwehr, G. Dorda: Solid State Commun. **14,** 387 (1974)
19. K. von Klitzing, G. Landwehr, G. Dorda: Solid State Commun. **15,** 489 (1974)
20. E. Bangert, K. von Klitzing, G. Landwehr: Proc. 12th Int. Conf. on the Physics on Semiconductors, Stuttgart 1974, M. H. Pilkuhn (ed.) (Teubner, Stuttgart 1974), p. 714
21. J. Ohkawa, Y. Uemura: Prog. Theor. Phys., Suppl. **57,** 164 (1975)
22. G. Landwehr, E. Bangert, K. von Klitzing, Th. Englert: Solid State Commun. **19,** 1031 (1976)
23. T. Neugebauer, K. von Klitzing, G. Landwehr: Solid State Commun. **17,** 295 (1975)
24. R. Kümmel: Z. Phys. B – Condensed Matter **22,** 223 (1975)
25. T. Ando: In: *The Application of High Magnetic Fields in Semiconductor Physics,* Lecture Notes, Phys. Inst. Univ. Würzburg 1976, p. 33
26. Y. Uemura: *ibid.,* p. 81
27. T. Ando, Y. Uemura: J. Phys. Soc. Jpn. **36,** 259 (1974)
28. J. Wakabayashi, S. Kawaji: Surf. Sci. **98,** 299 (1980)
29. T. Ando, Y. Uemura: J. Phys. Soc. Jpn. **37,** 1044 (1974)
30. Th. Englert, K. von Klitzing: Surf. Sci. **73,** 70 (1978)
31. R. J. Nicholas, K. von Klitzing, Th. Englert: Solid State Commun. **34,** 51 (1980)
32. Th. Englert, K. von Klitzing, R. J. Nicholas, G. Landwehr, G. Dorda, M. Pepper: Phys. Status Solidi (b) **99,** 237 (1980)
33. E. Braun, E. Staben, K. von Klitzing: PTB Mitt. **90,** 350 (1980)
34. L. Bliek, E. Braun, H. J. Engelmann, H. Leontief, F. Melchert, W. Schlapp, B. Stahl, P. Warnecke, G. Weimann: PTB Mitt. **93,** 21 (1983)
35. D. Thouless: Surf. Sci. **142,** 147 (1984)
36. T. Ando, Y. Matsumoto, Y. Uemura: J. Phys. Soc. Jpn. **39,** 279 (1975)

New Method for High-Accuracy Determination of the Fine-Structure Constant Based on Quantized Hall Resistance

K. v. Klitzing
Physikalisches Institut der Universität Würzburg, D-8700 Würzburg, Federal Republic of Germany, and Hochfeld-Magnetlabor des Max-Planck-Instituts für Festkörperforschung, F-38042 Grenoble, France

and

G. Dorda
Forschungslaboratorien der Siemens AG, D-8000 München, Federal Republic of Germany

and

M. Pepper
Cavendish Laboratory, Cambridge CB3 0HE, United Kingdom
(Received 30 May 1980)

Measurements of the Hall voltage of a two-dimensional electron gas, realized with a silicon metal-oxide-semiconductor field-effect transistor, show that the Hall resistance at particular, experimentally well-defined surface carrier concentrations has fixed values which depend only on the fine-structure constant and speed of light, and is insensitive to the geometry of the device. Preliminary data are reported.

PACS numbers: 73.25.+i, 06.20.Jr, 72.20.My, 73.40.Qv

In this paper we report a new, potentially high-accuracy method for determining the fine-structure constant, α. The new approach is based on the fact that the degenerate electron gas in the inversion layer of a MOSFET (metal-oxide-semiconductor field-effect transistor) is fully quantized when the transistor is operated at helium temperatures and in a strong magnetic field of order 15 T.[1] The inset in Fig. 1 shows a schematic diagram of a typical MOSFET device used in this work. The electric field perpendicular to the surface (gate field) produces subbands for the motion normal to the semiconductor-oxide interface, and the magnetic field produces Landau quantization of motion parallel to the interface. The density of states $D(E)$ consists of broadened δ functions[2]; minimal overlap is achieved if the magnetic field is sufficiently high. The number of states, N_L, within each Landau level is given by

$$N_L = eB/h, \quad (1)$$

where we exclude the spin and valley degeneracies. If the density of states at the Fermi energy, $N(E_F)$, is zero, an inversion layer carrier cannot be scattered, and the center of the cyclotron orbit drifts in the direction perpendicular to the electric and magnetic field. If $N(E_F)$ is finite but small, an arbitrarily small rate of scattering cannot occur and localization produced by the long lifetime is the same as a zero scattering rate, i.e., the same absence of current-carrying states occurs.[3] Thus, when the Fermi level is between

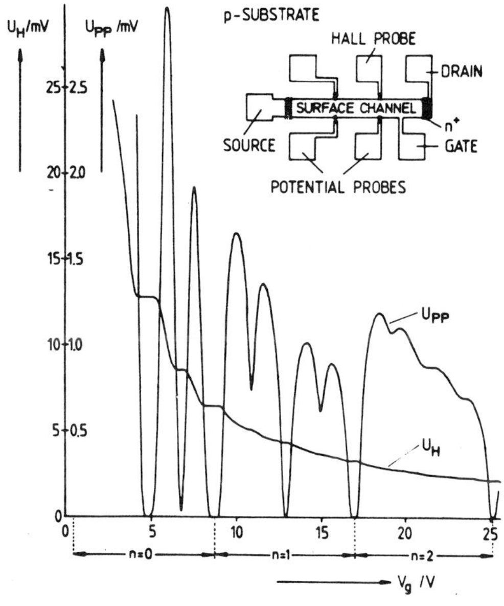

FIG. 1. Recordings of the Hall voltage U_H, and the voltage drop between the potential probes, U_{pp}, as a function of the gate voltage V_g at $T = 1.5$ K. The constant magnetic field (B) is 18 T and the source drain current, I, is 1 μA. The inset shows a top view of the device with a length of $L = 400$ μm, a width of $W = 50$ μm, and a distance between the potential probes of $L_{pp} = 130$ μm.

Landau levels the device current is thermally activated and the minima in σ_{xx}, $\sigma_{xx}^{\min}$, can be less than $10^{-7}\sigma_{xx}^{\max}$.[4] Increasing the magnetic field and decreasing the temperature, further decreases $\sigma_{xx}^{\min}$. The Hall conductivity σ_{xy}, which is usually a complicated function of the scattering process, becomes very simple in the absence of scattering and is given by[2]

$$\sigma_{xy} = -Ne/B, \tag{2}$$

where N is the carrier concentration.

The correction term to the above relation, $\Delta\sigma_{xy}$, is of the order of $\sigma_{xx}/\omega\tau$, where ω is the cyclotron frequency and τ is the relaxation time of the conduction electrons; $\omega\tau >> 1$ in strong magnetic fields. When the Fermi energy is between Landau levels, and $\sigma_{xx}^{\min} \sim 10^{-7}\sigma_{xx}^{\max}$, the correction $\Delta\sigma_{xy}/\sigma_{xy} < 10^{-8}$. Subject to any error imposed by $\Delta\sigma_{xy}$, when a Landau level is fully occupied and $N = N_L i$ ($i = 1,2,3,\ldots$), σ_{xy} is immediately given from Eqs. (1) and (2):

$$-\sigma_{xy} = e^2 i/h. \tag{3}$$

The Hall resistivity $\rho_{xy} = -\sigma_{xy}/(\sigma_{xx}^2 + \sigma_{xy}^2) \approx -\sigma_{xy}^{-1}$ is defined by E_H/j (E_H = Hall field, j = current density) and can be rewritten R_H/I, where R_H is the Hall resistance, U_H the Hall voltage and I the current. Thus, $R_H = h/e^2 i$, which may finally be written as[5]

$$R_H = \alpha^{-1}\mu_0 c/2i, \tag{4}$$

where μ_0 is the permeability of vacuum and exactly equal to $4\pi \times 10^{-7}$ H m^{-1}, c is the speed of light in vacuum and equal to 299 792 458 m s^{-1} with a current uncertainty[5] of 0.004 ppm and $\alpha \approx \frac{1}{137}$ is the fine-structure constant. It is clear from Eq. (4) that a high-accuracy measurement of the Hall resistance in SI units will give a value of α with essentially the same accuracy. Since resistances can be determined in SI units to a few parts in 10^8 by means of the so-called calculable cross capacitor by Thompson and Lampard,[6] the question of absolute units versus as-maintained units is much less of a problem than in the determination of e/h from the ac Josephson effect. Furthermore, the magnitude of R_H falls within a relatively convenient range: $R_H \approx (25\,813\ \Omega)/i$, with i typically between 2 and 8. Finally, we note that if α is assumed to be known from some other experiment (for example, from $2e/h$ and the proton gyromagnetic ratio γ_p), Eq. (4) may be used to derive a known standard resistance.

Two well-known corrections in the low-field Hall effect become unimportant. The first is the correction due to the shorting of the Hall voltage by the source and drain contacts.[7] This is important at low fields for samples with length-to-width ratio, L/W, less than 4, but becomes negligible when the Hall angle is 90°, i.e., $\sigma_{xx} = 0$.[8] The second correction which becomes unimportant is that due to an inexact alignment of the Hall probes, i.e., they are not exactly opposite: This is irrelevant, as the voltage drop along the sample vanishes when $\sigma_{xx} = 0$.[9]

The experiments were carried out on MOS devices with a range of oxide thicknesses (d_{ox} = 100 nm–400 nm), and length-to-width ratios ranging from $L/W = 25$ to $L/W = 0.65$. All the transistors were fabricated on the (100) surface orientation and, typically, the p-type substrate had room temperature resistivity of 10 Ω cm. The resistivity at helium temperature was higher than 10^{13} Ω cm, and no current flow between source and drain around the channel could be measured. The long devices ($L/W > 8$) had potential probes in addition to the Hall probes.

A typical recording of the measured Hall voltage U_H, and the voltage between the potential probes U_{pp}, as a function of the gate voltage is shown in Fig. 1. These results were obtained at a constant magnetic field of B = 18 T, a temperature of 1.5 K and a constant source drain current of I = 1 μA. Relevant device parameters were L = 400 μm, W = 50 μm and the distance between potential probes was about 130 μm.

The measured voltage U_{pp} is proportional to the resistivity component $\rho_{xx} = \sigma_{xx}[\rho_{xx}^2 + \rho_{xy}^2]$. At gate voltages where the E_F is in the energy gap between Landau levels, minima in both σ_{xx} and ρ_{xx} are observed.[9] Such minima are clearly visible, and are identified, in Fig. 1; the minima due to the lifting of the spin and the (twofold) valley degeneracy are also apparent. The Hall voltage clearly levels off at those values of carrier concentration where σ_{xx} and ρ_{xx} are zero. The values of U_H obtained in the regions are in good agreement with the predicted values, Eq. (4), if the error due to the 1-MΩ input impedance of the X-Y recorder is taken into account. It was found that the value of U_H in the "steps" was, for instance current, independent of sample geometry and direction of magnetic field, provided that σ_{xx} was zero.

An area of possible criticism of the theoretical basis of this experiment, is the role of carriers which are localized outside the main Landau level. Here we do not specify the localization mechanism, but the presence of localized carriers will invalidate both the relation $N = N_L$ and Eq. (4). However, the experimental results strongly suggest that such carriers do not invalidate Eq. (4). At present there is both theoretical and experimental investigation of this type of localization.[3,4,9-12] Ando[2] has suggested that the electrons in impurity bands, arising from short range scatterers, do not contribute to the Hall current; whereas the electrons in the Landau level give rise to the same Hall current as that obtained when all the electrons are in the level and can move freely. Clearly this process must be occuring but its range of validity must be carefully examined as an accompaniment to highly accurate measurements of Hall resistance.

For high-precision measurements we used a normal resistance R_0 in series with the device. The voltage drop, U_0, across R_0, and the voltages U_H and U_{pp} across and along the device was measured with a high impedance voltmeter ($R > 2\times10^{10}$ Ω). The resistance R_0 was calibrated by the Physikalisch Technische Bundesanstalt, Braunschweig, and had a value of $R_0 = 9999.69$ Ω at a temperature of 20 °C. A typical result of the measured Hall resistance $R_H = U_H/I = U_H R_0/U_0$, and the resistance, $R_{pp} = U_{pp}R_0/U_0$, between the potential probes of the device is shown in Fig. 2 ($B = 13$ T, $T = 1.8$ K). The minimum in σ_{xx} at $V_g = 23.6$ V corresponds to the minimum at $V_g = 8.7$ V in Fig. 1, because the thicknesses of the gate oxides of these two samples differ by a factor of 3.6. Our experimental arrangement was not sensitive enough to measure a value of R_{pp} of less than 0.1 Ω which was found in the gate-voltage region 23.40 V $< V_g <$ 23.80 V. The Hall resistance in this gate voltage region had a value of 6453.3 ± 0.1 Ω. This inaccuracy of ± 0.1 Ω was due to the limited sensitivity of the voltmeter. We would like to mention that most of the samples, especially devices with a small length-to-width ratio, showed a minimum in the Hall voltage as a function of V_g at gate voltage close to the left side of the plateau. In Fig. 2, this minimum is relatively shallow and has a value of 6 452.87 Ω at $V_g = 23.30$ V.

In order to demonstrate the insensitivity of the Hall resistance on the geometry of the device, measurements on two samples with a length-to-width ratio of $L/W = 0.65$ and $L/W = 25$, respectively, are plotted in Fig. 3. The gate-voltage scale

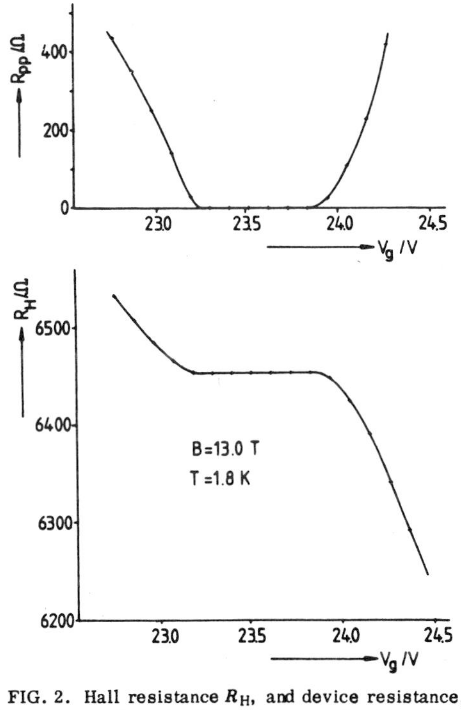

FIG. 2. Hall resistance R_H, and device resistance, R_{pp}, between the potential probes as a function of the gate voltage V_g in a region of gate voltage corresponding to a fully occupied, lowest ($n = 0$) Landau level. The plateau in R_H has a value of 6453.3 ± 0.1 Ω. The geometry of the device was $L = 400$ μm, $W = 50$ μm, and $L_{pp} = 130$ μm; $B = 13$ T.

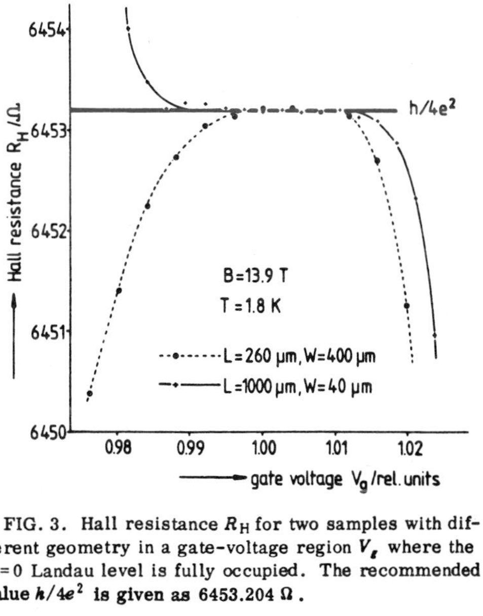

FIG. 3. Hall resistance R_H for two samples with different geometry in a gate-voltage region V_g where the $n = 0$ Landau level is fully occupied. The recommended value $h/4e^2$ is given as 6453.204 Ω.

is given in arbitrary units, and is different for the two samples because the thicknesses of the gate oxides are different. A gate voltage $V_g = 1.00$ corresponds, approximately, to a surface carrier concentration where the first fourfold-degenerate Landau level, $n = 0$, is completely filled. Within the experimental accuracy of 0.1 Ω, the same value for the plateau in the Hall resistance is measured. The value for $h/4e^2 = 6\,453.204 \pm 0.005$ Ω based on the recommended value for the fine-structure constant[5] is plotted in this figure, too. The decrease of the Hall resistance with decreasing gate voltage for the sample with $L/W = 0.65$ originates mainly from the shorting of the Hall voltage at the contacts. This effect is most pronounced when the Hall angle becomes smaller than 90°. In the limit of small Hall angles, the Hall voltage is reduced by a factor of 2 for the sample with $L/W = 0.65$.[7]

The mean value of the Hall resistance for all samples investigated was 6453.22 ± 0.10 Ω for measurements in the energy gap between the Landau levels $n = 0$ and $n = 1$ (corresponding to $i = 4$ in Eq. 4), 3226.62 ± 0.10 Ω for measurements in the energy gap between Landau levels $n = 1$ and $n = 2$ ($i = 8$), and $12\,906.5 \pm 1.0$ Ω for measurements in the energy gap between the spin split levels with $n = 0$ ($i = 2$). These resistances agree very well with the calculated values $h/e^2 i$ based on the recently reported[13] highly accurate value of $\alpha^{-1} = 137.035\,963(15)$ (0.11 ppm).

Measurements with a voltmeter with higher resolution and a calibrated standard resistor with a vanishing small temperature coefficient at $T = 25$ °C yield a value of $h/4e^2 = 6453.17 \pm 0.02$ Ω corresponding to a fine-structure constant of $\alpha^{-1} = 137.0353 \pm 0.0004$.

We would like to thank the Physikalisch Technische Bundesanstalt, Braunschweig, for experimental support and E. R. Cohen, Th. Englert, V. Kose, G. Landwehr, and B. N. Taylor for valuable discussions. One of us (M.P.) would like to thank the European Research Office of the U. S. Army for partial support.

[1]For a review see for example: F. Stern, Crit. Rev. Solid State Sci. 5, 499 (1974); G. Landwehr, in *Advances in Solid State Physics: Festkörperprobleme*, edited by H. J. Queisser (Pergamon, New York-Vieweg, Braunschweig, 1975), Vol. 15, p. 48.

[2]T. Ando, J. Phys. Soc. Jpn. 37, 622 (1974).

[3]H. Aoki and H. Kamimura, Solid State Commun. 21, 45 (1977).

[4]R. J. Nicholas, R. A. Stradling, and R. J. Tidey, Solid State Commun. 23, 341 (1977).

[5]E. R. Cohen and B. N. Taylor, J. Phys. Chem. Ref. Data 2, 633 (1973).

[6]A. M. Thompson and D. G. Lampard, Nature (London) 177, 888 (1956).

[7]I. Isenberg, B. R. Russel, and F. R. Greene, Rev. Sci. Instrum. 19, 685 (1948).

[8]R. F. Wick, J. Appl. Phys. 25, 741 (1954).

[9]Th. Englert and K. V. Klitzing, Surf. Sci. 73, 71 (1978).

[10]S. Kawaji and J. Wakabayashi, Surf. Sci. 58, 238 (1976).

[11]M. Pepper, Philos. Mag. 37B, 83 (1978).

[12]S. Kawaji, Surf. Sci. 73, 46 (1978).

[13]E. R. Williams and P. T. Olsen, Phys. Rev. Lett. 42, 1575 (1979).

PHYSICAL REVIEW B VOLUME 23, NUMBER 9 1 MAY 1981

Quantized Hall resistance and the measurement of the fine-structure constant

R. E. Prange
Department of Physics and Astronomy and Center for Theoretical Physics, University of Maryland, College Park, Maryland 20742
(Received 10 October 1980; revised manuscript received 17 February 1981)

An elementary, exact calculation of two-dimensional electrons in crossed electric and magnetic fields with a δ-function impurity is carried out in the quantum limit. A state localized on the impurity exists and carries no current. However, the remaining mobile electrons passing near the impurity carry an extra dissipationless Hall current exactly compensating the loss of current by the localized electron. The Hall resistance should thus be precisely h/e^2, as found experimentally by Klitzing *et al.* Other possible sources of deviation from this result are briefly examined.

In a recent Letter, v. Klitzing *et al.*[1] have reported a high-accuracy measurement of e^2/h (to one part in 10^5) which after improvements and together with the known value of the speed of light, c, potentially[2] would provide a measurement of the fine-structure constant of precision greater than that currently available (one part in 10^7). Their result is based on a measurement of the quantized Hall resistance in a two-dimensional electron gas, as realized in the inversion layer of a metal-oxide-semiconductor field-effect transistor. We here provide an elementary calculation which has a bearing on their result, and is a step in the direction of estimating theoretically the accuracy to which e^2/h is determined by the experiment.

Since free electrons which fill an integral number of Landau levels give a Hall resistance precisely an integral fraction of h/e^2, the problem is one of treating the imperfections, which might give rise to ordinary resistance, and/or to localized states which could cause the Hall resistance to deviate from its ideal value. (We do not treat the electrons as interacting, an omission which future work must remedy.) We here work out the instructive, elementary, and essentially exactly solvable case of two-dimensional electrons with a single δ-function impurity.

The main result is that (a) *a localized state exists,* which (b) carries *no current,* but (c) *the remaining nonlocalized states* carry an *extra Hall current* which *exactly compensates for that not carried* by the localized state. Thus, provided *all the nonlocalized states* of the appropriate Landau level are filled, the *total Hall current carried by the level is precisely the same as in the absence of impurities and localized states.*

We consider then free electrons of mass m ($=0.2m_e$ as appropriate for silicon) in the xy plane, subjected to an electric field E in the negative x direction, and a magnetic field B in the z direction. (The geometry is given in Fig. 1.) Spin and valley degeneracies are ignored, and attention is confined to the case in which only the ground Landau level is occupied. We choose m as the unit of mass, mc/eB as the unit of time, and $(hc/2\pi eB)^{1/2} \equiv l_B$ as the unit of length. (The cyclotron radius, $k_f hc/2\pi eB$, does not enter the problem.) The drift velocity, $cE/B = -e\Phi_H/W$ is denoted by v, where Φ_H is the Hall voltage

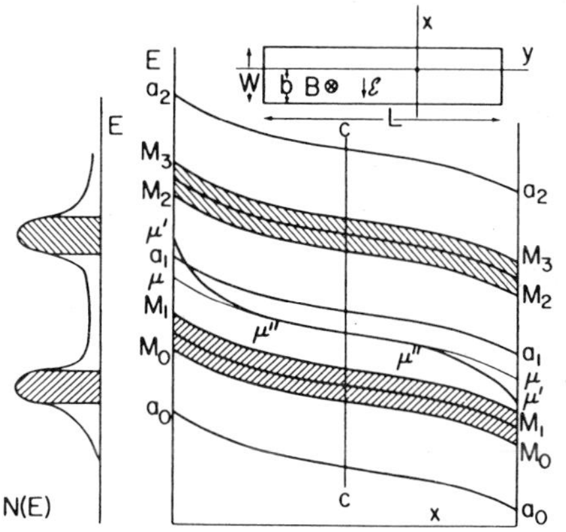

FIG. 1. The inset gives the geometry of the single-impurity problem. To the right below is the energy diagram of the lowest two Landau levels, as banded by impurities, and bent by the electric potential. The lines $a_i - a_i$ bound regions of energy states deriving from distinct levels. The lines $M_i - M_i$ are mobility edges separating localized and delocalized states. The hatched regions are those of delocalized states. The electrochemical potential is given by the line $\mu - \mu'' - \mu'' - \mu$. The position of this line depends on the total density of electrons, i.e., on the gate voltage. The states are occupied below the line $\mu' - \mu'' - \mu'' - \mu'$. The excess of electrons (holes) in the regions $\mu - \mu' - \mu''$ is the source of the Hall voltage. At the left is a schematic diagram of the density of states along the cut $c - c$.

tage, and W is the width of the device in the x direction. The Hamiltonian is $H = H_0 + 2\pi\lambda\delta^2(r)$, with $H_0 = \frac{1}{2}[-\partial^2/\partial x^2 + (p_y + x)^2] - vx$. The eigenstates of H_0 are

$$\psi_{np}(r) = \exp(ip_y y) H_n(x)\phi(x)/(2^n n! L)^{1/2} .$$

(The Landau gauge has been used.) The length of the sample in the y direction is L, H_n is the Hermite polynomial, $x = x + p$, $p = p_y - v$, and $\phi(x) = e^{-x^2/2}/\pi^{1/4}$. The eigenenergy corresponding to state ψ_{np} is $n + vp$. The values of p are $2\pi k/L$ with k integer and $-p$ ranges from $-b$ to $W - b$, where the impurity is at the origin of coordinates which is located a distance b above the lower edge of the sample. The total number of p values is thus $WL/2\pi$. This gives the degeneracy of the Landau level. For a sample as described in Ref. 1, we have in our units (with $B = 18$ T), $L = 7 \times 10^4$, $W = 8 \times 10^3$, and $v \sim 1/W \sim 10^{-4}$. We shall treat L and W as macroscopic and v as small, but it will be seen that $1/L$ is by no means the smallest number in the problem. The potential strength λ is of order unity, i.e., $\lambda >> 1/L$.

To find the full eigenstates of H, the state, denoted by ψ^α, is expanded in the "unperturbed" eigenstates of H_0 as $\psi^\alpha(r) = \sum c^\alpha_{np}\psi_{np}(r)$. It is easily seen that

$$c^\alpha_{np} = A^\alpha \psi_{np}(0)/(E_\alpha - n - pv) \tag{1}$$

with the eigenenergy E_α determined by

$$1 = 2\pi\lambda \sum_{np} \frac{|\psi_{np}(0)|^2}{E_\alpha - n - pv} \tag{2}$$

and with the amplitude determined by

$$(A^\alpha)^{-2} = \sum_{np} \frac{|\psi_{np}(0)|^2}{(E_\alpha - n - pv)^2} . \tag{3}$$

As is familiar, except for possible bound states breaking off above or below the bands of levels, the energies determined by (2) fall between the closely spaced levels of the successive p values. Thus, we may use as a label for the state α the nearest level of the system unperturbed by the impurity. For simplicity only levels belonging to the zeroth Landau level are considered.

We begin by making the approximation of retaining only the term $n = 0$ in the sums. This is suggested by the "strong magnetic field limit" which is usually taken in the literature.[3] The idea is that it is adequate to diagonalize the subspace consisting of the (nearly) degenerate states of one Landau level, if the levels are sufficiently separated. The current in this case may be obtained without the necessity of finding the explicit eigenfunctions and is completely independent of the form of the scattering potential. The (number) current operator in the x direction is $j_x = (1/i)(\partial/\partial x)$, and it is immediately seen that none of the states carries current along the electric field. In fact, for the exact solution to be found, the same is true. The current operator in the y direction is $j_y = p_y + x = p + x + v$. When this operator acts on a state ψ^α it gives $v\psi^\alpha$ plus a state orthogonal to ψ^α, because the operator $p + x$ changes the Landau level, and by assumption, there is only one Landau level in the sum defining ψ^α. Thus, all states belonging to the Landau level carry the same Hall current.

Since a localized state can certainly carry no current, we conclude that whenever the approximation is valid, there are no localized states. We shall see that for the case of the δ-function potential, there is a localized state and the approximation is not valid.

We therefore must solve the complete problem, keeping the admixture into the wave function of all Landau levels. An immediate difficulty arises because the sum on n in (2) does not converge if performed after the sum over p, at least if it is assumed that the latter sum may be approximated by an integral from $-\infty$ to ∞, using the largeness of L and W, and that, for large n, the term pv in the denominators may be neglected. However, the finiteness of the integration range will become a factor when the spatial extent of the wave functions ψ_{np} starts to become equal to the sample width, and there will be an effective cutoff at $n = M \sim W^2$. (The large magnitude of this cutoff is special to the δ-function potential. Finite-range potentials will have much smaller effective cutoffs.)

To evaluate (2) and (3), the sum over p is replaced by a principal value integral plus a contribution coming from the p values in the immediate neighborhood of the singular point. By introducing $k_\alpha - \delta_\alpha = LE_\alpha/2\pi v$, where k_α is an integer and $2|\delta_\alpha| < 1$, as well as $p_\alpha = 2\pi k_\alpha/L$ Eq. (2) may be rewritten as

$$1 = \lambda G(E_\alpha) + \lambda\phi(p_\alpha)^2\sigma_\alpha/v . \tag{2a}$$

Here $G(E)$ is the principal value integral, equal to $\sum_0^M (E - n)^{-1}$ for $|E - n| >> v$. (Since G is of order $\ln M$, the large but finite cutoff does not lead to an intolerably large G.) The discrete sum is convergent and is given by $\sigma_\alpha = -\pi\cot(\pi\delta_\alpha)$. In the same way the amplitudes are expressed as

$$(A^\alpha)^{-2} = -\left|\frac{\partial G}{\partial E_\alpha}\right| \Big/ 2\pi + \phi(p_\alpha)^2 \frac{L(\sigma_\alpha^2 + \pi^2)}{(2\pi v)^2} . \tag{3a}$$

Consider first states for which $p_\alpha >> 1$, that is, states which hardly overlap the impurity and for which $\phi(p_\alpha)^2 << 1/L$ is very small indeed. Except for the case that $G(E_\alpha) = 1/\lambda$, σ_α will necessarily be enormous, the second term on the right of (3a) will dominate, and the state ψ^α will differ insignificantly from the corresponding unperturbed state.

For the special energy E_R satisfying $G(E_R) = 1/\lambda$, however, $\sigma_\alpha \sim 1$, and the first term on the right-hand side of (3a) dominates. Under the circumstances, there is exactly one level for which this is true. This also subsumes the case for impurities near the sample edges, which may have bound states lying outside the quasicontinuum. (One may also take the thermodynamic limit, $L \to \infty$, in which case the state in question becomes an exceedingly narrow resonance, without changing the result.) This is a localized state, whose wave function is $\psi^R = \exp[-(x^2 + y^2 - 2ixy)/4]/\sqrt{2\pi}$ (for $v = 0$, and neglecting the interlevel mixing). It is a peculiarity of this system that the spatial extent of the localized state is controlled by the magnetic field when the potential fluctuation is short ranged and does not become large even when λ and E_R become small. (This may modify the theory of Anderson localization for weakly localized states.) The current of the localized state, j^R, of course vanishes as can be verified by direct computation which gives

$$j_y^R = v\left\{1 - \frac{(A^R)^2}{2\pi}\sum(n+1)[(E_R-n)^{-2} - (E_R-n-1)^{-2}]\right\} = 0 \ . \tag{4}$$

Next, the remaining class of states is considered, namely, those with energies E_α whose corresponding p_α is of order unity, that is, states made up of unperturbed eigenstates which have a significant overlap with the impurity. The eigenstates ψ^α do *not* overlap the impurity, of course, since they must be orthogonal to ψ^R which is localized at the site. The energies E_α are of order v, and thus $G(E_\alpha) \cong \mathrm{Re}G_0(p_\alpha)/v$ which is given by

$$G_0(p_\alpha) = \int dp \frac{\phi(p)^2}{p_\alpha - p + i\eta} \ . \tag{5}$$

Then, $G \sim 1/v$, $\phi(p_\alpha) \sim 1$, $\sigma_\alpha \sim 1$, and the second term on the right-hand side of (3a) dominates because L is large. The Hall current carried by such a state may be evaluated to leading order in v as

$$j_y^\alpha = v + (A^\alpha)^2/\pi v \ . \tag{6}$$

The sum over states α is smooth and may be replaced by an integral over p_α. It is found that

$$j^0 \equiv \sum_\alpha (j_y^\alpha - v) = v\left(\frac{2}{\pi}\right)\int dp \, \mathrm{Im}\left(\frac{1}{G_0(p)}\right) \ . \tag{7}$$

The integral is evaluated as $\pi \int dp \, p^2 \phi(p)^2 = \pi/2$, by recognizing its analytic properties. Thus, j^0, which is the excess current carried by the electrons which pass near the impurity, is just $j^0 = v$, exactly enough to compensate for the failure of the localized state to carry a current.

The case of two δ-function potentials may be studied as well. In this case, if the potential sites are separated by a distance $l \gg 1$, they do not interact, and the single-impurity problem gives the answer. (This may be extended to many well separated impurities.) On the other hand, if the separation l is small compared with unity, the potential acts as a single site. Only when $l \sim 1$ do the impurity levels interact and start to form an impurity band.

The preceding results, as well as the lore of Anderson localization due to Mott and others,[1,4,5] support the conjecture[1,3] that the following picture holds. In the absence of an electric field, and in the presence of a considerable number of impurities, defects or other potential fluctuations, and in a sufficiently strong magnetic field, a given Landau level will be broadened into a band. The central region of this band will be delocalized states. Beyond a mobility edge there will be localized states, which can only conduct current by hopping processes. The delocalized states of different bands will thus be separated by a region of localized states. The delocalized states of each Landau "band," however, collectively carry a total Hall current I, in the presence of an electric field or potential gradient, which is

$$I = (-ev/L)LW/2\pi = e^2\Phi_H/2\pi (= e^2\Phi_H/h) \ . \tag{8}$$

Our calculation thus tends to confirm the expectation that the quantum Hall resistance is h/e^2.

This result applies to a dilute system of δ-function impurities. Ando *et al.*[3] also obtained a quantized Hall current within the framework of their approximations, which are the assumption of "high magnetic field," the single-site approximation with the effects of scattering taken into account self-consistently, and the assumption that there are gaps between the resulting "impurity bands." The spirit of the "high-field" approximation seems to be the same as discussed earlier, but in detail it is different since the elegant formalism of Kubo *et al.*[6] is used. (This method breaks the current correlation expression for the conductivity into two parts, one of which is treated exactly, and the other approximately.) Although the conditions for the validity of the high-field approximation have not been spelled out, it seems likely that it will be valid when the field is so great that the potential hardly varies on the scale of l_B. The δ-function results on the other hand ought to be qualitatively valid when the potential is confined to a small region compared with l_B. Given the value of l_B appropriate for the experiment (70 Å), it is unlikely that either approximation is *a priori* very good. The experiment, however, is evidence that the results which have been obtained in these limiting cases must be valid under very general conditions.

Aside from these questions which somehow must

receive a favorable answer if the experiment is to be explained at all, there are several other considerations which might lead to effects at the part in 10^6 level. In particular, W is not so large that corrections of order $1/W$ can be tolerated. Thus, the edge effects must be carefully investigated. It is known from the theory of the Landau diamagnetism, for example, that the surface Landau levels tend to carry current in a direction counter to that of the bulk. Perhaps this kind of correction can be avoided along the lines suggested by the foregoing, namely, that any surface states or other anomalous states will be compensated for by an increase in the Hall current of the delocalized states. Another small effect needing investigation is the nonparabolicity of the energy bands.

It is interesting to ask whether the quantum Hall current is a supercurrent in the sense of the theory of superconductivity, and whether a persistent current can be set up. There is no dissipation connected with the Hall current *per se,* since it is perpendicular to the field. It is a supercurrent in the sense that the wave function is locked into place by an energy gap, and it is because of the vector potential that the current exists.

Thus the more interesting question is whether there will be a small current parallel to the electric field, that is, whether in this direction, the system is a perfect insulator. In our model, such a current can come only by taking into account inelastic processes so far neglected, which give rise to a change in occupation of the states. When an entire level of current-carrying states are filled there is no possibility of changing their occupation without large energy cost. The localized states also are activated so they too are perfectly insulating at sufficiently low temperature. Thus we expect that the Hall potential can be maintained without dissipation and that a persistent current can be set up.

This raises the possibility of photoinducing a potential drop in the y direction. If light falls on the junction, it can excite electrons into the unfilled delocalized states of a higher Landau level, and these electrons will provide a current in the x direction which will in turn cause a shift in the direction of the Hall current and lead to a potential drop along the length of the sample. By controlling the frequency of the radiation, something about the position of the mobility edges may be inferred.

The approximation of constant electric field must also be examined. It is not known where the charge which gives rise to this field resides, at least in the case that the diagonal component of the conductivity, σ_{xx}, vanishes, and the actual electric field configuration may depend on how the Hall current is set up. If the charge is localized toward the edges of the sample, it will attract an equal and opposite image charge in the facing metal a few hundred nm away. The potential of such a line of dipoles will vary most pronouncedly in the first few hundred nm away from the edge. If this is the case, most of the Hall current will flow along the edges of the sample, and the inner part will carry practically no current. This will increase the effective value of the local υ and corrections of order υ^2 could start to play a role. There is, fortunately, no evidence thus far that the electric field is far from constant.[7] The situation might arise, however, in a Corbino (disk with center hole) geometry where a Hall voltage could be applied by moving up external charges.

In the actual experimental configuration,[1] the primary charge giving rise to the Hall potential presumably resides in the localized states. We thus envisage a situation which is schematically shown in the figure. The charges in these localized states will be unable to relax toward equilibrium at low temperature since they require thermally activated inelastic processes to change their state. The Landau band will then bend to follow the potential, and a picture as in the figure will result.

ACKNOWLEDGMENTS

I wish to thank Dennis Drew for introducing me to this problem and to express gratitude to him and Victor Korenman for fruitful discussions on the subject. This work was supported in part by NSF Grant No. DMR 7908819.

[1]K. v. Klitzing, G. Dorda, and M. Pepper, Phys. Rev. Lett. 45, 494 (1980).

[2]Several laboratories have already apparently achieved 1-ppm precision by this method. E. Braun, E. Staben, and K. v. Klitzing, PTB-Mitteilungen 90, 350 (1980); B. Taylor (private communication).

[3]T. Ando, Y. Matsumoto, and Y. Uemura, J. Phys. Soc. Jpn. 39, 279 (1975), and preceding papers.

[4]R. J. Nicholas, R. A. Stradling, and R. J. Tidley, Solid State Commun. 23, 341 (1977).

[5]H. Aoki and H. Kanimura, Solid State Commun. 21, 45 (1977).

[6]R. Kubo, S. J. Miyake, and N. Hashitsume, in *Solid State Physics*, edited by F. Seitz and D. Turnbull (Academic, New York, 1965), Vol. 17, p. 269.

[7]K. v. Klitzing (private communication).

PHYSICAL REVIEW B VOLUME 23, NUMBER 10 15 MAY 1981

Quantized Hall conductivity in two dimensions

R. B. Laughlin
Bell Laboratories, Murray Hill, New Jersey 07974
(Received 20 January 1981)

It is shown that the quantization of the Hall conductivity of two-dimensional metals which has been observed recently by Klitzing, Dorda, and Pepper and by Tsui and Gossard is a consequence of gauge invariance and the existence of a mobility gap. Edge effects are shown to have no influence on the accuracy of quantization. An estimate of the error based on thermal activation of carriers to the mobility edge is suggested.

There has been considerable interest in the remarkable observation made recently by von Klitzing, Dorda, and Pepper[1,2] and by Tsui and Gossard[2] that, under suitable conditions, the Hall conductivity of an inversion layer is quantized to better than one part in 10^5 to integral multiples of e^2/h. The singularity of the result lies in the apparent total absence of the usual dependence of this quantity on the density of mobile electrons, a sample-dependent parameter. As it has been proposed[1] to use this effect to define a new resistance standard or to refine the known value of the fine-structure constant, an important issue at present is to what accuracy the quantization is exact, particularly in the regime of high impurity density. Some light has been shed on this question by the renormalized weak-scattering calculations of Ando,[3] who has shown that the presence of an isolated impurity does not affect the Hall current. A similar result has been obtained recently by Prange,[4] who has shown that an isolated δ-function impurity does not affect the Hall conductivity to lowest order in the drift velocity $v = cE/H$, even though it binds a localized state, because the remaining delocalized states carry exactly enough extra current to compensate for its loss. The exactness of these results and their apparent insensitivity to the type or location of the impurity suggest that the effect is due, ultimately, to a fundamental principle. In this communication, we point out that it is, in fact, due to the long-range phase rigidity characteristic of a supercurrent, and that quantization can be derived from gauge invariance and the existence of a mobility gap.

We consider the situation illustrated in Fig. 1, of a ribbon of two-dimensional metal bent into a loop of circumference L, and pierced everywhere by a magnetic field H_0 normal to its surface. The density of states of this system, also illustrated in Fig. 1, consists, in the absence of disorder, of a sequence of δ functions, one for each Landau level. These broaden, in the presence of disorder, into bands of extended states separated by tails of localized ones. We consider the disordered case with the Fermi level in a mobility gap, as shown.

We wish to relate the total current I carried around the loop to the potential drop V from one edge to another. This current is equal to the adiabatic derivative of the total electronic energy U of the system with respect to the magnetic flux ϕ through the loop. This may be obtained by differentiating with respect to a uniform vector potential A pointing around the loop, in the manner

$$I = c\frac{\partial U}{\partial \phi} = \frac{c\,\partial U}{L\,\partial A} \quad . \tag{1}$$

This derivative is nonzero only by virtue of the phase coherence of the wave functions around the loop. If, for example, all the states are localized then the only effect of A is to multiply each wave function by $\exp(ieAx/\hbar c)$, where x is the coordinate around the loop, and the energy change and current are zero. If a state is extended, on the other hand, such a gauge transformation is illegal unless

$$A = n\frac{hc}{eL} \tag{2}$$

In the case on noninteracting electrons, phase coherence enables a vector potential increment to

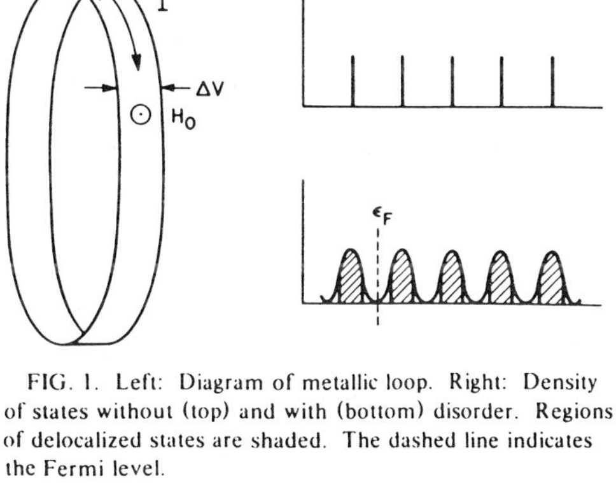

FIG. 1. Left: Diagram of metallic loop. Right: Density of states without (top) and with (bottom) disorder. Regions of delocalized states are shaded. The dashed line indicates the Fermi level.

change the total energy by forcing the filled states toward one edge of the ribbon. Specifically, if one adopts the usual isotropic effective-mass Hamiltonian,

$$H=\frac{1}{2m^*}\left(\vec{p}-\frac{e}{c}\vec{A}\right)^2+eE_0y \ , \tag{3}$$

where E_0 is the electric field across the ribbon, and adopts Landau gauge

$$\vec{A}=H_0y\hat{x} \ , \tag{4}$$

then the wave functions, given by

$$\psi_{k,n}=e^{ikx}\phi_n(y-y_0) \ , \tag{5}$$

where ϕ_n is the solution to the harmonic-oscillator equation

$$\left[\frac{1}{2m^*}p_y^2+\frac{1}{2m^*}\left(\frac{eH_0}{c}\right)^2y^2\right]\phi_n=(n+\tfrac{1}{2})\hbar\omega_c\phi_n \ , \tag{6}$$

and y_0 is related to k by

$$y_0=\frac{1}{\omega_c}\left(\frac{\hbar k}{m^*}-\frac{cE_0}{H_0}\right) \ , \tag{7}$$

are affected by a vector potential increment $\Delta A\hat{x}$ only through the location of their centers, in the manner

$$y_0\rightarrow y_0-\frac{\Delta A}{H_0} \ . \tag{8}$$

The energy of the state, still given by

$$\epsilon_{n,k}=(n+\tfrac{1}{2})\hbar\omega_c+eE_0y_0+\tfrac{1}{2}m^*(cE_0/H_0)^2 \tag{9}$$

thus changes linearly with ΔA. This gives rise to the derivative in Eq. (1), which may be conveniently evaluated via the substitution

$$\frac{\partial U}{\partial\phi}\rightarrow\frac{\Delta U}{\Delta\phi} \tag{10}$$

with $\Delta\phi=hc/e$ a flux quantum. Since, by gauge invariance (2), adding $\Delta\phi$ maps the system back into itself, the energy increase due to it results from the net transfer of n electrons (no spin degeneracy) from one edge to the other. The current is thus

$$I=c\frac{neV}{\Delta\phi}=\frac{ne^2V}{h}. \tag{11}$$

We now consider the dirty interacting system. As in the ideal case, gauge invariance is an exact symmetry forcing the addition of a flux quantum to result only in excitation or deexcitation of the original system. Also as in the ideal case, there is a gap, although the gap now exists between the electrons and holes affected by the perturbation, those contiguous about the loop, rather than in the density of states. Since adiabatic change of the many-body Hamiltonian cannot excite quasiparticles across this gap, it can only produce an excitation of the charge-transfer variety discussed in the ideal case, although the number of electrons transferred need not be the ideal number, and can be zero, as is the case for most systems with gaps. Therefore, Eq. (11) is always true, as a bulk property, for some integer n whenever the local Fermi level lies everywhere in a gap in the extended-state spectrum.

At the edges of the ribbon, the effective gap collapses and communication between the extended states and the local Fermi level is reestablished. Particles injected into this region rapidly thermalize to the Fermi level, in the process losing all memory of having been mapped adiabatically. This would be a significant source of error in Eq. (11) were it not for the fact that *isothermal* differentiation with respect to ϕ, the thermodynamically correct procedure for obtaining I, is equivalent to adiabatic differentiation in the sample interior and is reversible. Thus, slow addition of $\Delta\phi$ physically removes a particle from the local Fermi level at one edge of the ribbon and injects it at the local Fermi level of the other, acting as a pump. Since the Fermi energy is defined as the change in V resulting from the injection of a particle, and since eV is defined to be the Fermi-level difference, edge effects are not a source of error in Eq. (11).

Several other sources remain to be investigated, including possible ϕ dependence, the effect of substituting the ring geometry of Fig. 1 for the usual strip geometry, and effects of tunneling. However, we find it intuitively appealing that the quantum effect should go hand in hand with the persistence of currents, and thus that the physically significant source of error should be thermal activation of carriers to the mobility edge. These carriers produce a large, but finite, normal resistance per square R, which in the steady-state strip geometry, results in a Hall resistance too small in the amount

$$\left|\frac{\Delta R_H}{R_H}\right|=\left(\frac{R_H}{R}\right)^2 \ . \tag{12}$$

In summary, we have shown that the quantum Hall effect is intimately related to the extended nature of the states near the center of the disorder-broadened Landau level, and that edge effects do not influence the accuracy of the quantization. We speculate that the only significant source of error is thermal activation of carriers to the mobility edge.

ACKNOWLEDGMENTS

I am grateful to P. A. Lee, D. C. Tsui, R. E. Prange, and H. Störmer for helpful discussions.

[1]K. V. Klitzing, G. Dorda and M. Pepper, Phys. Rev. Lett. 45, 494 (1980).
[2]Identical behavior has been seen in GaAs-Al$_x$Ga$_{1-x}$As heterostructures. D. C. Tsui and A. C. Gossard (unpublished).
[3]T. Ando, J. Phys. Soc. Jpn. 37, 622 (1974).
[4]R. E. Prange (unpublished).

Quantized Hall conductance, current-carrying edge states, and the existence of extended states in a two-dimensional disordered potential

B. I. Halperin
Lyman Laboratory of Physics, Harvard University, Cambridge, Massachusetts 02138
(Received 21 August 1981)

When a conducting layer is placed in a strong perpendicular magnetic field, there exist current-carrying electron states which are localized within approximately a cyclotron radius of the sample boundary but are extended around the perimeter of the sample. It is shown that these quasi-one-dimensional states remain extended and carry a current even in the presence of a moderate amount of disorder. The role of the edge states in the quantized Hall conductance is discussed in the context of the general explanation of Laughlin. An extension of Laughlin's analysis is also used to investigate the existence of extended states in a weakly disordered two-dimensional system, when a strong magnetic field is present.

I. INTRODUCTION

In a recent paper Laughlin has given a very elegant and general explanation of the phenomenon that under appropriate conditions, for a two-dimensional sample in a strong magnetic field, at $T=0$, the Hall conductance is quantized in *exact* multiples of the unit e^2/h.[1-4] The purpose of the present paper is to discuss some curious properties of electronic states in a magnetic field that are implied by Laughlin's analysis, and, incidentally, to clarify some details of Laughlin's argument. In particular, it is shown in Secs. II and III below that states at the perimeter of the sample are quasi-one-dimensional states which carry a current, and which do not become localized in the presence of a disordered potential of moderate strength. The perimeter states play an important role in the Hall measurement, if the Fermi levels are different at two edges of the sample.

Following the method of Laughlin,[1] we consider a film of annular geometry, in a magnetic field perpendicular to the plane of the film. In this case, the currents at the inner and outer edge are in opposite directions, and they contribute no net current around the annulus if the Fermi levels are the same at the two edges. If the two Fermi levels differ by an amount $e\Delta$, however, we find that the edge states contribute a net current δI around the ring given by

$$\delta I = ne^2\Delta/h \ , \tag{1}$$

where n is an integer. This contribution is consistent with the quantized Hall conductance, as the chemical potential difference Δ is included, along with any electrostatic potential present, in the potential difference that would be measured by a voltmeter connected between the inner and outer edges of the ring. Of course, the edge current and the quantity Δ are taken into account automatically in the general analysis of Laughlin.[1]

In Sec. IV below, we use an extension of Laughlin's analysis to investigate the question of whether extended states can exist in principle in the interior of a two-dimensional disordered system. We conclude that there must exist a band of extended states in the vicinity of the Landau energy, or at least an energy at which the localization length diverges, if the random potential is weak compared to the cyclotron energy $\hbar\omega_c$.

II. IDEAL SAMPLE

Let us first consider a collection of noninteracting electrons, confined in an ideal uniform film of annular geometry, with a uniform magnetic field $\vec{B}_0$ perpendicular to the plane of the sample. (See Fig. 1.) We assume in addition that there is a magnetic flux Φ, confined to the interior of a solenoid magnet threading the hole in the annulus, and we shall be able to vary the flux Φ without changing the magnetic field in the region where the electrons are confined. (This is a slight modification of the cylinder geometry considered by Laughlin.) We shall assume that no electric field is present so that the electrostatic potential seen by the electrons is constant in the interior of the film, and we assume that the dimensions of the annulus are very large compared to the cyclotron radius r_c for electrons in the magnetic field. We adopt the gauge where the vector potential $\vec{A}$ points in the

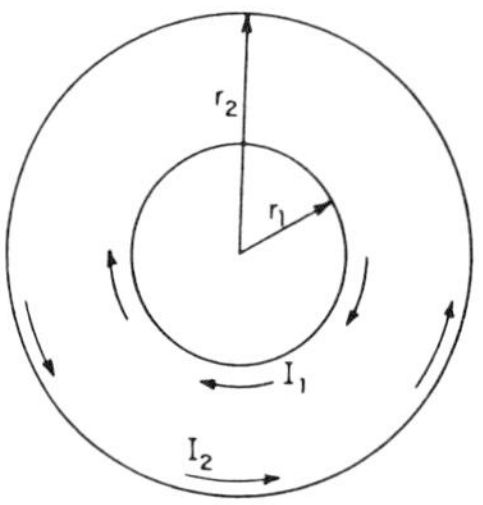

FIG. 1. Geometry of sample. Annular film, in region $r_1 < r < r_2$ is placed in uniform magnetic field B_0, pointing out of the page. Additional magnetic flux Φ is confined to region $r < r_1$. Curved arrows show direction of currents I_1 and I_2 at the boundaries of film.

azimuthal (θ) direction, and the magnitude of $\vec{A}$ depends only on the distance from the center of the annulus:

$$A = \tfrac{1}{2} B_0 r + \Phi / 2\pi r \ . \tag{2}$$

Away from the edges of the film, the electronic states in this geometry have the form

$$\psi_{m,\nu}(\vec{r}) \simeq \text{const} \times e^{im\theta} f_\nu(r - r_m) \ , \tag{3}$$

where m and ν are integers, with $\nu \geq 0$, f_ν is the $\nu+1$ eigenstate of a one-dimensional harmonic oscillator, and the radius r_m is determined by

$$B_0 \pi r_m^2 = m\Phi_0 - \Phi \ . \tag{4}$$

Here Φ_0 is the flux quantum, hc/e. The width of f is of order r_c, where r_c is the cyclotron radius. Of course, Eq. (3) is only applicable if r_m is in the range $r_1 < r_m < r_2$, with $r_m - r_1$ and $r_2 - r_m$ large compared to r_c. We shall assume throughout that r_c is small compared to r_1 and $r_2 - r_1$. The energies of the states (3) are given by the Landau formula

$$E_{m,\nu} = \hbar\omega_c(\nu + \tfrac{1}{2}) \ , \tag{5}$$

where ω_c is the cyclotron frequency determined by the field B_0 and the carrier effective mass m^*:

$$\omega_c = | eB_0 | / m^* c \ . \tag{6}$$

The electron density $| \psi_{m,\nu}(r) |^2$ associated with Eq. (3) is symmetric about the radius r_m, and decays rapidly for $| r - r_m | / r_0 >> 1$. The current carried by the state is given by

$$I_{m,\nu} = \frac{e}{m^*} \int_0^\infty dr \, | \psi_{m,\nu}(\vec{r}) |^2 \left| \frac{m\hbar}{r} - \frac{eA(r)}{c} \right|$$

$$\simeq \frac{e^2 B_0}{m^* c} \int_0^\infty dr \, | \psi_{m,\nu} |^2 (r_m - r) \ . \tag{7}$$

The integral may be taken over the radial coordinate r, at any fixed value of θ. The net current vanishes for states in the interior of the annulus, since the probability densities of the harmonic oscillator states are symmetric about the point $r = r_m$.

The situation is very different when r_m is closer than a few times r_c to an edge of the sample. Then the condition that the wave function vanish at the edges of the sample will shift the energies of the eigenstates away from the Landau energies (5).

Let us focus our attention on the behavior near the outer edge of the annulus, and let us continue to use the index ν to label the number of nodes in the radial wave function. We may then write the electronic wave functions as

$$\psi_{m,\nu}(\vec{r}) = \text{const} \times e^{im\theta} g_\nu(r - r_m, r_2 - r_m) \ , \tag{8}$$

where $g_\nu(x,s)$ is a wave function which is defined in the region $-\infty < x < s$ and has ν nodes, which vanishes for $x \to s$ and $x \to -\infty$, and which obeys the eigenvalue equation

$$\left[-\frac{\hbar^2}{2m^*} \frac{d^2}{dx^2} + \frac{B_0^2 e^2 x^2}{2m^* c^2} \right] g_\nu = E g_\nu \ . \tag{9}$$

Now it is clear that the eigenvalue $E_{m,\nu}$ will approach the value $E_\nu = \hbar\omega_c(\nu + \tfrac{1}{2})$, for $r_2 - r_m >> r_c$. The energy $E_{m,\nu}$ will increase monotonically as r_m increases, passing through the value $E_{m,\nu} = \hbar\omega_c(2\nu + \tfrac{3}{2})$, when $r_m = r_2$, and increasing eventually as $(r_m - r_2)^2 e^2 B_0^2 / 2m^* c^2$ for $r_m - r_2 > r_c$. The energy curve is sketched in Fig. 2.

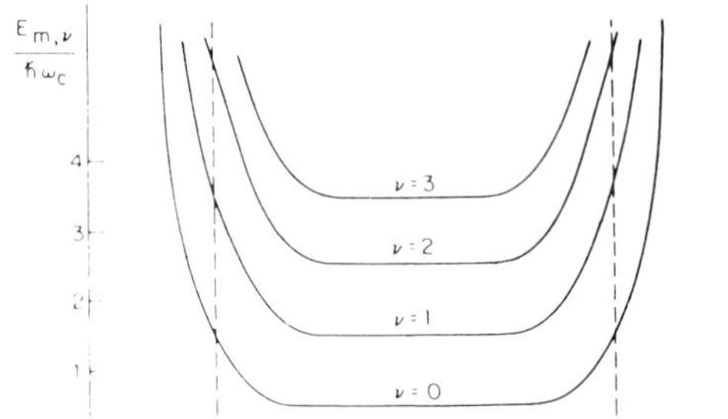

FIG. 2. Energy levels of nonrandom system, in units of $\hbar\omega_c$, as a function of the parameter r_m. The latter quantity is determined by the azimuthal quantum number m, according to Eq. (4), and it is the radius at which the azimuthal current density vanishes for quantum number m. The radius r_m is the center of the wave function $\psi_{m\nu}$ provided that r_m is not too close to the boundary r_1 or r_2.

Since the density $|\psi_{m,v}(\vec{r})|^2$ is no longer symmetric about $r=r_m$, we no longer expect that $I_{m,v}=0$. In fact, it is readily established that

$$I_{m,v}=-c\frac{\partial E_{m,v}}{\partial \Phi}=\frac{e}{h}\frac{\partial E_{m,v}}{\partial m} . \quad (10)$$

For $B_0>0$, we find that $I_{m,v}>0$, for $r_m \simeq r_2$, while $I_{m,v}<0$, near the inner edge $r_m \simeq r_1$.

Note that the quantity $|\partial E_{m,v}/\partial m|$ is just the energy separation between adjacent energy levels for a given quantum number v. Thus the total current carried by states of a given v in a small-energy interval δE is equal to $(e/h)\delta E$ at the outer edge of the sample, and $-(e/h)\delta E$ at the inner edge. (We neglect here any spin or valley degeneracy of the carriers.)

Let us suppose that the Fermi level lies in between the energies E_v of two Landau levels $v=n-1$ and $v=n$, in the interior of the sample. Suppose also that near r_2 and r_1 there are Fermi levels $E_F^{(2)}$ and $E_F^{(1)}$, respectively, which differ from each other, but still lie in the interval between E_{n-1} and E_n. Then the total current carried by the edge states between $E_F^{(2)}$ and $E_F^{(1)}$ is clearly given by $neh^{-1}(E_F^{(2)}-E_F^{(1)})$, in agreement with Eq. (1).

In a real experiment, the measured Hall potential eV is the sum of an electrostatic potential eV_0 and the difference in Fermi levels $E_F^{(2)}-E_F^{(1)}$. The edge current is then only a *fraction* of the total Hall current, given by $(E_F^{(2)}-E_F^{(1)})/eV \approx \alpha n r_c \hbar\omega_c C/e^2$ where C is the capacitance per unit length of the edge states, and α is a number of order unity.

III. DISORDERED SAMPLE

Now we must show that the edge currents are not destroyed by a moderate amount of disorder in the sample. Let us consider the effect of a weak random potential $V(\vec{r})$, with $|V(\vec{r})| << \hbar\omega_c$. Let us consider for simplicity a situation where the Fermi level E_F lies midway between the unperturbed Landau energies E_0 and E_1. It is then clear that there will be no energy eigenstates with E near E_F in the interior of the sample, but there will remain two bands of states with E near E_F which are radially localized near r_2 and r_1, respectively.

Consider an energy eigenstate ψ from the band at r_2, and write the state as superposition of the eigenstates ψ_{mv} of the nonrandom system:

$$\psi(\vec{r})=\sum_{mv} c_{mv}\psi_{mv}(\vec{r}) . \quad (11)$$

The expansion coefficient c_{mv} will be relatively large for $v=0$ and r_m near to, but slightly smaller than r_2. The coefficient c_{mv} will be smaller by a factor of order $V(\vec{r})/\hbar\omega_c$ for $v\geq 1$, and c_{mv} will be "exponentially small" for $|r_2-r_m| >> r_c$.

The azimuthal current carried by the state ψ is given by

$$\langle I\rangle=\sum_{mvv'} c_{mv}^* c_{mv'} I_{mvv'} , \quad (12)$$

where

$$I_{mvv'}\equiv\frac{e}{2\pi m^*}\int\int dr\,d\theta\,\psi_{mv}^*(\vec{r})\psi_{mv'}(\vec{r}) \times\left[\frac{m\hbar}{r}-\frac{eA(r)}{c}\right] . \quad (13)$$

Note that azimuthal current must be independent of θ, since current conservation requires $\vec{\nabla}\cdot\langle\vec{j}(\vec{r})\rangle=0$, where $\langle\vec{j}(\vec{r})\rangle$ is the current density carried by any exact eigenstate of the Hamiltonian. We see that $I_{mvv'}$ is identical to $I_{mv'}$ when $v=v'$. Furthermore, for r_m near r_2, I_{m01} is of the same order as I_{m0}, namely of order $e\omega_c r_c/r_2$. It follows then that the off-diagonal contribution $(v\neq v')$ to Eq. (12) cannot cancel the positive diagonal contribution $(v=v'=0)$, when $V(\vec{r})/\hbar\omega_c$ is small; hence the current $\langle I\rangle$ is nonzero. It follows also from current conservation that the eigenstate ψ is not localized azimuthally in any region of θ, but must be spread more or less uniformly around the annulus.

It is clear, physically, that the situation is unaltered if there are a few isolated regions with $V(\vec{r}) >> \hbar\omega_c$. Although there may be localized bound states or resonances in the regions of strong potential, the current-carrying edge states will simply be displaced, locally, to go around these regions. Of course, if the random potential becomes sufficiently strong that electron scattering rate is large compared to ω_c, it is no longer useful to employ the Landau levels as starting points and the arguments given here breakdown.

The arguments given above can be extended, with little difficulty, to the case where E is midway between the $v=1$ and $v=2$ Landau levels, etc. In this case there will be several values of v for which the expansion coefficients c_{mv} can be large. The contributions of the off-diagonal terms $(v\neq v')$ in Eq. (12) to the current carried by the state ψ are nevertheless small for $V(\vec{r})/\hbar\omega_c << 1$, because the matrix element $I_{mvv'}$ is diagonal in m, while the largest values of c_{mv} occur at different values of m

for different oscillator levels ν. The reasoning clearly breaks down, on the other hand, if E_F is too close to an unperturbed energy E_ν.

Our argument that $\langle I \rangle \neq 0$ for an edge state in a weakly disordered system did not show that the current carried satisfies Eq. (1) exactly in this case. The validity of this equation may be most easily established by considering what happens as one adiabatically increases the threading flux Φ by one flux quantum, in the manner described by Laughlin, in Ref. 1. We shall not repeat that analysis here in detail, but we shall mention some essential features in the following section.

IV. DO EXTENDED STATES EXIST IN TWO DIMENTIONS?

A starting point of Laughlin's analysis of the quantized Hall conductance is the assumption that for a collection of noninteracting electrons in an infinite two-dimensional sample with weak disorder, in a strong perpendicular magnetic field, there exist energy bands of extended states ("Landau levels") separated by energy regions of localized states and/or energy gaps where there are no states at all. Laughlin shows that if the Fermi energy occurs at a position outside the bands of extended Landau states, and if the flux Φ threading the hole of an annular sample is increased adiabatically by one flux quantum Φ_0, then the net effect will be to transfer an integral number n electrons from the Fermi level at the outer edge to states at the Fermi level of the inner edge of the sample. Since the net change in the energy of the sample is $-neV$, where V is the voltage difference between the outer and inner edge of the sample, and since the work done in the flux change is equal to $-c^{-1}I\Phi_0$, where I is the current around the loop, Laughlin has established that $I/V = nec/\Phi_0$.

It is natural to identify the integer n with the number of bands of extended states below the Fermi energy (multiplited by the spin and valley degeneracy of the carriers), and it is natural to suppose that for weak disorder this number will be the same as the number of Landau levels that would occur below E_F in the absence of disorder. If the disorder is sufficiently strong, however, so that all states below E_F are localized, then we would obtain the integer $n=0$, and the quantized Hall conductance would not be observed.

It is now generally believed that in the absence of a magnetic field or other mechanisms to break the time-reversal invariance of the Schrödinger equation, the electronic states in a two-dimensional random potential are *always* localized, in principle.[5] When time-reversal symmetry is broken, the leading term responsible for localization in the renormalization-group equations is known to be absent; nevertheless, it has remained an open question whether extended states can exist in a two-dimensional system under these conditions.[5,6]

If two-dimensional states were actually always localized there would seem to be a serious problem, in principle, with the starting point of Laughlin's theory. One could take the point of view that the experimental existence of a nonzero, quantized Hall conductance is sufficient evidence for the existence of extended states, and that further discussion of this point is unnecessary.[7] For the sake of intellectual completeness, however, it seems worthwhile to note that the existence of extended states and of nonzero Hall conductance can actually be demonstrated theoretically, at least in the case of a weakly disordered sample in a strong magnetic field, by an extension of Laughlin's arguments, which will be given below. (Actually, we cannot rule out the possibility that the energy regions of extended states have vanishing width in the limit of an infinite sample, but this possibility would still be compatible with a nonzero quantized Hall conductance.) In addition, the theoretical argument can be applied directly to the theoretically important case of noninteracting electrons, whereas the electron-electron interactions could *a priori* be important in the experimental systems.[8] In the discussion below, we shall in fact confine ourseives to the noninteracting case, although a small modification of the arguments also confirms that the electron-electron interaction, if it is not too strong, will not destroy a nonzero, quantized Hall conductance.[9]

We begin by generalizing the annular geometry of Fig. 1 as follows. We divide the sample into three concentric regions, bounded by radii $r_1 < r_1' < r_2' < r_2$. For $r_1 < r < r_1'$ and $r_2' < r < r_2$, we assume the potential $V(\vec{r})=0$. For $r_1' < r < r_2'$, we assume a weak random potential $V(\vec{r}) << \hbar\omega_c$. There is no macroscopic electrostatic field present, and we assume infinite reflecting walls at r_1 and r_2 as before. We shall assume the dimensions of the sample to be arbitrarily large compared to any microscopic length.

The electronic energy levels in this geometry are indicated in Fig. 3. In the border regions $r_1 < r < r_1'$ and $r_2' < r < r_2$, the analysis of Sec. II applies, and the electronic states are well under-

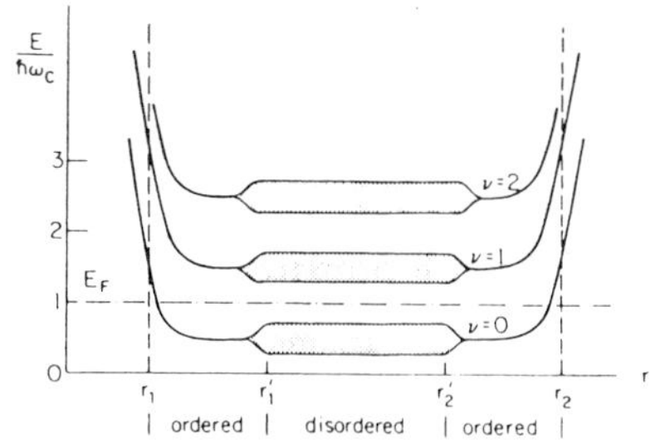

FIG. 3. Energy bands, as a function of position for the inhomogeneous geometry described in Sec. IV. Regions $r_1 < r < r_1'$ and $r_2' < r < r_2$ contain ordered "ideal" conductor, while region $r_1' < r < r_2'$ contains a weak random potential $V(\vec{r})$.

stood. The states have energies $E_{m\nu}$ which are given by the Landau formula $E_\nu = \hbar\omega_c(\nu + \frac{1}{2})$, except at the boundaries r_1 and r_2, where they are pushed upward in energy as in Fig. 2. Now, in the interior disordered region $r_1' < r < r_2'$, we expect that the states will occur in a series of energy bands of finite width, centered about the energies E_ν. If the potential $V(\vec{r})$ is sufficiently weak there should be no states in the region midway between two Landau levels. (Alternatively, if there are a small number of strong impurities, there may be a small density of isolated impurity levels in the mid-gap region; these states will be localized on a scale of order r_c, however, and will not be important for our argument.)

We may now choose one of two hypotheses:

(a) The states in the disordered region are *localized at all energies* with a finite energy-dependent localization length $\lambda(E)$.

(b) The states near the center of each magnetic energy band are *delocalized*, or at least $\lambda(E) \to \infty$ for some energy E in the band.

We shall adopt hypothesis (a), and see that this leads to a contradiction.

Assume that, initially, all electron states in the sample are filled up to a Fermi level E_F, which we choose to lie at the energy $\hbar\omega_c$, midway between the $\nu=0$ and $\nu=1$ Landau levels. Let $\lambda_{\max}$ be the maximum value of $\lambda(E)$, for $E < E_F$, and choose $r_2' - r_1' >> \lambda_{\max}$. Let us now increase adiabatically the flux Φ through the hole in the annulus, by one flux quantum Φ_0. Since, initially, there was no net current flowing in the sample, there is no work done in this process, or, more accurately, the work $-c^{-1}\int I\, d\Phi$ is inversely proportional to the size of the system, as the induced current will be small for large r. We also know that the electronic wave functions in the ordered regions will contract slightly during the flux change so that at the end there is one state unoccupied just below E_F, at $r \approx r_2$, and one new state occupied just above E_F, at $r \approx r_1$.

This change in occupation costs no energy in the limit of a large sample. If, however, in the disordered region the states below the Fermi surface are all localized, there will be no way to transport an electron across this region, since, as discussed by Laughlin, localized states remain unchanged during the flux increase, except for an uninteresting phase factor $e^{i\theta(\vec{r})}$. Then the electron removed from $r \approx r_2$ must be "transferred" to a new occupied state at $r \approx r_2'$, and the new electron at $r \approx r_1$ must be associated with a hole near $r \approx r_1'$. However, by construction, there are no states in the interior of the sample with energy near E_F (except perhaps for some strongly localized impurity states, whose occupation cannot change during the flux increase). It follows that the required change of occupation must cost an energy of order $\hbar\omega_c$, which would be a violation of conservation of energy. Therefore, there must be at least some delocalized states below the Fermi level, even in the disordered region of the sample.

It is interesting to ask what happens to the above argument when the random potential is sufficiently strong that all states below the Fermi energy are localized in the disordered region. It seems that the bands of extended states do not disappear, but rather are pushed upwards in energy as the disorder is increased, and that the Hall conductance ceases when the lowest extended band rises above the Fermi energy.[10] In an inhomogeneous geometry such as that considered above, there will be current-carrying states at the Fermi level near the boundaries of the disordered region (radii r_1' and r_2') analogous to the current-carrying states at the edge of the sample. Under these circumstances, the addition of a flux quantum will transfer an electron from a state at r_2 to a state at r_2' and another electron from a state at r_1' to a state at r_1, so that the Laughlin argument cannot be applied to the sample as a whole. The Hall current will then be determined by the voltage drops across the nondisordered regions only.

As a final remark, we note that by using the geometry described above, we have put Laughlin's derivation of the exact quantization of the Hall conductance in a form which does not require any *a priori* assumption about the behavior of extended states in the disordered region, during the adiabatic change of Φ. We have only made use of the known behavior of the wave functions in the ordered boundary regions, and the relatively trivial behavior of any localized states at the Fermi level during the change of Φ. The transfer of charge through the disordered region, and the quantized relation for I/V, are then implied by conservation of energy and particle number.

ACKNOWLEDGMENTS

The author has benefited from stimulating discussions with numerous colleagues, among whom James Black, P. A. Lee, S. Hikami, and S. J. Allen deserve special mention. Particular thanks are due to G. Lubkin and B. Schwarzschild for directing the author's attention to this subject and for telling him about the work of Laughlin. The author is also grateful to R. B. Laughlin for helpful comments on the manuscript. This work was supported in part by the National Science Foundation through Grant No. DMR77-10210.

---

[1]R. B. Laughlin, Phys. Rev. B 23, 5632 (1981).

[2]The quantized Hall conductance was predicted originally, on the basis of an approximate calculation of a simple model, by T. Ando, Y. Matsumoto, and Y. Uemura, J. Phys. Soc. Jpn. 39, 279 (1975). Subsequently, the quantization was observed to hold with great precision experimentally by K. V. Klitzing, G. Dorda, and M. Pepper, Phys. Rev. Lett. 45, 494 (1980), and by others. Recent experimental results from a number of laboratories are reported in the Proceedings of the Fourth International Conference on Electronic Properties of Two-Dimensional Systems, New London, New Hampshire, 1981 [Surf. Sci. (in press)].

[3]See also the discussion of R. E. Prange, Phys. Rev. B 23, 4802 (1981), and references therein, and H. Aoki and T. Ando, Solid State Commun. 38, 1079 (1981).

[4]D. J. Thouless [J. Phys. C 14, 3475 (1981)] has taken an approach quite different from Laughlin's, and has presented an argument, based on the convergence of perturbation theory, that the quantization of Hall conductance remains exact in the presence of a random potential smaller than $\frac{1}{2}\hbar\omega_c$.

[5]See P. A. Lee and D. S. Fisher, Phys. Rev. Lett. 47, 882 (1981), and references therein.

[6]P. A. Lee (private communication); S. Hikami (private communication).

[7]Ando and Aoki, Ref. 3, and also Thouless, Ref. 4, have also emphasized that a nonzero Hall conductance at $T=0$ implies the existence of some extended states below the Fermi level.

[8]It has been suggested by H. Fukuyama, P. M. Platzman, P. A. Lee, and P. W. Anderson that the electron-electron interaction must be taken into account explicitly if one wishes to understand experiments in which the carrier density is varied and the Fermi level passes through a region of extended states [H. Fukuyama (private communication); H. Fukuyama and P. M. Platzman (unpublished)].

[9]One way to consider electron-electron interactions is via a *Gedanken* experiment, where the electron-electron interaction applies only when the electrons are inside the disordered region $r_1' < r < r_2'$ of Fig. 3. If the interaction is not too strong, there must remain an energy gap between the first and second Landau levels. If necessary, we may add a constant background potential in the disordered region to keep the Fermi level in this gap. The requirements of conservation of energy and of particle number then imply that a nonzero integral number of electrons is transferred through the disordered region, when the flux Φ is increased by one flux quantum, just as in the noninteracting case, analyzed in Sec. IV. If there is a finite density of localized states at the Fermi level in the noninteracting case, then we must make the additional reasonable assumption that these states remain localized (i.e., nonconducting) in the presence of the electron-electron interaction.

[10]This suggestion was also made by R. B. Laughlin (private communication).

J. Phys. C: Solid State Phys., **15** (1982) L717–L721. Printed in Great Britain

LETTER TO THE EDITOR

Theory of quantised Hall conductivity in two dimensions

P Středa
Institute of Physics, Czechoslovak Academy of Sciences, Prague, Czechoslovakia

Received 10 December 1981

Abstract. On the basis of linear response theory, the Hall conductivity is expressed as a sum of two contributions: one corresponding to the classical Drude–Zener formula, and a second which has no classical analogy. The developed theory is applied to the Hall effect, thermopower and thermal conductivity in two-dimensional systems. The periodic potential is taken into account.

Several recent experiments made by von Klitzing, Dorda and Pepper (1980) and by Tsui and Gossard (1981) lead to the remarkable conclusion that the Hall conductivity of a two-dimensional system in its quantum limit is quantised to better than one part in 10^5 to integral multiples of e^2/h. Theoretical explanation of this effect was originally given by the renormalised weak-scattering calculations of Ando (1974). Later, several papers appeared in which the more advanced derivations were presented (Prange 1981, Thouless 1981). Nevertheless all these theories are based on simple model calculations or only particular aspects of the problem are taken into account. The apparent insensitivity of experimental results (von Klitzing *et al* 1980, Tsui and Gossard 1981) to the type or location of impurities and also to the type of the host material suggests that the effect is due to a fundamental principle, especially due to the long-range phase rigidity characteristic of a supercurrent as was pointed out by Laughlin (1981). Since the results are simple and general, the existence of an equilibrium quantity might be intuitively supposed, which would allow us to deduce them in a straightforward manner. In this Letter we shall prove that the Hall conductivity is closely connected to the derivative of the number of electrons with respect to the magnetic field taken at Fermi energy. The thermopower and thermal conductivity will also be discussed.

To derive the expression for the conductivity, we shall use the linear response theory based on the following assumptions:

(i) The electron system can be described as a Fermi–Dirac assembly of independent quasiparticles.

(ii) Only elastic scattering is admissible.

(iii) A two-dimensional solid is formed by a layer in the (x, y) plane and magnetic field $\boldsymbol{B}$ is perpendicular to the layer ($\boldsymbol{B} \equiv (0, 0, B)$).

(iv) There are gaps in the energy spectrum of the one-electron Hamiltonian describing the system

$$H = (1/2m)[\boldsymbol{P} - (e/c)\boldsymbol{A}]^2 + V(x, y) \qquad (1)$$

where m and e are electron mass and charge respectively, $\boldsymbol{A}$ is the vector potential ($\boldsymbol{B} = \mathrm{curl}\, \boldsymbol{A}$) and $V(x, y)$ is an arbitrary fixed potential.

(v) An electric field $\boldsymbol{E}$ established in the solid results in an electric current $\boldsymbol{I}$ linearly related to the field through Ohm's law

$$\boldsymbol{I} = \boldsymbol{\sigma}\boldsymbol{E} \tag{2}$$

where $\boldsymbol{\sigma}$ is the conductivity tensor.

Under the above assumptions, the components of the conductivity tensor $\boldsymbol{\sigma}$ are given by the expressions derived for example by Smrčka and Středa (1977):

$$\sigma_{ij}(T) = -\int \frac{\mathrm{d}\rho_0(\eta)}{\mathrm{d}\eta} \sigma_{ij}(\eta, 0)\, \mathrm{d}\eta \tag{3}$$

where $\rho_0(\eta)$ is the equilibrium Fermi–Dirac distribution function and $\sigma_{ij}(E_\mathrm{F}, 0)$ are components of the conductivity tensor at zero temperature T (E_F denotes the Fermi energy). The diagonal components are given by the following expression (Kubo *et al* 1965):

$$\sigma_{ii} \equiv \sigma_{ii}(E_\mathrm{F}, 0) = \pi\hbar e^2 \,\mathrm{Tr}[v_i \delta(E_\mathrm{F} - H) v_i \delta(E_\mathrm{F} - H)] \tag{4}$$

and non-diagonal components by (Bastin *et al* 1971)

$$\sigma_{ij} \equiv \sigma_{ij}(E_\mathrm{F}, 0) = e^2 \int_{-\infty}^{E_\mathrm{F}} A_{ij}(\eta)\, \mathrm{d}\eta \tag{5}$$

$$A_{ij}(\eta) = \mathrm{i}\hbar \,\mathrm{Tr}[v_i (\mathrm{d}G^+/\mathrm{d}\eta)\, v_j \delta(\eta - H) - v_i \delta(\eta - H)\, v_j \,\mathrm{d}G^-/\mathrm{d}\eta] \tag{6}$$

where the Green function is defined by

$$G^{\pm}(\eta) = (\eta - H \pm \mathrm{i}0)^{-1} \qquad \delta(\eta - H) = -\frac{1}{2\pi\mathrm{i}}(G^+ - G^-) \tag{7}$$

and velocity operator is given by the commutation relation

$$v_i = \frac{1}{\mathrm{i}\hbar}[r_i, H] = -\frac{1}{\mathrm{i}\hbar}[r_i, G^{-1}] = \frac{1}{m}\left(p_i - \frac{e}{c}A_i\right). \tag{8}$$

To proceed further we split the Hall conductivity into two parts using an expression originally derived by Smrčka and Středa (1977):

$$A_{ij}(\eta) = \frac{1}{2}\frac{\mathrm{d}}{\mathrm{d}\eta} B_{ij}(\eta) + \frac{1}{2}\mathrm{Tr}\frac{\mathrm{d}\delta(\eta - H)}{\mathrm{d}\eta}(r_i v_j - r_j v_i) \tag{9}$$

$$B_{ij}(\eta) = \mathrm{i}\hbar \,\mathrm{Tr}[v_i G^+(\eta) v_j \delta(\eta - H) - v_i \delta(\eta - H) v_j G^-(\eta)]. \tag{10}$$

The second term on the right-hand side of equation (9) can be rewritten into the more convenient form

$$\frac{1}{2}\mathrm{Tr}\frac{\mathrm{d}\delta(\eta - H)}{\mathrm{d}\eta}(r_x v_y - r_y v_x) = \frac{c}{e}\frac{\partial}{\partial B}\mathrm{Tr}\, \delta(\eta - H) \tag{11}$$

if the definition of the Hamiltonian (1), the expression (8) and commutation relation

$$[r_x, v_y] = [r_y, v_x] = 0$$

are used. The expression (11) is valid for arbitrary choices of vector potential $\boldsymbol{A}$, nevertheless the simplest derivation is obtained if the circular gauge centred at the origin is used; $A \equiv \frac{1}{2}B(-y, x, 0)$. Introducing expressions (6), (9) and (11) into (5) we get immediately useful expression for the Hall conductivity

$$\sigma_{ij} \equiv \sigma^{\mathrm{I}}_{ij} + \sigma^{\mathrm{II}}_{ij}$$

$$\sigma^{\mathrm{I}}_{ij} = \frac{e^2}{2} \mathrm{i}\hbar \,\mathrm{Tr}[v_i G^+(E_\mathrm{F}) v_j \delta(E_\mathrm{F} - H) - v_i \delta(E_\mathrm{F} - H) v_j G^-(E_\mathrm{F})] \tag{12}$$

$$\sigma^{\mathrm{II}}_{xy} = -\sigma^{\mathrm{II}}_{yx} = ec \frac{\partial N(E)}{\partial B}\bigg|_{E=E_\mathrm{F}}$$

where $N(E)$ is the number of states below the energy E defined by

$$N(E) = \int_{-\infty}^{E} \mathrm{Tr}\, \delta(\eta - H)\, \mathrm{d}\eta \tag{13}$$

and the derivative with respect to the magnetic field B is taken at the Fermi energy.

The formula (12) satisfies the Onsager relations and is valid for the two-dimensional layer as well as for the three-dimensional substances.

The first term σ^{I}_{xy} depends on the structure of solid, crystallographic orientation and of course on the potential $V(\boldsymbol{r})$. The free electron model with arbitrary energy-dependent self-energy $\Sigma(E)$ leads to the classical Drude–Zener result if the vertex corrections are omitted:

$$\sigma^{\mathrm{I}}_{xy} = -\omega\tau\sigma_{xx} \tag{14}$$

where $\omega = |e|B/mc$ is the cyclotron frequency; the lifetime τ is equal to $\hbar/2\Gamma$ ($\Gamma = -\mathrm{Im}\,\Sigma(E_\mathrm{F})$).

The second term $\sigma^{\mathrm{II}}_{xy}$ has no classical analogy. In the classical limit the density of states is not a field-dependent quantity and $\sigma^{\mathrm{II}}_{xy}$ vanishes. The results for the free-electron model were obtained earlier (Středa and Smrčka 1975). Generally, $\sigma^{\mathrm{II}}_{xy}$ depends on material constants only through the number of particles. It does not depend on the crystallographic orientation and the type of scattering.

The general formulae (4) and (12) will be used to explain quantised Hall effect in the two-dimensional system. Since the influence of the bound states was recently extensively studied (Prange 1981, Thouless 1981) we shall concentrate our attention mainly on the influence of the periodic potential. According to Baraff and Tsui (1981) we shall suppose that ionised donors outside the layer serve as a reservoir of electrons. This reservoir which can produce the plateaus in the oscillations of conductivity does not contribute to the density of electron states in the two-dimensional layer.

To derive the expression for the quantised Hall conductivity, we shall employ the assumption (iv) mentioned above, namely that there are gaps in the electron energy spectrum in the magnetic field and that the Fermi energy is lying just within a gap, where the density of states is zero. Since δ functions in expressions (4) and (12) describe contributions to the density at Fermi energy, diagonal elements of the conductivity tensor σ and the classical term of the Hall conductivity σ^{I}_{xy} are equal to zero and we get

$$\sigma_{xx} = \sigma_{yy} = 0, \qquad \sigma_{xy} = -\sigma_{yx} = ec(\partial N(E)/\partial B)|_{E=E_\mathrm{F}}. \tag{15}$$

For simplicity, we limit ourselves to one electron band and to the so-called 'rational' magnetic field, namely

$$\frac{|e|}{hc}\boldsymbol{B}\cdot\boldsymbol{a}_1 \times \boldsymbol{a}_2 = \frac{1}{J} \tag{16}$$

where a_1 and a_2 are elementary lattice vectors and J is an integer. The electron energy structure of this system is well described for example by Rabinovitch (1969). The one-electron band is separated into J narrow bands (Landau levels) with $(J^2|a_1 \times a_2|)^{-1}$ states. Each state is J-times degenerate. The Landau levels are very narrow and well separated at the periphery of the electron band and broadening is increased as the centre is approached. The width of gaps is rising with increasing field intensity. Any impurities and deformations cannot change the number of states at a single Landau level without the disappearance of the gap.

We should like to point out that the limitation imposed by condition (16) to the field intensities is not essential. For example, in the case of a two-dimensional interface of GaAs–$Al_xGa_{1-x}As$ heterojunctions, J approximately equals 4000 at magnetic field 10 T. The change of J by one corresponds to the change of B by 0.0025 T. The points selected by the condition (16) form quite a dense quasicontinuum at least near the edges of an electron band. Moreover, the density of states corresponding to the 'rational' field (16) is numerically indistinguishable from that of a slightly different field which does not fulfil equation (16), although very different analytically (Wannier *et al* 1979).

Let us suppose that n narrow bands are lying below E_F. Since the magnetic field is supposed to be perpendicular to the two-dimensional layer, the number of states is given by

$$N = n\frac{1}{J^2|a_1 \times a_2|}J = n\frac{|e|}{hc}B. \qquad (17)$$

At the bottom of the band, where broad gaps exist, the small changes in magnetic field do not change the number of Landau levels below Fermi energy and we get

$$\sigma_{xy} = -\frac{e^2}{h}n \qquad (18)$$

using expressions (15) and (17). This is the result of recent experiments (von Klitzing *et al* 1980, Tsui and Gossard 1981) and can also be obtained for free electrons in a magnetic field (Ando 1974, Laughlin 1981). If the band is fully occupied, the number of states with and without magnetic field coincides and the Hall conductivity is just zero as expected. For the nearly occupied band the expression for the hole Hall conductivity is obtained:

$$\sigma_{xy}^{(h)} = \frac{e^2}{h}n \qquad (19)$$

where n is now the number of empty Landau levels.

From the theoretical point of view the very interesting region is the middle of the band, where also open orbits exist. In this region the change of the magnetic field from one rational value to another can change the number of Landau levels below the Fermi energy, and simple arguments leading to expressions (18) and (19) are no longer valid. This problem will be treated in a separate paper.

The other transport coefficients also have interesting features. The thermopower S and the electronic part of the thermal conductivity K are given by following expressions

$$S = (1/T)(L_{12}L_{11}^{-1} - (E_F/e)\,\mathbf{1}) \qquad (20)$$

$$K = (1/T)(L_{22} - L_{21}L_{11}^{-1}L_{12}) \qquad (21)$$

where L_{ij} are the phenomenological transport coefficients (for definition see e.g. Smrčka and Středa 1977) and **1** denotes the identity matrix. The assumptions mentioned at the beginning of this Letter lead to the expressions derived by Smrčka and Středa (1977):

$$L_{11} = \sigma \qquad L_{12} = (E_F/e)\sigma \qquad L_{22} = (E_F/e)^2\sigma. \tag{22}$$

Substituting equation (22) into equations (20) and (21) we arrive, after some algebra, at the conclusion that all components of both tensors S and K are equal to zero. This formal result reflects the obvious fact that no thermal gradient can be established in the electron gas when the Fermi energy is lying within a gap. Note also that this is the missing thermopower which makes it possible to measure the Hall voltage with unusually high accuracy.

The presented new expression (12) for the Hall conductivity is quite general and it is composed of two parts: the term σ_{xy}^{I} which corresponds to the classical expression and the purely quantum contribution σ_{xy}^{II}. It should be stressed here that it differs substantially from the well known expression derived by Kubo *et al* (1965), where the motion of centres of orbits and the relative motion of electrons are described by separate terms. Their approach can be successfully applied, e.g. in the case of closed orbits, but on the other hand it cannot explain in a simple way why a fully occupied band does not give any Hall current. It is also difficult to apply in the quasiclassical limit of weak magnetic fields.

References

Ando T 1974 *J. Phys. Soc. Japan* **37** 622–30
Baraff G A and Tsui D C 1981 *Phys. Rev.* **B24** 2274–77
Bastin A, Lewinner C, Betbeder-Matibet O and Nozières P 1971 *J. Phys. Chem. Solids* **32** 1811–24
von Klitzing K, Dorda G and Pepper M 1980 *Phys. Rev. Lett.* **45** 494–7
Kubo R, Miyake S I and Hashitsume N 1965 *Solid State Phys.* **17** 269–364
Laughlin R B 1981 *Phys. Rev.* **B23** 5632–3
Prange R E 1981 *Phys. Rev.* **B23** 4802–5
Rabinovitch A 1969 *Physics of Solids in Intense Magnetic Fields* ed E D Haidemenakis (New York: Plenum) pp 337–43
Smrčka L and Středa P 1977 *J. Phys. C: Solid State Phys.* **10** 2153–61
Středa P and Smrčka L 1975 *Phys. Stat. Solidi* (b) **70** 537–48
Thouless D J 1981 *J. Phys. C: Solid State Phys.* **14** 3475–80
Tsui D C and Gossard A C 1981 *Appl. Phys. Lett.* **38** 550–2
Wannier G H, Obermair G M and Ray 1979 *Phys. Status Solidi* (b) **93** 337–42

PHYSICAL REVIEW B VOLUME 33, NUMBER 2 15 JANUARY 1985

Localization and scaling in the quantum Hall regime

H. P. Wei and D. C. Tsui
Department of Electrical Engineering and Computer Science, Princeton University, Princeton, New Jersey 08544

A. M. M. Pruisken
The Institute for Advanced Study, School of Natural Sciences, Princeton, New Jersey 08544
*and Pupin Physics Laboratory, Columbia University, New York, New York 10027**

(Received 29 July 1985)

A new experimental study of the quantum Hall effect is presented. Detailed measurements are made of the temperature and magnetic field dependences in the electronic transport coefficients, σ_{xx} and σ_{xy}, in $In_xGa_{1-x}As$-InP heterostructures for the $n=0$ and 1 Landau levels. The results are studied in the context of the (two-parameter) renormalization-group theory of the integral quantum Hall effect.

The concept of scaling has provided an important framework in recent years to understand the nature of quantum transport in impure metals. The discovery of weak localization,[1] followed by its field-theoretic interpretation,[2] has established a variety of universality classes in the metal-insulator problem and, furthermore, the unique role of dimensionality $D=2$ as a lower critical dimension.[3] The quantum Hall effect[4,5] (QHE) exhibits the richest and most complex manifestation of scaling behavior in two-dimensional (2D) electronic transport problems.[6-11] Recent theoretical work[6] has predicted a two-parameter scaling of the integral QHE (IQHE), which means that both the dissipative conductance σ_{xx} and the Hall conductance σ_{xy} vary with length scale L. This theory results from a field-theoretic representation of the problem of localization in a magnetic field B in terms of a nonlinear σ model in the presence of the so-called θ term,[8] a topological issue which originally arose in quantum chromodynamics several years ago.[9,10] We believe that an experimental observation of this scaling behavior will provide a test of the localization theory and also a possible experimental realization of the field-theoretic concept of θ vacuum.

In this paper we report the first experimental study of scaling in the IQHE, in the context of the theory of Ref. 6. We first recapitulate some of the essential points in the theory. Recall that the renormalization-group parameters, σ_{xx} and σ_{xy}, are dimensionless numbers measuring the conductance in units of e^2/h and defined for a length scale L. The effect of the renormalization-group functions, $d\sigma_{\eta\nu}/d\ln L=\beta_{\eta\nu}(\sigma_{xx},\sigma_{xy})$, is illustrated by the flow lines in the inset of Fig. 1. The fact that the flow lines are directed toward the fixed points $(\sigma_{xx},\sigma_{xy})=(0,n)$ means that any initial set of system parameters $(\sigma_{xx}^0,\sigma_{xy}^0)$ will renormalize to the quantized values $(0,n)$ after successive length-scale transformations. In addition to these "localization fixed points," which describe localized wave functions of the electrons near the Fermi energy E_F, there are intermediate-coupling fixed points denoted by $\otimes$ on $\sigma_{xy}=n+\frac{1}{2}$. The latter describe the singular behavior in the renormalized transport coefficients, corresponding microscopically to the occurrence of a diverging localization length. These "delocalization" fixed points are associated with the extended states at E_F. An important aspect of the theory is the identification of the starting point $(\sigma_{xx}^0,\sigma_{xy}^0)$ for scaling, which contains the microscopics of the system, such as the details of the impurity potential and the information on the electron spin. Generally speaking, these parameters have a meaning in terms of the kinetic theory (Boltzmann equation), and the length scale in the problem is set by the length over which the wave functions lose phase coherence. They can be obtained microscopically from the self-consistent Born approximation (SCBA).[12] The dot-dashed lines indicate the loci of $(\sigma_{xx}^0,\sigma_{xy}^0)$ calculated from the SCBA for short-ranged scatterers in the strong-B limit, which is of interest to us. We have considered the effective g factor as an adjustable parameter and given the results for completely resolved as well as completely unresolved spin splitting of the Landau levels.

We next mention that the theory contains the characteristic aspects of "weak" and "strong" localization, which are well recognized in experiments on the IQHE, as limiting ex-

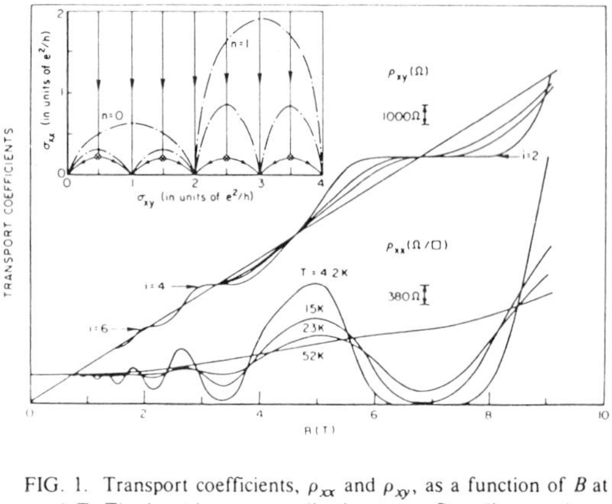

FIG. 1. Transport coefficients, ρ_{xx} and ρ_{xy}, as a function of B at several T. The inset is a renormalization-group flow diagram showing two-parameter scaling of the integral quantum Hall effect for the $n=0$ and $n=1$ Landau levels. Here, σ_{xx} and σ_{xy} are in units of e^2/h. The dot-dashed curves are the loci of $(\sigma_{xx}^0,\sigma_{xy}^0)$ calculated from SCBA for short-ranged scatterers of completely unresolved spin splitting (upper curve) and completely resolved spin splitting (lower curve).

amples. Notice hereto that the maximum of σ_{xx}^0 varies linearly with the Landau-level index within the SCBA. The regime $\sigma_{xx}^0 \gtrsim 1$ is "quasimetallic" in the usual sense of 2D localization; i.e., the functional dependence of the localization length ξ can be obtained from the weak-localization result and is $\xi \sim e^{(4\pi\sigma_{xx}^0)^2}$. This extremely strong dependence on the classical conductance becomes effective near the center of higher-index Landau levels and causes the localization length to exceed the sample size. In this situation, localization effects are completely suppressed and the experimentally observed $\log T$ variation in σ_{xx} close to the center of the higher-index Landau levels may be explained as due to interaction effects.[13-16] On the other hand, the loci of the "starting points" for scaling, which are close to the "localization" fixed points $(0,n)$, correspond to the Landau-level band tails. Via scaling, these points represent strongly localized states near E_F (ξ will be on the order of the magnetic length, ~ 100 Å, which for all practical purposes can be considered as microscopic). At sufficiently low T, σ_{xx} is known to show variable-range-hopping behavior,[15-17] typical for 2D localized wave functions.[18] Finally, the intermediate regime $0 < \sigma_{xx}^0 \lesssim 1$ is the regime of actual interest. Although the theory is developed for $T=0$ and the length-scale transformations are, in principle, accomplished by varying the sample size, in practice, however, the experiment is carried out at finite T and the effective sample size is varied by varying T. At present there is no microscopic theory on inelastic scattering and our translation of length scale L into T is purely conceptual.[19]

Our experiment was carried out on an $In_xGa_{1-x}As$-InP heterojunction[20] with a 2D electron density $n = 3.4 \times 10^{11}$ cm^{-2}, a mobility $\mu = 35\,000$ $cm^2/V\,s$ and an effective mass $m^* = 0.047m$. The magnetoresistance coefficients ρ_{xx} and ρ_{xy} were measured on a Hall bridge, using an ac lock-in technique at 34 Hz and a current of 10^{-8} A, in the T range between 50 and 0.5 K and for B up to 10 T. The temperature was measured and controlled, respectively, by a calibrated carbon glass resistor and a capacitance sensor, which are insensitive to the magnetic field. Within a given run, the data are reproducible to better than 4%.

In Fig. 1 we show the characteristic features of the IQHE in ρ_{xx} and ρ_{xy} as a function of B, measured at several temperatures. The T dependence of the data predominantly shows up as an increase in the ρ_{xy} plateaus and a narrowing of the ρ_{xx} peaks with decreasing T. The quantized plateaus, given by $\rho_{xy} = h/ie^2$, are developed around $B \simeq 6.7$ and 3.3 T for $i=2$ and 4, respectively. The former plateau corresponds to E_F being in the band tails of the $n=1$ and $n=0,\downarrow$ Landau levels; the latter the $n=1$ and 2 Landau levels. The absence of the $i=3$ plateau indicates that spin splitting, although resolved for the $n=0$ Landau level at higher $B \sim 14$ T,[21] is not resolved for the $n=1$ Landau level. The quantum oscillations observed for $B \lesssim 4$ T correspond to E_F in the $n \gtrsim 2$ Landau levels. The peak value of σ_{xx} from these oscillations is known to follow the $\log T$ variation due to interaction effects.[15,16] For the reasons discussed above, we have restricted ourselves to detailed measurements on the $n=1$ and the $n=0,\downarrow$ Landau levels in the B range from 4 to 10 T (covering $\sigma_{xy} = 1$–4). In order to facilitate direct comparison with the renormalization-group flow diagram, we have converted, for a given B, the measured $\rho_{xx}(T)$ and $\rho_{xy}(T)$ into $\sigma_{xx}(T)$ and $\sigma_{xy}(T)$. The results are plotted as the T-driven flow lines, shown in Figs. 2(a) and 3.

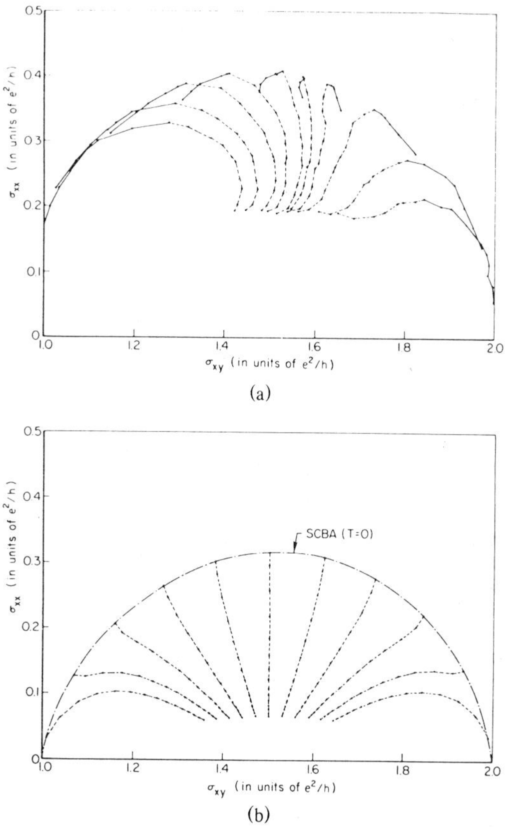

FIG. 2. (a) Experimental $\sigma_{xx}(T)$ and $\sigma_{xy}(T)$ plotted as T-driven flow lines from $T=50$ to 1.5 K. Each line corresponds to a fixed B. The dashed lines are from 50 to 4.2 K and the solid lines from 4.2 to 1.5 K. (b) T-driven flow lines of $\sigma_{xx}^0(T)$ and $\sigma_{xy}^0(T)$, calculated for the spin-resolved $n=0,\downarrow$ Landau level using Eq. (1).

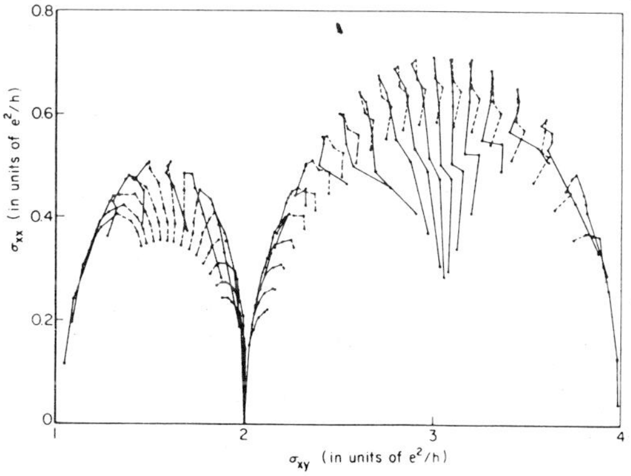

FIG. 3. Experimental $\sigma_{xx}(T)$ and $\sigma_{xy}(T)$ plotted as T-driven flow lines from $T=10$ to 0.5 K. The dashed lines are from 10 to 4.2 K and the solid lines from 4.2 to 0.5 K.

The data of Fig. 2(a) are from the center of the $n=0, \downarrow$ Landau level for $T=50$–1.5 K. The initial rise in σ_{xx} with decreasing T is the same phenomenon observed in Fig. 1 at the ρ_{xx} peaks. The main point which we want to make here is that this is not due to scaling. In fact, it can simply be understood from the T dependence of the Fermi-Dirac distribution f, which causes the "classical" conductance to behave according to

$$\sigma^0_{\mu\nu}(T)=\int dE\,\frac{\partial f(T)}{\partial E}\sigma^0_{\mu\nu}(T=0) \quad . \tag{1}$$

Such a functional dependence is expected to dominate when k_BT is comparable to the Landau-level broadening due to impurities. Within the SCBA, the characteristic temperature estimated for this effect is a few degrees kelvin. Moreover, we have plotted in Fig. 2(b) the $\sigma^0_{\mu\nu}(T)$ as calculated from Eq. (1) with $\sigma^0_{\mu\nu}(T=0)$ approximated in the SCBA and the assumption of an independent $n=0, \downarrow$ Landau level. At $T=0$, the locus of $(\sigma^0_{\mu\nu}, \sigma^0_{\mu\nu})$ in Fig. 2(b) reduces to that indicated by the dot-dashed lines in the inset of Fig. 1. For $T\neq 0$, Fig. 2(b) reproduces the functional behavior of the high-temperature data shown in Fig. 2(a). The lack of symmetry in the data about the line $\sigma_{xy}=1\frac{1}{2}$ indicates that the influence of the other Landau levels, which has been neglected in the calculation, is not negligible; in particular, the spin splitting of the $n=0$ Landau level is not completely resolved.

In the regime $T\lesssim 4.2$ K, Fig. 2(a) deviates from the classical behavior in that there is a tendency in the data to flow out toward the fixed points (0,1) and (0,2). This aspect is somewhat more pronounced in the data of Fig. 3, which covers the T range 10–0.5 K. We attribute this behavior in the range 4.2–0.5 K (indicated by the full lines in Fig. 3) to genuine scaling of the conductances. This behavior follows the flow lines in the inset of Fig. 1 in the regime $1<\sigma_{xy}<2$, starting (roughly) from the indicated classical result of a semicircle. We have investigated the 0.5-K data in more detail and found that there is a remarkable symmetry about the line $\sigma_{xy}=1\frac{1}{2}$. The data of Fig. 3 were measured two months later than those of Fig. 2(a) on the same sample. The slightly different results in the overlapping T range in Figs. 2(a) and 3 are attributed to slight changes in the microscopic details of the random potential, seen by the 2D electron gas, as a result of the difference in cooling down the sample.

A very interesting aspect of the data on the $n=1, \uparrow$ and $\downarrow$ Landau levels is apparent in Fig. 3 ($\sigma_{xy}=2$–4). The data indicate that although the spin splitting is completely unresolved in this T range, there is a definite flow toward the $\sigma_{xy}=3$ fixed point, i.e., the formation of the $i=3$ Hall plateau at lower T. This phenomenon has been attributed to exchange enhancement of the electron g factor with decreasing T. This would mean that the lower-T data of Fig. 3 interpolate predominantly between the classical results for the $n=1$ Landau level with vanishing spin splitting and that with complete spin splitting (see Fig. 1). We stress here that this exchange effect can be considered as "irrelevant" in the context of renormalization-group theory. That is, the formation of the $i=3$ plateau will occur in any case, even with a vanishing electron g factor. Indeed, in view of the lower-temperature measurements,[15,16] where the formation of the $i=3$ plateau is observed in the 50 mK regime, together with the variable-range-hopping behavior of σ_{xx}, it seems plausible that the flow toward $\sigma_{xy}=3$ is predominantly a localization effect with spin splitting only playing a minor role.

It is worth mentioning that the flow diagram is only meaningful in an asymptotic sense. This is because only the "relevant" parts are considered in the effective Lagrangian description of Ref. 8, which is what is needed in order to reveal the singular behavior in the electronic transport with the aid of the renormalization group. The effect of "irrelevant" operators will become noticeable, especially in the regime of interest, $\sigma^0_{xx}\lesssim 1$. The exchange-enhanced spin splitting and the T dependence in the Fermi-Dirac distribution are examples. The part of the flow diagram which should survive after many length-scale transformations (i.e., as $T\to 0$), is in effect the renormalization-group trajectory which connects the "localization" fixed points $(0,n)$ with the "delocalization" fixed points on $\sigma_{xy}=n+\frac{1}{2}$. This trajectory forms the only set of allowed values of the conductance parameters in the thermodynamic limit. It is of course understood that for increasing length scales (or decreasing T) the density of points becomes increasingly dilute near the "delocalization" fixed points, in favor of the density near the quantization fixed points. This corresponds physically to broadening of the quantized Hall plateaus and narrowing of the step regions in between the plateaus. In other words, the step region becomes more and more difficult to observe with increasing length scale ($T\to 0$). In particular, the sample homogeneity becomes increasingly important, since ρ_{xx} and ρ_{xy} are measured on different parts of the sample, and this requirement will impose practical limitations on the observability of the delocalization fixed point.

In summary, we can say that in the temperature range studied (50–0.5 K) our data are consistent with the two-parameter scaling theory of the IQHE. Except for the notion that the inelastic scattering length plays the role of an effective sample size, many-body effects are found to be irrelevant. Furthermore, the data are consistent with the existence of an intermediate-coupling, "delocalization" critical point, which controls the singular behavior in the transport parameters in between the Hall plateaus. Clearly, more research is needed in order to penetrate deeper into the critical regime. This requires lower temperatures and optimal sample homogeneity and work along these lines is currently in progress. As a final remark, we stress the importance of short-range correlated randomness in our interpretation of the experiments. Despite the fact that little is known on the microscopic details, the modeling in terms of a white-noise random potential, approximated in SCBA, accounts to a large extent for the temperature dependence of the transport parameters. On the other hand, it is known that the classical diffusion decreases strongly with an increasing correlation length in the random potential.[12] Since only the value of σ^0_{xx} enters into the scaling picture, it seems that the assumption of a slowly varying random potential cannot explain the large σ^0_{xx} in the data. However, smaller σ^0_{xx} and, hence, increasing spatial correlation of the randomness may be a prerequisite for the observation of the fractional QHE.[5,7]

We are grateful to Professor E. Abrahams, Professor P. W. Anderson, and Professor R. Dashen for stimulating discussions and Dr. A. M. Chang and Dr. M. Razeghi for their contributions to this work. The work at Princeton University is supported by the National Science Foundation through Grant No. DMR-8212167. A.M.M.P. has been supported by the U.S. Department of Energy under Grant No. DE-AC02-76ER02220 and by the Office of the Naval Research under Grant No. N00014-80-C-0657.

---

*Present address.

[1]E. Abrahams, P. W. Anderson, D. C. Licciardello, and T. V. Ramakrishnan, Phys. Rev. Lett. **42**, 673 (1979).

[2]F. Wegner, Z. Phys. B **38**, 207 (1979).

[3]*Anderson Localization,* edited by Y. Nagaoka and H. Fukuyama, Springer Series in Solid State Sciences, Vol. 39 (Springer-Verlag, Berlin, 1982).

[4]K. von Klitzing, G. Dorda, and M. Pepper, Phys. Rev. Lett. **45**, 494 (1980).

[5]D. C. Tsui, H. L. Stormer, and A. C. Gossard, Phys. Rev. Lett. **48**, 1559 (1982).

[6]A. M. M. Pruisken, Phys. Rev. B **32**, 2636 (1985).

[7]R. B. Laughlin, M. L. Cohen, M. J. Kosterlitz, H. Levine, S. B. Libby, and A. M. M. Pruisken, Phys. Rev. B **32**, 1311 (1985).

[8]A. M. M. Pruisken, Nucl. Phys. **B235**FS[11], 277 (1984).

[9]A. M. M. Pruisken, in *Localization, Interaction and Transport Phenomena,* edited by B. Kramer, G. Bergmann, and Y. Bruynseraede, Springer Series in Solid State Sciences, Vol. 61 (Springer-Verlag, Berlin, 1985), p. 188, and references therein.

[10]H. Levine, S. B. Libby, and A. M. M. Pruisken, Phys. Rev. Lett. **51**, 1915 (1983); Nucl. Phys. **B240**FS[12], 30 (1984); **B240**FS[12] 49 (1984); **B240**FS[12], 71 (1984), and references in these articles.

[11]D. E. Khmel'nitzkii, Pis'ma Zh. Eksp. Teor. Fiz. **38**, 454 (1983) [JETP Lett. **38**, 552 (1983)].

[12]T. Ando, Y. Matsumoto, and Y. Uemura, J. Phys. Soc. Jpn. **39**, 279 (1975); **36**, 959 (1974).

[13]S. M. Girvin, M. Johnson, and P. A. Lee, Phys. Rev. B **26**, 1651 (1982).

[14]M. A. Palaanen, D. C. Tsui, and A. C. Gossard, Phys. Rev. B 25, 5566 (1982).

[15]D. C. Tsui, H. L. Stormer, and A. C. Gossard, Phys. Rev. B 25, 1405 (1982).

[16]A. Briggs, Y. Guldner, J. P. Vieren, M. Voos, J. P. Hirtz, and M. Razeghi, Phys. Rev. B **27**, 6549 (1983).

[17]G. Ebert, K. von Klitzing, C. Probst, E. Schubert, K. Ploog, and G. Weigmann, Solid State Commun. **45**, 625 (1983); B. Tausendfreund and K. von Klitzing, Surf. Sci. **142**, 220 (1984).

[18]Y. Ono, J. Phys. Soc. Jpn. **51**, 237 (1982).

[19]D. J. Thouless, Phys. Rev. Lett. **39**, 1167 (1977); E. Abrahams, P. W. Anderson, P. A. Lee, and T. V. Ramakrishnan, Phys. Rev. B **24**, 6783 (1981).

[20]H. P. Wei, D. C. Tsui, and M. Razeghi, Appl. Phys. Lett. **45**, 666 (1984).

[21]D. C. Tsui, H. L. Stormer, and M. Razeghi (unpublished).

Quantization of coupled orbits in metals

By A. B. Pippard, F.R.S.
Cavendish Laboratory, University of Cambridge

(*Received* 12 *April* 1962)

The failure of semi-classical quantization of electron orbits in metals in the presence of a strong enough magnetic field (magnetic breakthrough) is discussed in an elementary fashion by means of first-order perturbation theory. The interference effects, which arise when orbits about different centres are coupled, are reproduced in a simple network analogue. Exact analysis of the network shows how the energy levels are broadened by the coupling and eventually reform into a different set of levels. Fourier analysis of the level density reveals what might be observed in the de Haas–van Alphen effect when magnetic breakthrough is significant, and it is concluded that in principle the whole evolution of the level system should be observable.

According to the semi-classical theory which is the basis of most treatments of the de Haas–van Alphen and related effects, indeed of most phenomena involving the influence of magnetic fields on metallic behaviour, an electron moves in k-space round orbits which are the intersections of the surfaces of constant energy by planes normal to the magnetic field, H. Correspondingly (Onsager 1952) in real space the electrons describe similar orbits, oriented at $\frac{1}{2}\pi$ to the k-orbits in the plane normal to H. If the orbits are closed curves the motion is periodic and may be quantized by semi-classical methods; the permitted orbits have areas in k-space which (apart from a small phase correction) are integral multiples of $2\pi eH/\hbar$. An equivalent statement is that the permitted energies of motion in the plane normal to H form a set of degenerate discrete levels separated by $\hbar\omega_c$, ω_c being the cyclotron frequency of the orbits concerned. The validity of this treatment of the problem has been more or less established by several investigators (see Blount 1962), but the methods they have used are far from elementary and not always easy to apply to a detailed study of particular points where the semi-classical theory fails. One such point was noted by Cohen & Falicov (1961), who suggested that when an electron in its k-orbit approaches a Brillouin zone boundary, it may not suffer Bragg reflexion but may instead pass through the boundary, jumping onto another orbit, even though this involves tunnelling through a region where no classically permitted energy surface exists. They remark that if the energy discontinuity ϵ across the zone boundary is smaller than the level separation $\hbar\omega_c$ this effect may be expected to occur, but Blount (1962) has shown that breakthrough is much easier to achieve than this criterion suggests, and has given an improved criterion that breakthrough of the electrons at the Fermi surface is significant when ϵ is of the order of $(E_F\hbar\omega_c)^{\frac{1}{2}}$, where E_F is the Fermi energy. Since in practice E_F may well be $10^4\hbar\omega_c$, breakthrough may be expected in magnetic fields much less than what Cohen & Falicov suggested, and need not be confined to exceptional situations in which ϵ happens to be very small.

Blount's argument concerns itself with the probability that an electron on one orbit in real space may jump onto another similar orbit having a different centre.

Cohen & Falicov discussed the removal of the degeneracy of the semi-classical energy levels by a weak lattice potential. These treatments are of course different aspects of the same phenomenon, line broadening due to coupling of previously independent degenerate modes. We shall begin by carrying through the calculation left incomplete by Cohen & Falicov, and shall derive Blount's criterion as the condition for significant line broadening. In so doing we shall note an interference effect which leads us to develop a model of coupled electron orbits, of such simplicity that we may follow the process of magnetic breakthrough in detail and see how the energy-level structure is modified thereby.

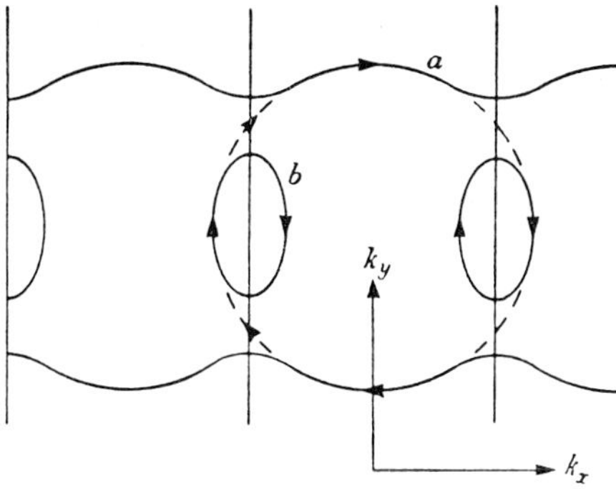

FIGURE 1. Orbits for two-dimensional metal with one Fourier component of lattice potential.

To make the argument as transparent as possible we treat a two-dimensional metal in which the periodic lattice potential has but one component,

$$V(x, y) = V_g \cos gx;$$

the Brillouin zone structure consists of an array of parallel lines $k_x = \frac{1}{2}ng$, where n is a non-zero integer. If V_g is small, a typical energy contour passing through a zone boundary will be that shown in figure 1. The diagram has been extended periodically to show that, according to the semi-classical argument, in a magnetic field normal to the paper two types of orbit are possible, (a) an open orbit, involving Bragg reflexion at each zone boundary, and (b) a closed orbit, also involving Bragg reflexion; there is, of course, no discontinuity in the electron wave function on reflexion. The occurrence of breakthrough is indicated by the dotted lines, and becomes a likely effect if H is large and the energy gap ϵ at the zone boundary is small. Under these conditions the normal path of an electron may be taken as a circular orbit, such as would obtain if the electrons were strictly free, and the level broadening results from the coupling of neighbouring orbits by occasional Bragg reflexion. The levels themselves have a separation $\hbar\omega_c$ determined by the cyclotron frequency of a free electron. In smaller fields, when breakdown is a rarity, the levels are determined by the small orbits (b) and are farther apart; there is also a continuum of levels due to electrons in the open orbits (a). Both in large and small fields perturbation theory serves to elucidate the level broadening, but is not so valuable in plotting out the transition from one set of narrow levels (high field) to another set of different spacing and a continuum (low field), and here we invoke our model. But first we carry through the perturbation treatment in the simplest case, which is the high field limit, and for which first-order perturbation theory is adequate.

The electron orbits in a high field are the circular free electron orbits if break-through always occurs, and we can write the wave functions of the electrons exactly under these conditions, in effect supposing $V_g = 0$. We can do the same when there is no magnetic field, so that plane waves describe the electrons, and thus we can compare the level broadening in a field, due to V_g, with the energy gap at the zone boundary. It is desirable to choose a representation for the unperturbed wave functions such that the perturbation does not couple two degenerate levels, and this is easily achieved in both cases. When $H = 0$ we take for an electron on the zone boundary

$$\psi = e^{ik_y y} \cos \tfrac{1}{2}gx \quad \text{or} \quad e^{ik_y y} \sin \tfrac{1}{2}gx.$$

Correspondingly the first-order perturbation energy is $\pm \frac{1}{2}V_g$, so that the band gap ϵ is just V_g.

In a magnetic field derived from the vector potential $(0, Hx, 0)$ the free-electron wave function takes the form

$$\psi = e^{ik_y y} X(x),$$

and X obeys the oscillator equation

$$(\hbar^2/2m)X'' + [E - \tfrac{1}{2}m\omega_c^2 x^2]X = 0, \tag{1}$$

where x has its origin at $\hbar k_y/(eH)$ (i.e. different for every k_y), and ω_c is the cyclotron frequency eH/m. In real metals we usually are concerned with electron orbits many de Broglie wavelengths in perimeter, and for such orbits the W.K.B. solution of (1) is adequate:

$$X = \left(\frac{2eH}{\pi\hbar k_x}\right)^{\frac{1}{2}} \cos\phi(x) \quad \text{or} \quad \left(\frac{2eH}{\pi\hbar k_x}\right)^{\frac{1}{2}} \sin\phi(x), \tag{2}$$

in which ϕ is the phase integral

$$\int_0^x k_x(x)\,\mathrm{d}x \quad \text{and} \quad k_x^2 = \frac{2m}{\hbar^2}[E - \tfrac{1}{2}m\omega_c^2 x^2].$$

This result has an immediate pictorial interpretation. The electron may be imagined moving round a circular orbit in real space of radius $(2mE)^{\frac{1}{2}}/(eH)$, and round a circular orbit in k-space of radius $(2mE)^{\frac{1}{2}}/\hbar$, which we shall call k_0. The two orbits are traversed in quadrature, so that

$$\phi = \int_0^x k_x \mathrm{d}x = -\frac{\hbar}{eH}\int k_x \mathrm{d}k_y = \frac{\hbar}{eH}A, \tag{3}$$

where A is the shaded area in figure 2. This diagram is of course much more than a geometrical construction—by appropriate gauging of the vector potential the wave function can be represented as a wave of uniform amplitude closely confined to the circle and running round continuously, so that the quantum condition $E = n\hbar\omega_c$† is seen as a prescription for the single-valuedness of the wave function.

† We do not concern ourselves with such refinements as lead in the correct theory to the condition $E = (n + \frac{1}{2})\hbar\omega_c$.

1-2

Moreover, k_x at any part on the orbit in real space is merely the x-component of k_0 (which is directed along the orbit).

Consider now the first-order contribution of the lattice potential to the energy which, since k_x varies with x, arises from only a small region of the wave function where $k_x \doteqdot \frac{1}{2}g$. This contrasts with the behaviour of the plane wave, of which all regions contribute. Thus the perturbation of the oscillator energy is smaller than of the plane wave. There are four points on the circular orbit where interaction occurs, or two points in real space for the standing wave representation (2), at $\pm x'$ say, and we consider the region near $+x'$ first, writing an approximate form of k_x near x',

$$k_x \doteqdot \tfrac{1}{2}g + \alpha(x-x'), \quad \text{where} \quad \alpha = \mathrm{d}k_x/\mathrm{d}x.$$

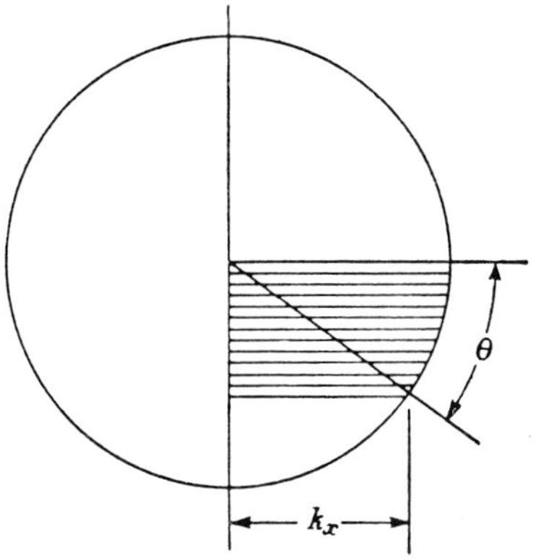

FIGURE 2. Graphical representation of phase integral.

Correspondingly, from (2), we write for X^2, needed to evaluate the perturbation energy,

$$X^2 \doteqdot \frac{2eH}{\pi\hbar g}\{1 \pm \cos[g+2\alpha(x-x')]x\}.$$

When evaluating the integral $\int \psi^* V \psi \,\mathrm{d}\tau$, i.e. $\int X^2 V \,\mathrm{d}x$, we must remember that for each different k_y the origin of x is different and thus the cosine term in X^2 (the only term contributing to the perturbation energy) may have any phase with respect to V ($= V_g \cos gx$). It is this that results in line broadening, and the overall broadening is obtained by choosing that phase difference which maximizes the integral. An equivalent procedure is to express the oscillatory part of X^2 in complex notation and evaluate the amplitude of the resulting complex perturbation integral; i.e. the overall line breadth is Δ where

$$\tfrac{1}{2}\Delta = \frac{2eHV_g}{\pi\hbar g}\left|\int_{-\infty}^{\infty} \exp\{\mathrm{i}[g+2\alpha(x-x')]x\}\cos gx \,\mathrm{d}x\right|,$$

or

$$\Delta = \frac{2eHV_g}{\pi\hbar g}\left|\int_{-\infty}^{\infty} \exp[2\mathrm{i}\alpha(x-x')x]\,\mathrm{d}x\right|,$$

if we concern ourselves only with that part of the integral due to regions near x', where it is permissible to replace $\cos gx$ by $\frac{1}{2}\mathrm{e}^{-\mathrm{i}gx}$. Thus

$$\Delta = \frac{eHV_g}{\pi\hbar g}(2\pi/\alpha)^{\frac{1}{2}}. \tag{4}$$

This expression gives us the range over which the energy levels are spread, and it is worth noting that the spread is not uniform but concentrated towards the edges of the range, for there is considerable latitude in the placing of the wave function on the lattice if something near the maximum perturbation is to be achieved. In fact the density of states within a broadened level varies as $\sec[\pi(E-\bar{E})/\partial\Delta]$, where $\bar{E}$ is the mean energy of the broadened level.

Returning to (4) we note that α is dk_x/dx at the point where $k_x = \frac{1}{2}g$, and this is conveniently expressed in terms of the angle θ (figure 2) representing position round the orbit—$\cos\theta = k_x/k_0$. Remembering that the band gap ϵ is the same as V_g we reach the result

$$\Delta/\epsilon = \left\{\frac{2eH}{\pi\hbar k_0^2 \sin 2\theta}\right\}^{\frac{1}{2}}.$$

This can be further simplified by putting $\hbar^2k_0^2/2m = E$, the energy of the electron, and $eH/m = \omega_c$; then

$$\Delta/\epsilon = \frac{1}{\beta}\left(\frac{\hbar\omega_c}{E}\right)^{\frac{1}{2}}, \quad \text{where} \quad \beta = (\pi \sin 2\theta)^{\frac{1}{2}}. \tag{5}$$

Now the condition that magnetic breakthrough shall occur readily is just that the lattice potential is too weak to cause significant perturbation of the energy levels. Since in the free-electron gas these levels are spaced $\hbar\omega_c$ apart, the condition for easy breakthrough is that $\Delta \ll \hbar\omega_c$, i.e.

$$\epsilon \ll \beta(\hbar\omega_c E)^{\frac{1}{2}}. \tag{6}$$

This result is of the same form as Blount's (1962), derived by a different argument.

In this treatment we have considered only the contribution to Δ from the two parts of the circular orbit at which x' is positive and $k_x = \frac{1}{2}g$. There are also two parts at negative x', and the total line broadening will depend on the phase relationship between the two perturbations. If at positive x' the maxima of V lie on maxima of X^2, broadening will be enhanced when the same occurs at negative x'; i.e. when the phase changes between x' and $-x'$ for V and for X^2 differ by a multiple of 2π. This condition for maximum broadening may be displayed geometrically, as in figure 3. Following (3) and the construction of figure 2, we represent the phase change of X from the centre of the orbit to the point x' by the lower right quarter of the horizontally shaded region, and similarly the phase change between $-x'$ and the centre by the upper right quarter. Consequently the right half of the horizontally shaded region represents (apart from the factor $\hbar/eH$) the phase change between $\pm x'$ of X, and the whole horizontally shaded region the phase change of the oscillatory part of X^2. The phase change of the lattice potential, which has constant wave number g, is represented by the vertically shaded rectangle. The condition for maximum broadening 2Δ is therefore that the area A_b of the two segments which are horizontally but not vertically shaded, in effect the area of the orbit b in figure 1, shall be such that

$$\hbar A_b/eH = 2n\pi. \tag{7}$$

For weak lattice perturbation, then, the level system of the free electrons is periodically broadened, as shown in figure 4.

The spacing of the centroids of the broadened levels is determined by A_0, the area in k-space of the unperturbed circular orbit, and the periodicity of the broadening is determined by A_b. With a weak perturbation we can hardly speak of the orbit b as being quantized in the sense of contributing its own set of levels, but we might expect such an effect to appear when the perturbation is stronger than can

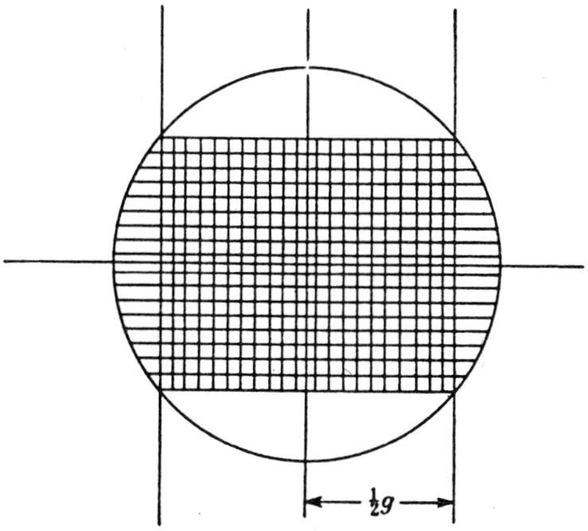

FIGURE 3. Graphical representation of condition for level broadening.

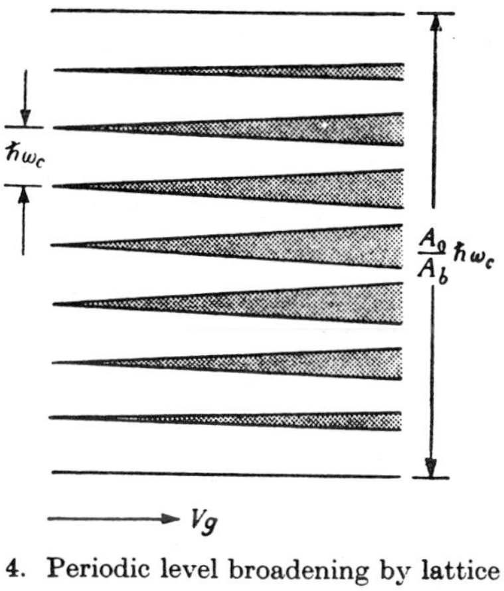

FIGURE 4. Periodic level broadening by lattice potential.

be treated by first-order perturbation theory. For as the likelihood of breakthrough diminishes, the orbits b become more and more the normal trajectories whose sharply quantized levels are broadened by coupling to the open orbits a. The treatment given by Blount cannot without extension cover this situation, for it concerns itself only with breakthrough at one point in the orbit and not with the phase relationships with other points at which breakthrough is possible. Since these are the key to a fuller understanding we shall make them the central feature in what we now present, a somewhat heuristic approach to the problem which, if perhaps not exact, does seem in its outcome to provide a plausible picture of the changeover from one set of quantized levels to the other as the probability of breakthrough varies from unity to zero.

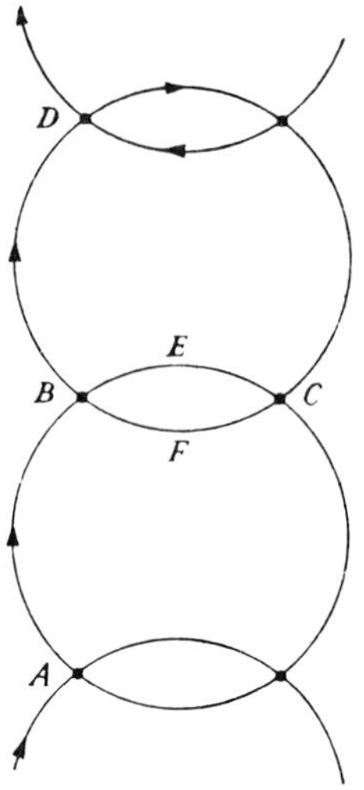

FIGURE 5. Network representation of coupled orbits.

The fact that the wave function of an electron in a magnetic field may be represented as a wave travelling on, and rather closely confined to, a well-defined path in real space, suggest that we may ignore everything about it except the form of the path and the phase of the wave at any point on it. Thus a free electron is schematized as a circular loop on which travels a scalar wave at constant velocity. This is not like a wave on a string, for propagation is here possible only in one direction, given by the sense of the electron's motion in its orbit, and, for example, at a discontinuity there is no possibility of a reflected wave. The discontinuities with which we are concerned are created at certain points on the path by the possibility of breakthrough. When the electron is at such a point that it may take one of two paths, as shown in figure 1, either following the continuous trajectory or breaking through to another, we have in our network analogue reached a junction of two separate networks, which are coupled together to a degree determined by the probability of breakthrough. Remembering that the trajectory of an electron in real space has the same form as its trajectory in k-space, but turned through $\frac{1}{2}\pi$, we have no difficulty in constructing, as in figure 5, an appropriate network to represent the structure of figure 1, an infinite chain of linked circular loops, around each of which clockwise propagation only is possible, and for which certain relations must exist between the amplitudes of the waves in different arms meeting at a junction.

To determine these relations consider a single junction, B say, and suppose a wave of unit amplitude to arrive from A; then waves will leave along BD and BEC, but not of course along BFC and BA. Let the wave along BEC have amplitude and phase at B represented by p, and that along BD be represented by q. Clearly

$pp^* + qq^* = 1$ to conserve particles. We may similarly consider a wave arriving along CFB, and splitting into p' along BD and q' along BEC. By analyzing the system when arbitrary waves arrive along both paths together it is readily shown that $qp'^* + pq'^* = 0$, and hence that $|p| = |p'|$, $|q| = |q'|$, and p and q are in phase quadrature. Since we shall not be concerned with small phase errors but only with the general structure of the solutions of the problem, we shall take q to be real and $p = \mathrm{i}(1-q^2)^{\frac{1}{2}}$.

We now proceed to find the conditions under which the network can carry waves. We shall see that solutions only exist for certain ranges of the variables ξ and χ, which are the phase lengths of the paths BC and AB; this has an immediate interpretation in terms of the energy level diagram of the electrons, for these phase paths are related to the areas of certain orbits in k-space, and the permitted energies are those that give rise to appropriate areas. From (3) it is clear that 2ξ is $\hbar/eH$ times the area of the orbit b in figure 1, while $2(\xi + \chi)$ is $\hbar/eH$ times the area of the

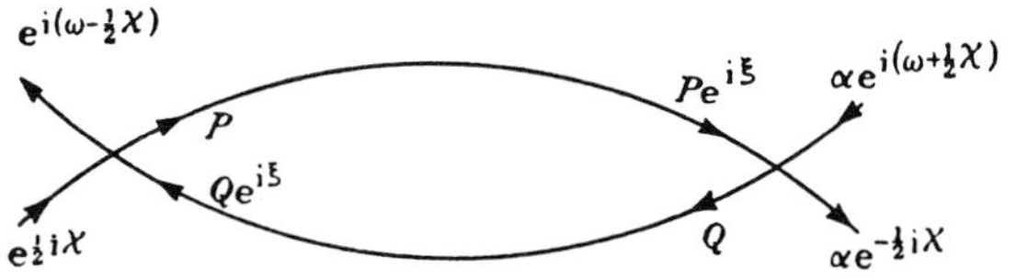

FIGURE 6. Assumed amplitudes at junction of orbits.

unperturbed circular orbit. Following the usual procedure for periodic lattices we look for solutions in which the amplitudes in each cell of the network are the same, and there is a phase difference ω between successive cells. We therefore ascribe amplitudes and phases to the various paths at the junctions B and C, as shown in figure 6, and write down the relations between them as follows:

$$\left.\begin{aligned} \mathrm{e}^{\mathrm{i}(\omega-\frac{1}{2}\chi)} &= q\,\mathrm{e}^{\frac{1}{2}\mathrm{i}\chi} + pQ\,\mathrm{e}^{\mathrm{i}\xi}, \\ P &= p\,\mathrm{e}^{\frac{1}{2}\mathrm{i}\chi} + qQ\,\mathrm{e}^{\mathrm{i}\xi}, \\ \alpha\,\mathrm{e}^{-\frac{1}{2}\mathrm{i}\chi} &= q\alpha\,\mathrm{e}^{\mathrm{i}(\omega+\frac{1}{2}\chi)} + pP\,\mathrm{e}^{\mathrm{i}\xi}, \\ Q &= p\alpha\,\mathrm{e}^{\mathrm{i}(\omega+\frac{1}{2}\chi)} + qP\,\mathrm{e}^{\mathrm{i}\xi}. \end{aligned}\right\} \tag{8}$$

Eliminating P, Q and α from these equations, we find an equation for ω in the form

$$\cos\omega = \frac{\sin(\xi+\chi) + q^2 \sin(\xi-\chi)}{2q\sin\xi}. \tag{9}$$

Since ω is real, (9) contains the energy level structure of the system in the form of the range of values of ξ and χ which allow $|\cos\omega| \leqslant 1$.

Before proceeding to discuss this solution we remark that it might be more realistic to think of the individual quantized states as standing waves on the network rather than travelling waves. When we construct such standing waves by superposing solutions having the same amplitude and $\pm\omega$ for phase difference, we do not get (as on a simple transmission line) a real solution with sinusoidal amplitude variation along the line, but because on each side we compound a wave of unit amplitude with one of amplitude α, the amplitude varies above the line as $1+|\alpha|\,e^{2in\omega}$, with

$$|\alpha|^2 = \frac{1+q^2-2q\cos(\omega-\chi)}{1+q^2-2q\cos(\omega+\chi)}.$$

This, however, is a detail of no significance to what follows.

The meaning of (9) is conveniently displayed geometrically, as in figure 7. It is clear that $PN = \sin(\xi+\chi)+q^2\sin(\xi-\chi)$ and $PT = 2q\sin\xi$. Hence $|\cos\omega|$ is less than unity only if the line PN does not cut the circle, and the range of values of χ for which this is so measures the extent to which the quantized levels were broadened

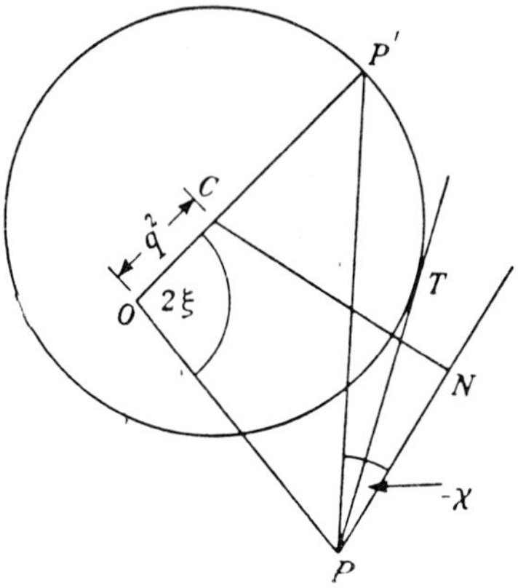

FIGURE 7. Geometrical representation of (9). $OP = OP' = 1$; the circle has centre C and radius $1-q^2$; angle $CNP = \frac{1}{2}\pi$.

by interaction between neighbouring orbits. This diagram indicates without the trouble of computation the conditions for more or less broadening, and shows that there are always gaps in the spectrum, the broadening never being so great as to cause crossing of lines from different unbroadened levels. It is, however, a simple matter to compute (9), and in figure 8 are displayed contours of constant ω on a $\chi-\xi$ plot, for the case $q^2 = \frac{1}{2}$. The diagram repeats continuously in χ and ξ with a repetition length of π in each direction. When q is small the contours are confined to the regions near the lines $\xi+\chi = n\pi$, while as q tends to unity they spread until they fill the whole square. Contraction to a point occurs for all q when ξ and χ are multiples of π. It is clear from the diagram that considerable bunching of contours occurs near the edges of the allowed regions, as we have already noted in the perturbation treatment.

By means of such contour plots we may construct a diagram showing how the levels shift with change of q. It is convenient to draw not an energy level diagram but, what is virtually equivalent, a chart of the values of $1/H$ at which permitted levels exist for an electron of given energy. When this energy is the Fermi energy the chart relates directly to the de Haas–van Alphen and other oscillatory phenomena studied experimentally. Since the areas of the orbits in phase space are constant for constant energy, it is clear that as $1/H$ is changed χ and ξ maintain the same proportion, and the value of either (we shall choose ξ) is a measure of $1/H$.

In constructing figure 9 we have arbitrarily chosen the orbit areas such that $\chi/\xi = \frac{11}{8}$, and followed the variation with q of the value of ξ at which ω takes a given value. The reason for this is that ω is a quantum number playing the role of k in a continuum; for a chain of N orbits the permitted values of ω are integral multiples of π/N. The number of states between each pair of neighbouring lines in the diagram is the same. When $q = 0$ each circular orbit is independent and there are N degenerate solutions† for each field value which makes $\chi + \xi = n\pi$, so that the orbit perimeter is an integral number of wavelengths. As q increases the degeneracy

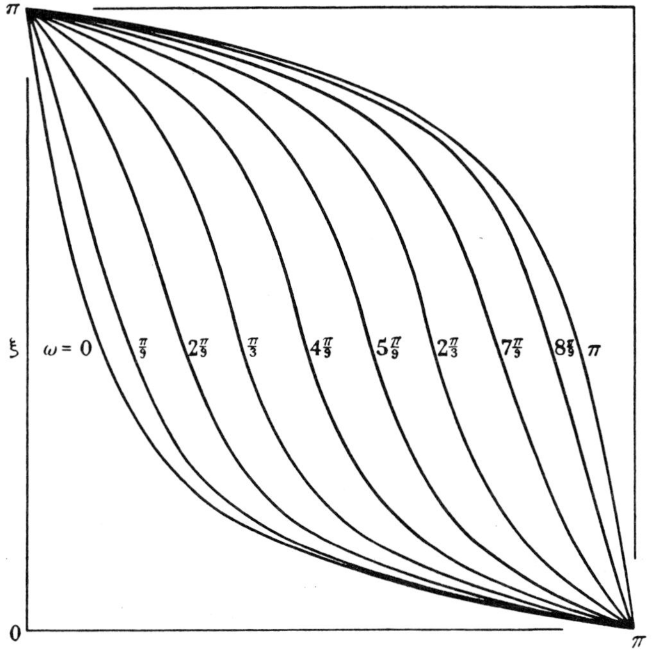

FIGURE 8. Solutions of (9) for $q^2 = \frac{1}{2}$; contours of $\omega(\xi, \chi)$.

is removed, and ultimately there is a concentration of some levels in groups of N towards the points $\xi = n\pi$, while the rest spread out evenly. This corresponds to the quantized levels of the N orbits (b) and the unquantized open orbits (a). It will be observed that for q nearly unity the continuum of levels due to the open orbits is periodically split, the gaps being wide where a resonance of b-orbits is responsible ($\xi = n\pi$), and rather less wide where non-resonant b-orbits couple open orbits having $\chi = n\pi$. In the latter case the waves running in opposite directions along two neighbouring open orbits have the same phase relationship at each successive junction. However weak the excitation of the intervening b-orbits, the matrix element

† It should be remarked that there will be in the sample of metal many identical replicas of the chain of N rings, overlapping but not interacting with one another. The reason for the absence of interaction is that breakthrough cannot occur at any arbitrary point in the orbit, but only at such points that k is changed on breakthrough by a reciprocal lattice vector.

coupling the degenerate open orbits will not vanish, and the degeneracy will be removed, just as at a band edge in the nearly-free electron theory of metals. This analogy has already been pointed out by Blount.

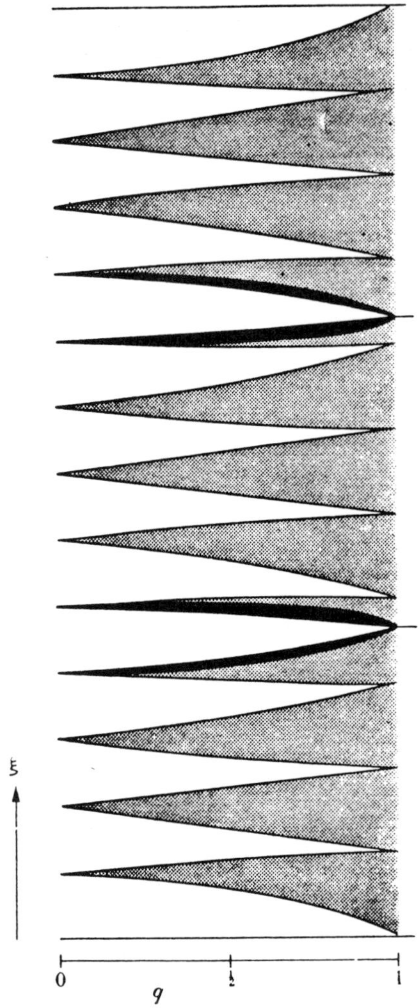

FIGURE 9. Evolution of level diagram as q varies from 0 to 1, for the case $\chi = \frac{11}{3}\xi$. The grey shading indicates the development of a continuum of levels; the black shading shows the coalescence of N levels into a degenerate level. The diagram covers a range 3π in ξ.

The energy level structure may be explored experimentally by the de Haas–van Alphen effect, which picks out periodicities in density of levels. Usually all that can be observed, because of the effects of temperature and collisions, is the lowest Fourier component of each distinct periodicity. The calculations presented here enable a Fourier analysis of the level density to be carried out, and this has been done approximately to reveal the main features, which seem to offer no surprises.

When $q = 0$, only one periodicity is present, that due to the circular orbits, whose area we take as 14. Since the frequency of the oscillations is proportional to the area of the k-orbit, we label this frequency 14 also, and call the amplitude unity. When $q = 1$, the b-orbits contribute a frequency 3, with amplitude $\frac{3}{14}$ or 0·215. There is also a continuum of states due to open orbits. In between, the behaviour is intermediate, as shown in figure 10 which gives the amplitude of all frequency components up to 30 for the case $q^2 = \frac{1}{2}$.† The prominent frequencies can all be ascribed to specific orbits, as the diagram indicates. What is perhaps worthy of comment is that while the frequency 14 is just over one-quarter of the maximum value (when $q = 0$), the frequency 3 is one-half its maximum (when $q = 1$). Thus

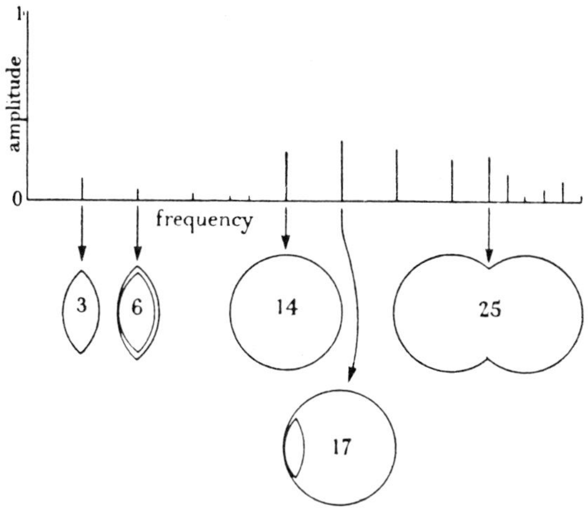

FIGURE 10. Spectrum of energy levels for $q^2 = \frac{1}{2}$, showing the orbits responsible for certain prominent components.

there is no intermediate coupling range at which the levels become so smeared out that all structure is lost; rather is there a gradual merging of one structure into another, and one might hope under favourable conditions to observe the whole process experimentally. It is rather unlikely that in any real metal so simple an example of breakthrough will be found. More commonly the Brillouin zone structure will be such as to produce a two-dimensional array of interlocked orbits rather than the simple chain we have considered. But the results of an analysis of such a system (which is probably a tedious matter suitable only for a digital computer) are likely to be qualitatively similar. At any rate we do not propose to investigate any such example here.

One final point may be made. We have introduced the parameter q^2 as a measure of the probability that an electron will follow the semiclassical path at a zone boundary, but we have not related its magnitude to the parameters of the metal. This can clearly be done in the range of small q^2 by means of our perturbation argument. It is readily shown from (9) that when q^2 is small the maximum level

† I am indebted to Mrs Margaret Mutch for the numerical calculations.

broadening is a fraction $4q/\pi$ of the separation between successive levels. In terms of Δ as used in (5) this means that

$$\frac{2\Delta}{\hbar\omega_c} \approx \frac{4q}{\pi},$$

or

$$q^2 \approx \frac{\pi^2}{4\beta^2}\frac{\epsilon^2}{\hbar\omega_c E} = \frac{\pi\epsilon^2}{4\hbar eH\,|v_x v_y|}, \tag{10}$$

where v_x and v_y are the velocity components of the free electron at the zone boundary. Now Blount has shown by direct calculation that when the field is small, so that q^2 is near unity, the probability of breakthrough

$$1-q^2 = |p|^2 \approx \exp\left\{\frac{-\pi\epsilon^2}{4\hbar eH\,|v_x v_y|}\right\}. \tag{11}$$

The fact that this formula, when applied to the case $q^2 \ll 1$, agrees with (10), indicates that (11) may be a reasonably good approximation (or even exact) for all values of q^2. Thus the parameters to be used in the network analogue are determined at least approximately.

I am most grateful to Dr V. Heine for his critical advice, particularly his insistence that the original form of this analysis was unnecessarily complicated.

References

Blount, E. I. 1962 *Phys. Rev.* **126**, 1636.
Cohen, M. H. & Falicov, L. 1961 *Phys. Rev. Lett.* **7**, 231.
Onsager, L. 1952 *Phil. Mag.* **43**, 1006.

"Quantization of Coupled Orbits in Metals," A.B. Pippard, Proc.Roy. Soc. A, **270** pp. 1–13 (1962).

Energy levels and wave functions of Bloch electrons in rational and irrational magnetic fields*

Douglas R. Hofstadter†
Physics Department, University of Oregon, Eugene, Oregon 97403
(Received 9 February 1976)

An effective single-band Hamiltonian representing a crystal electron in a uniform magnetic field is constructed from the tight-binding form of a Bloch band by replacing $\hbar\vec{k}$ by the operator $\vec{p} - e\vec{A}/c$. The resultant Schrödinger equation becomes a finite-difference equation whose eigenvalues can be computed by a matrix method. The magnetic flux which passes through a lattice cell, divided by a flux quantum, yields a dimensionless parameter whose rationality or irrationality highly influences the nature of the computed spectrum. The graph of the spectrum over a wide range of "rational" fields is plotted. A recursive structure is discovered in the graph, which enables a number of theorems to be proven, bearing particularly on the question of continuity. The recursive structure is not unlike that predicted by Azbel', using a continued fraction for the dimensionless parameter. An iterative algorithm for deriving the clustering pattern of the magnetic subbands is given, which follows from the recursive structure. From this algorithm, the nature of the spectrum at an "irrational" field can be deduced; it is seen to be an uncountable but measure-zero set of points (a Cantor set). Despite these features, it is shown that the graph is continuous as the magnetic field varies. It is also shown how a spectrum with simplified properties can be derived from the rigorously derived spectrum, by introducing a spread in the field values. This spectrum satisfies all the intuitively desirable properties of a spectrum. The spectrum here presented is shown to agree with that predicted by A. Rauh in a completely different model for crystal electrons in a magnetic field. A new type of magnetic "superlattice" is introduced, constructed so that its unit cell intercepts precisely one quantum of flux. It is shown that this cell represents the periodicity of solutions of the difference equation. It is also shown how this superlattice allows the determination of the wave function at nonlattice sites. Evidence is offered that the wave functions belonging to irrational fields are everywhere defined and are continuous in this model, whereas those belonging to rational fields are only defined on a discrete set of points. A method for investigating these predictions experimentally is sketched.

I. INTRODUCTION

The problem of Bloch electrons in magnetic fields is a very peculiar problem, because it is one of the very few places in physics where the difference between rational numbers and irrational numbers makes itself felt.[1,2] Common sense tells us that there can be no physical effect stemming from the irrationality of some parameter, because an arbitrarily small change in that parameter would make it rational — and this would create some physical effect with the property of being everywhere discontinuous, which is unreasonable. The only alternative, then, is to show that a theory which apparently distinguishes between rational and irrational values of some parameter does so only in a mathematical sense, and yields physical observables which are nevertheless continuous. It is the purpose of this paper to present a method which effects such a reconciliation of "rational" and "irrational" magnetic fields. The method is illustrated in a maximally simple model of the physical situation, but the ideas which arise are, it is to be hoped, applicable to more realistic models of the physical situation.

II. DERIVATION OF THE DIFFERENCE EQUATION

Briefly, then, the model involves a two-dimensional square lattice of spacing a, immersed in a uniform magnetic field H perpendicular to it. We restrict our considerations to what happens to a single Bloch band when the field is applied. This is one strong simplifying feature of the model; the next is that we postulate the following tight-binding form for the Bloch energy function:

$$W(\vec{k}) = 2E_0(\cos k_x a + \cos k_y a).$$

Perhaps the most difficult step to justify on physical grounds is the following one, which I shall refer to as the "Peierls substitution"[3]: we replace $\hbar k$ in the above function by the operator $\vec{p} - e\vec{A}/c$ ($\vec{A}$ being the vector potential), to create an operator out of $W(\vec{k})$, which we then treat as an effective single-band Hamiltonian. Work to justify this substitution has been done.[4-7]

When this substitution is made, the effective Hamiltonian is seen to contain translation operators $\exp(ap_x/\hbar)$ and $\exp(ap_y/\hbar)$. Depending on the gauge chosen, there are, in addition, certain phase factors dependent on the magnetic field strength, which multiply the translation operators. If the Landau gauge — $\vec{A}=H(0,x,0)$ — is chosen, then only the translations along y are multiplied by phases. From now on, we assume this gauge. Now when the effective Hamiltonian is introduced into a time-independent Schrödinger equation with a two-dimensional wave function, the following eigenvalue equation results:

$$E_0[\psi(x+a,y)+\psi(x-a,y)+e^{-ieHax/\hbar c}\psi(x,y+a) + e^{+ieHax/\hbar c}\psi(x,y-a)]=E\psi(x,y).$$

Note how the wave function at (x,y) is linked to its four nearest neighbors in the lattice. It is convenient to make the substitutions

$$x=ma\,,\quad y=na\,,\quad E/E_0=\epsilon\,.$$

It is furthermore reasonable to assume plane-wave behavior in the y direction, since the coefficients in the above equation only involve x. Therefore, we write

$$\psi(ma,na)=e^{i\nu n}g(m)\,.$$

Finally we introduce the parameter about which all the fuss is made.

$$\alpha=a^2H/2\pi(\hbar c/e)\,.$$

Notice that α is dimensionless, being the ratio of flux through a lattice cell to one flux quantum. (The author is indebted to Professor F. Bloch for pointing out that this parameter can be interpreted as the ratio of two characteristic periods of this problem: one is the period of the motion of an electron in a state with crystal momentum $2\pi\hbar/a$, which is $a^2m/2\pi\hbar$; the other is the reciprocal of the cyclotron frequency eH/mc.) A value of $\alpha=1$ implies an enormous magnetic field (on the order of a billion gauss), if the lattice spacing is typical of real crystals (on the order of 2 Å). Despite this, we are going to be interested in the results for such values of α (for a treatment of smaller values of α in this same equation, see Ref. 8.)

With all these substitutions, our Schrödinger equation turns into a one-dimensional difference equation:

$$g(m+1)+g(m-1)+2\cos(2\pi m\alpha-\nu)g(m)=\epsilon g(m)\,. \tag{1}$$

This equation is sometimes called "Harper's" equation, and has been studied by a number of authors.[8-11]

III. CALCULATION OF THE SPECTRUM AND THE RATIONALITY CONDITION

Another way of writing Eq. (1) is

$$\begin{pmatrix} g(m+1)\\ g(m)\end{pmatrix}=\begin{pmatrix}\epsilon-2\cos(2\pi m\alpha-\nu) & -1\\ 1 & 0\end{pmatrix}\begin{pmatrix} g(m)\\ g(m-1)\end{pmatrix}.$$

The 2×2 matrix is called "$A(m)$." When a product of m successive A matrices is multiplied with the two vector $\langle g(1),g(0)\rangle$, the result is the two vector $\langle g(m+1),g(m)\rangle$. The physical condition which must be imposed on the wave function (i.e., the function g) is boundedness, for all m. This translates into a condition on the products of successive A matrices. Now if the A matrices are periodic in m (which they may very well be, since m enters only under a cosine), then long products of A matrices consist essentially in repetitions of one block of A matrices, whose length is the period in m. Let us assume that the A matrices are indeed periodic in m, with period q. This is a requirement on α, namely that there should exist an integer p such that

$$2\pi\alpha(m+q)-\nu=2\pi\alpha m-\nu+2\pi p\,.$$

Algebra reveals the fact that this condition on α is precisely that of rationality[1]:

$$\alpha=p/q\,.$$

We now proceed, making full use of this somewhat bizarre ansatz. (Presently, we will consider the case when α is irrational.) The product of q successive A matrices will be called "Q." The condition of physicality is now transferred from the g's to the matrix Q. It can be shown without trouble that the correct condition on Q is that its two eigenvalues be of unit magntiude. That condition can then be shown to be equivalent to requiring its trace to be less than or equal to 2, in absolute value. Hence, a concise test for the boundedness of the g's is the following:

$$|\mathrm{Tr}Q(\epsilon;\nu)|\le 2\,.$$

Trace conditions of this type have been found by other authors.[2,12] This one was discovered by Professor G. Obermair, and extensively used by the author. Now it can be shown that the only way that ν affects the value of TrQ is additively, i.e., that as ν changes, the shape of the graph of TrQ, plotted against ϵ, is unchanged — it merely moves as a whole, up and down. (A proof of essentially this fact can be found in Ref. 2.) Therefore $\mathrm{Tr}Q(\epsilon;\nu)=\mathrm{Tr}Q(\epsilon)+2f(\nu)$, where $f(\nu)$ is a periodic function of unit amplitude, and $Q(\epsilon)$ is defined as $Q(\epsilon;1/2q)$. We are interested in all values ϵ which, for some ν, yield bounded g's. (Such values will be called "eigenvalues" of the difference equation.) Therefore, we want to form the union of all eigenvalues ϵ, as ν varies. Since $2f(\nu)$ ranges between $+2$ and -2, the condition on the trace can be rewritten as follows:

$$|\mathrm{Tr}Q(\epsilon)|\le 4\,.$$

The trace of Q is always a polynomial of degree

q; hence one might expect the above condition to be satisfied in roughly q distinct regions of the ϵ axis (one region centered on each root). This is indeed the case, and is the basis for a very striking (and at first disturbing) fact about this problem: when $\alpha = p/q$, the Bloch band always breaks up into precisely q distinct energy bands. Since small variations in the magnitude of α can produce enormous fluctuations in the value of the denominator q, one is apparently faced with an unacceptable physical prediction. However, nature is ingenious enough to find a way out of this apparent anomaly. Before we go into the résolution, however, let us mention certain facts about the spectrum belonging to any value of α. Most can be proven trivially: (i) Spectrum(α) and spectrum ($\alpha + N$) are identical. (ii) Spectrum(α) and spectrum($-\alpha$) are identical. (iii) ϵ belongs to spectrum(α) if and only if $-\epsilon$ belongs to spectrum(α). (iv) If ϵ belongs to spectrum (α) for any α, then $-4 \leq \epsilon \leq +4$. The last property is a little subtler than the previous three; it can be proven in different ways. One proof has been published.[13]

From properties (i) and (iv), it follows that a graph of the spectrum need only include values of ϵ between $+4$ and -4, and values of α in any unit interval. We shall look at the interval $[0, 1]$. Furthermore, as a consequence of properties, the graph inside the above-defined rectangular region must have two axes of reflection, namely the horizontal line $\alpha = \frac{1}{2}$, and the vertical line $\epsilon = 0$. A plot of spectrum(α), with α along the vertical axis, appears in Fig. 1. (Only rational values of α with denominator less than 50 are shown.)

IV. RECURSIVE STRUCTURE OF THE GRAPH

This graph has some very unusual properties. The large gaps form a very striking pattern somewhat resembling a butterfly; perhaps equally striking are the delicacy and beauty of the fine-grained structure. These are due to a very intricate scheme, by which bands cluster into groups, which themselves may cluster into larger groups, and so on. The exact rules of formation of these hierarchically organized clustering patterns (Π's) are what we now wish to cover. Our description of Π's will be based on three statements, each of which describes some aspect of the structure of the graph. All of these statements are based on extremely close examination of the numerical data, and are to be taken as "empirically proven" theorems of mathematics. It would be preferable to have a rigorous proof but that has so far eluded capture. Before we present the three statements, let us first adopt some nomenclature. A "unit cell" is any portion of the graph located between successive integers N and $N+1$—in fact we will call that unit cell the Nth unit cell. Every unit cell has a "local variable" β, which runs from 0 to 1; in particular, β is defined to be the fractional part of α, usually denoted as $\{\alpha\}$. At $\beta = 0$ and $\beta = 1$, there is one band which stretches across the full width of the cell, separating it from its upper and lower neighbors; this band is therefore called a "cell wall." It turns out that certain rational values of β play a very important role in the description of the structure of a unit cell; these are the "pure cases"

FIG. 1. Spectrum inside a unit cell. ϵ is the horizontal variable, ranging between $+4$ and -4, and $\beta = \{\alpha\}$ is the vertical variable, ranging from 0 to 1.

$1/N$ and $1-1/N$ $(N \geq 2)$;

and the "special cases"

$N/(2N+1)$ and $(N+1)/(2N+1)$ $(N \geq 2)$

(of the special cases, those with numerator N are the "lower" special cases, and those with numerator $N+1$ are the "upper"). The spectra belonging to these rational values form a "skeleton" on which the rest of the graph is hung. Figure 2 shows that skeleton; in it are shown the bands belonging to pure cases (up to $N=37$); in addition, one out of the $2N+1$ bands per special case is included, the centermost (i.e., the $N+1st$, counting from either end). The rest of the graph can be built up from this skeleton by a recursive process. Roughly, that process amounts to compressing the skeleton down to a small fraction of its size, distorting its vertical and horizontal scale in the process, and inserting this shrunken skeleton in between neighboring "ribs" of the large skeleton. When appropriately shrunken skeletons have been inserted between each pair of ribs, then the process is reiterated on the next level down; and this must continue indefinitely. Our goal is to turn this picturesque description into a precise description, and then to extract physical consequences from this weird structure. For this, we need the three statements:

Statement I. At the height inside any cell where its local variable equals a pure case $1/N$ or $1-1/N$, there are N bands between the left and right borders of the cell. (In unit cells, when N is even, there seem to be only $N-1$ bands, because the two centermost bands touch in the middle, where $\epsilon=0$.) As N goes to infinity, the ratio of band size to gap size goes to zero (in other words, the bands become negligibly thin, compared to the gaps). Furthermore, the pure-case bands are distributed in such a way that the entire length of each cell roof and cell floor is approached, in the limit that N goes to infinity. Moreover, the number of pure-case bands per unit energy interval is a slowly varying and roughly constant function; that is, there is no clustering of the bands belonging to a pure case.

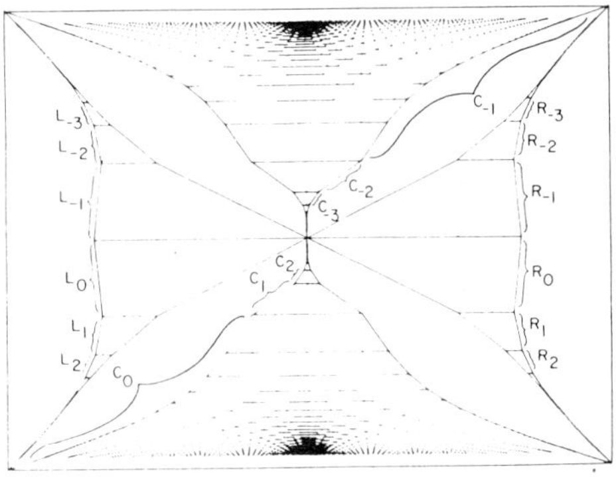

FIG. 2. Unit cell, shown with a "skeleton": the spectra belonging to pure cases $\beta=1/N$ and $\beta=1-1/N$, as well as the center band belonging to special cases $\beta=N/(2N+1)$ and $\beta=(N+1)/(2N+1)$. The L chain, C chain, and R chain are shown, all consisting of subcells formed by joining bands in the "skeleton" by straight-line segments. The labeling scheme for subcells in the three chains is indicated.

At heights where the local variable equals a special case, there is a set of bands, of which only the centermost is of interest here. The width of these center bands approaches zero as N goes to infinity. When upper special cases are considered, these bands approach a limit point, which is the inner edge of one of the two bands at $\frac{1}{2}$; when lower special cases are considered, the limit-point is the inner edge of the other band at $\frac{1}{2}$.

The next two statements involve the concept of "subcells," which are at the core of the recursive description of the graph's structure; but the concept of subcells can only be defined after the "skeleton" has been introduced (statement I). This is the reason that the following definition has been sandwiched between statements. It is best understood with the help of Fig. 2.

Rules for Subcell-Construction. The L and R subcells of any cell are formed as follows: Connect the edges of the outermost bands of neighboring pure cases by straight lines. The trapezoidal boxes thus created form the "L chain" and the "R chain" (on the left- and right-hand sides of the cell).

The C subcells of any cell are formed as follows: Connect the outer edges of the next-to-outermost bands of pure cases with $N>2$ by straight lines. This will produce two large boxes whose sides are unions of infinitely many straight-line segments. The remaining C subcells are formed by joining the centermost bands of neighboring special cases by straight lines. All the C subcells taken together form the "C chain." Each subcell has a unique label; the labeling scheme is shown in Fig. 2. We now affirm the existence of large empty swaths crossing the graph.

Statement II. The regions of a cell outside its subcells are gaps (contain no bands or portions of bands).

Finally, statement III contains the essence of the recursive nature of this graph.

Statement III. Each subcell of any cell can be given its own local variable, defined in terms of the local variable of the parent cell. (See below.) Each subcell, when indexed by its own local variable, is a cell in its own right, in that it satisfies statements I, II, and III.

The subcell's local variable is defined as follows: Let β be the "outer" local variable (i.e., that of the parent cell), and β' be the "inner" local variable. Assume first that $\beta \leq \frac{1}{2}$. Then let N be defined by

$$N = [1/\beta] .$$

(Note: The notation $[x]$ stands for the greatest integer less than or equal to x; it follows that N is the denominator of the pure case just above β.)

If the subcell is of L type or R type, then the equation relating β and β' is

$$\beta = (N + \beta')^{-1} .$$

(Note how this forces β' to lie between 0 and 1.) Let us denote the function of β which yields this value of β' by "$\Lambda(\beta)$."

If the subcell is of C type, then the relation between inner and outer local variables is

$$\beta = (2 + 1/\alpha')^{-1} ,$$

$$\beta' = \{\alpha'\} \quad \text{(fractional part of } \alpha') .$$

Let us denote this function of β by "$\Gamma(\beta)$."

Finally, if β is between $\frac{1}{2}$ and 1, then β' is equal to the value of β' which belongs to $1 - \beta$.

The statements are a little startling; they need evidence. In Figs. 3 and 4 are plotted two "rectangularized" subcells of a unit cell, namely L_2 and C_2. A "rectangularized" cell is made from the cell itself by a family of one-dimensional linear transformations. There is a linear stretching at each height, which makes the effective width of the cell be the same at every height (like a unit cell); and the bands as stretched in that way are then plotted using the cell's own local variable, rather than that of its parent cell, as the vertical axis. The characteristic butterfly pattern of the large gaps is very obvious in the rectangularized graphs. Note, however, that pure cases with even denominators inside the L cell do not possess the "degeneracy" property (of having two bands which "kiss" at the center).

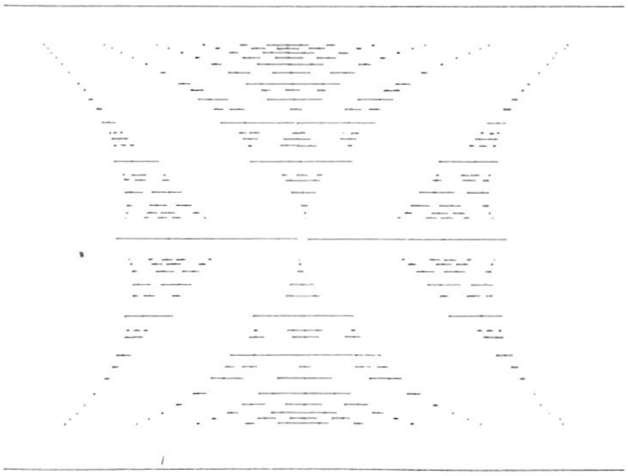

FIG. 3. Rectangularization of L_2.

FIG. 4. Rectangularization of C_2. The number of bands calculated was much smaller, which explains why so little detail is visible. All the bands shown belong to the pure-case part of the skeleton of this subcell. (Compare Fig. 2.)

The recursive structure as here presented confirms in the main (but differs in detail with) the important but extremely difficult article by Azbel´,[10] which states that the spectrum is entirely determined by the continued fraction of α. The connection is through the Λ function. If the local variable function Λ is iterated, one obtains the following representation for β:

$$\beta = \cfrac{1}{N_1 + \cfrac{1}{N_2 + \cfrac{1}{N_3 + \ddots}}} ,$$

which is unique, and will terminate for any rational α. Azbel´ predicts that spectrum(α) will consist of N_1 bands, each of which breaks up into N_2 subbands, each of which breaks up into N_3 subbands, and so on. This is approximately the same as our result, when all of the N's are large. Our prediction is that the L and R cells will each contain N_1 bands, but the number inside the C cell is not given by this expansion. As the nesting continues, N_2 subbands are indeed found in the L and R subcells of each of the L and R cells, but in the C subcells, once again there is no simple prediction based on the continued-fraction expansion. Qualitatively, though, Azbel´'s prediction contains the essence of the structure, and is very intuitively appealing.

From this recursive breakdown of the graph there

follow a number of theorems, most of which involve somewhat tedious topological reasoning (the proofs in complete detail are worked out in the author's thesis[14]). First of all it is important to be able to pinpoint any particular cell, no matter how deeply it is nested inside other cells. A simple notation will do this for us: the outermost cell is written first, followed by successively shrinking cells inside it . . . . For example, "$U_7L_{-2}C_0R_3L_1$" stands for a cell-in-a-cell-in-a-cell-in-a-cell-in-a-cell. The subscripts are to be interpreted as shown in Fig. 2. ("U_N" stands for the unit cell where $[\alpha]=N$. However, the notation for the unit cell is usually omitted, since all unit cells are identical.)

A result which is quite difficult to establish is the simple fact that all cells are (nearly) homeomorphic to each other. (Homeomorphisms are the topological version of isomorphisms: a homeomorphism is a one-to-one continuous mapping between two manifolds whose inverse is also continuous.) The "nearly" has to be included since there is a feature which could not be preserved under a continuous mapping, and that is the "degeneracy" at rationals with even denominators which exists in unit cells, but not in L or R cells. This means that there is a "branch cut" across which the homeomorphism does not carry. To be precise, each cell can be cut into two pieces — a left and a right half. For unit cells, the dividing line is merely the vertical line at $\epsilon=0$; for other cells, the dividing line can be defined in terms of the center bands of rationals with odd denominators. The left and right halves of any cell, as determined by its dividing line, are homeomorphic to each other and to the halves of every other cell as well. However, the homeomorphism can only be extended over the line in case both cells are of the same type, in the sense that they share the property of degeneracy, or share the property of its absence.

V. HOW THE BANDS ARE CLUSTERED

We now can make a precise definition of the cluster patterns. Suppose we wish to describe the distribution of bands at the value $\alpha=p/q$. Let $\beta=\{\alpha\}$, so that β is the local variable for the unit cell to which α belongs. The recursive decomposition tells us that the spectrum at β consists of three parts, which must be separated by gaps: one inside an L subcell, one inside a C subcell, and one inside an R subcell. Furthermore, the L and R subcells contain bands at that height with a Π belonging to $\Lambda(\beta)$, and the C subcell contains bands at the height with a Π belonging to $\Gamma(\beta)$. In other words, the Π at α consists of three Π's, from right to left, belonging to $\Lambda(\beta)$, $\Gamma(\beta)$, $\Lambda(\beta)$, respectively. Let us take the example of the value $\alpha=\frac{5}{17}$. Its spectrum is shown below:

-- · --   -- --- --   -- · --

A suggestive symbolic representation for the cluster pattern is

$$(2\text{-}1\text{-}2)\text{-}(2\text{-}3\text{-}2)\text{-}(2\text{-}1\text{-}2).$$

The five bands on either side are located inside the L and R chains; the central seven are located inside the C chain. The reason the breakdown is 5-7-5 is explained recursively as follows:

For the L and R subcells, the local variable is given by

$$\tfrac{5}{17}=(N+\beta')^{-1}=(3+\tfrac{2}{5})^{-1},$$

so that $\beta'=\Lambda(\beta)=\frac{2}{5}$. The demoninator is 5, hence we expect to see 5 bands inside L_1 and R_1.

For the C subcell, the local variable is given by

$$\tfrac{5}{17}=(2+1/\alpha')^{-1}=(2+1/\tfrac{5}{7})^{-1},$$

which yields

$$\beta'=\{\alpha'\}=\{\tfrac{5}{7}\}=\tfrac{5}{7}.$$

The analysis then "predicts" that the spectrum at $\alpha=\frac{5}{17}$ will consist of a set of five bands belonging to the local variable $\frac{2}{5}$; then a gap; then a set of seven bands belonging to the local variable $\frac{5}{7}$; then another gap; then another set of five bands belonging to the local variable $\frac{2}{5}$. But the analysis can be carried further, because the very same operations can be carried out inside the subcells, starting with their local variables and deriving local variables which are even more local. For $\frac{2}{5}$ and $\frac{5}{7}$, this gives

$$\Pi(\tfrac{2}{5})=\Pi(\tfrac{1}{2})\Pi(0)\Pi(\tfrac{1}{2}),$$

$$\Pi(\tfrac{5}{7})=\Pi(\tfrac{1}{2})\Pi(\tfrac{1}{3})\Pi(\tfrac{1}{2}).$$

It is useful to adopt the notation "N" as shorthand for "$\Pi(1/N)$," because, according to statement I, the bands belonging to $1/N$ are smoothly spread out across the cell to which they belong, with no clustering. And $\Pi(0)$ is denoted "1" because at $\beta=0$ there is only one band. With this shorthand, then, we can write

$$\Pi(\tfrac{2}{5})=2\text{-}1\text{-}2,$$

$$\Pi(\tfrac{5}{7})=2\text{-}3\text{-}2.$$

And these Π's can then be stuffed back into the original Π for $\frac{5}{17}$:

$$\Pi(\tfrac{5}{17})=(2\text{-}1\text{-}2)\text{-}(2\text{-}3\text{-}2)\text{-}(2\text{-}1\text{-}2).$$

This coincides with what our eye told us. There is a guarantee that this recursive analysis of Π's will come to an end, because the two operations which produce new local variables always reduce the numerator or denominator of the input fraction. In the end, one must eventually wind up with pure cases, or zero. The number of levels which one must descend before this process terminates is, however, rather difficult to predict.

VI. SPECTRA BELONGING TO IRRATIONAL FIELDS

The only case in which it is easy to predict what will happen is if you begin with an irrational value of α. In that case, the two operations yield new irrational values, which in turn yield irrational values, etc., ad infinitum. This leads to the very interesting question, "What is left — if anything — in the spectrum of an irrational field, according to this process?" Readers who are familiar with the pathology of point sets may already be anticipating the answer: there is indeed something left, and it is homeomorphic to the Cantor set. (The Cantor set is an uncountable yet measure-zero set of reals in an interval; see Ref. 15 for a detailed exposition of its fundamental properties.)

To demonstrate this starting from the three statements, one looks at the sequences of nested cells which are created by the repeated recursion in statement (iii). That is, given the original irrational α, one knows that its spectrum is confined to some particular unit cell. Statement (iii) says that the confinement can be further specified, as being inside three particular subcells of that cell. Reapplication of statement (iii) creates more deeply nested confining cells; for rational α the process terminates, but for irrational α the end product is uncountably many different infinite sequences of nested cells. It is a well-known theorem of topology that any nested sequence of closed intervals whose lengths tend to zero contains a unique limit point; its two-dimensional generalization to closed sets whose maximum dimension shrinks to zero is immediate, and that theorem is what tells us that the spectrum belonging to any irrational value of α consists of uncountably many points, between any pair of which there is a finite gap. Rigorous topological analysis establishes that the spectrum is indeed homeomorphic to the Cantor set.

It is legitimate to question whether these values of ϵ are actually eigenvalues of the difference equation, i.e., whether, in fact, the wave function $g(m)$ does remain bounded as m goes to infinity. Numerical work suggests that the answer is yes: such values really are the eigenvalues. It would be highly interesting to see a rigorous proof of this fact, or a refutation. Until proven wrong, however, we shall adopt this construction via recursion as the definition of the spectrum belonging to an irrational field value.

VII. MAGNETIC FIELD FLUCTUATIONS CREATE A "BLURRED GRAPH"

When this is done, we have finally achieved an important result: we have found a spectrum for every single value of α. Now the crucial question is, "How physical is this spectrum?" After all, it still remains true that the spectrum at a rational p/q consists of q bands, and q is still a highly fluctuating function of p/q. One can still feel suspicious of the graph. Despite the intellectual misgivings, though, the eye sees something rather continuous. There is something to this visual insight, and it can be stated formally in the following continuity theorem, which has been proven in the author's thesis: For any α, as α' approaches α, then all points of spectrum(α) are approached by points belonging to spectrum(α'); furthermore, only the points of spectrum(α) are so approached.

This theorem confirms the eye's assessment, that vertical motion along the graph is "continuous," in some sense; yet there is something discontinuous about vertical motion as well. Define $M(\alpha)$ to be the Lebesgue measure of spectrum(α). For all rational α, $M(\alpha)$ is positive, since every rational has bands of positive length. But for all irrational α, $M(\alpha)$ is zero. Therefore M has very peculiar behavior: at rational values, M is discontinuous, since there are irrationals arbitrarily near any rational; yet at irrational values, M is continuous. [The proof of this latter statement can be found in the author's thesis; it depends on a careful examination of how spectrum(α') is determined by sequences of nested cells, when α' is taken to be arbitrarily close to irrational values of α.] The function M is continuous at all irrationals, discontinuous at all rationals. This is a direct consequence of our recursive picture, and once again makes one wonder whether the graph is physically meaningful, or not.

Fortunately, there is a very simple resolution to this problem, consisting in the observation that every physical parameter has an experimental uncertainty in it, which smears it over some interval. Thus, the magnetic field, no matter how carefully controlled, has some fluctuations, which may be terribly small. This suggests the following concept: form a union of the spectra of all α within a "window" of height $\Delta\alpha$. This can be thought of as a blurred version of the graph, created by rapid up-down jiggling, where the amplitude of the jiggling is given by $\frac{1}{2}\Delta\alpha$. A blurred graph is shown in Fig. 5, using $\Delta\alpha = \frac{1}{100}$. As you can see, the result of the smearing-process yields a graph with a radically simplified appearance. It can be proven that the number of bands in any smeared graph is bounded by the constant $1/\Delta\alpha + 1$, for all α, and that the band edges change smoothly with

FIG. 5. One quadrant of the smeared graph created by using $\Delta\alpha = \frac{1}{100}$.

α. This establishes a totally continuous behavior for all magnetic field values, as a consequence of the imprecision of the field value. This also gets rid, of course, of the measure anomaly. As $\Delta\alpha$ approaches zero, the fine structure of the graph is bit by bit recovered; the infinitely fine-grained detail never returns (for positive $\Delta\alpha$), but more and more of it is revealed by decreasing the uncertainty $\Delta\alpha$. Of course, at the unphysical value $\Delta\alpha = 0$, the entire graph returns.

VIII. CORRESPONDENCE WITH RAUH'S LANDAU-LEVEL APPROACH

One unexpected feature of the recursive nature of the graph is how it corroborates a picture set forth by Rauh concerning the broadening of Landau levels when a periodic potential is "turned on."[11,13] In Rauh's work, the simplest possible two-dimensionally periodic potential, $V(x,y) = 2V_0(\cos kx + \cos ky)$, is chosen as a perturbing potential acting on an electron in an initially pure Landau state. The same difference equation arises, with a totally different interpretation: $g(m)$ represents the amplitude of a Landau state of fixed principal quantum number, whose center of localization along the axis of square integrability is ma/α (with α as we have defined it), and ϵ is proportional to the energy splitting. More interesting is the interpretation of α. Instead of measuring the flux in flux quanta, it measures the reciprocal of that quantity. Therefore, a large α in Rauh's equation means a small field. One can use the equations linking inner and outer variables to establish a link between Rauh's conclusions and our graph. This is done as follows. Observe the way the L chain turns into a very thin line as it approaches the Bloch band; it is so thin that it resembles a single level, split by a perturbation. Therefore, we choose to identify the leftmost band with a Landau level, perturbed by the periodic potential of the crystal. The splitting of the band is present in our picture, since in reality the line is composed of very thin L cells. At height β (assuming $\beta < \frac{1}{2}$), the structure inside each of those cells is given by spectrum (x), where

$$\beta = (N + x)^{-1}.$$

Here, N is integral, and x is between 0 and 1. Now by the first symmetry property of the graph,

$$\text{spectrum}(x) = \text{spectrum}(N + x) = \text{spectrum}(1/\beta).$$

Notice that this says that the split-up of the lowest-lying Landau level is given by the same eigenvalue equation, but with parameter $1/\beta$ instead of β. This is completely consistent with Rauh's work. Moreover, one can identify other chains in the graph with Landau levels, and under this identification, it turns out that each one of them splits up in a pattern given by spectrum $(1/\beta)$. The natural candidate for the 2nd-lowest Landau level is the L chain located inside C_0; the 3rd lowest is the L chain inside C_0C_0, and so on. The number of such levels is essentially $1/\beta$; half of them are L chains inside nested C cells, and the other half are their symmetric counterparts: R chains inside nested C cells. To determine how any one of them is split, we must iterate the formation of local variables. In particular, to derive the splitting of the nth Landau level, we must begin with β, apply the Γ function $n-1$ times to it, and finish by taking Λ of the result. As before, let

$$\beta = (N + x)^{-1},$$

with N integral, and x between 0 and 1. Further, assume N is at least 4. Then by definition, $\Lambda(\beta) = x$. Simple calculation shows also that

$$\Gamma(\beta) = [(N-2) + x]^{-1}.$$

From this expression, we can directly read off $\Lambda(\Gamma(\beta))$: it is also x. Now if $N-2$ is also at least 4, then we can immediately get

$$\Gamma(\Gamma(\beta)) = [(N-4) + x]^{-1},$$

and Λ of this is, once again, x. So it will go, with 2 being subtracted from the integer in the denominator over and over again, as long as that integer stays 4 or more. When β is small (i.e., when N is big), then the number of Γ's which can be iterated before the integer ceases to satisfy that condition is roughly $\frac{1}{2}N$. This implies that there are roughly $\frac{1}{2}N$ Landau levels to the left of center, and symmetrically, $\frac{1}{2}N$ to the right of center, all N of which are split according to the pattern of spectrum(x) —but as before, spectrum(x) and spectrum$(1/\beta)$ are identical. Therefore all the Landau levels do split in a similar way. And we have shown that their number is roughly N, which is to say $1/\beta$. Therefore the separation between Landau levels is roughly $8\beta E_0$. This corresponds to the spacing which one can calculate using an effective-mass approximation at the edges of the band. Altogether, the Landau-level-based theory and the Bloch-band-based theory achieve in this way a satisfying harmony.

IX. WAVE FUNCTIONS AND IRRATIONAL FIELDS

In certain other approaches to this problem, notably those based on the magnetic translation group,[1,2] the rationality of α is forced if one seeks representations of the magnetic translation group by the Frobenius method, which involves finding an invariant subgroup. For some subgroup of magnetic translations to be invariant, all of its members must commute, and this in turn forces certain phase factors, involving the flux through the parallelogram defined by the two translations, to be unity. The end result is that one must choose a rational value p/q for α, and the invariant subgroup consists of "superlattice" translations, where the superlattice consists of lattice points separated from each other by q lattice spacings in both x and y directions. That way, the amount of flux is always an integer, the phase factors are always unity, and the subgroup of magnetic translations is indeed invariant. The problem with this whole approach is that such superlattices can only be defined in the case of rational fields, and there seems to be no obvious way to extend the results to irrational fields.

An alternative type of "superlattice" can be formulated, however, which comes up naturally in the context of our difference equation. One begins with the observation that there are solutions of the Bloch-Floquet type to the difference equation — that is, solutions with the property that

$$g(m+nP)=e^{inkP}g(m),$$

where n is any integer, P is a constant, and k is a wave number. One's first guess might well be that P must be an integer, corresponding to moving through an integral number of lattice spacings. This assumption is erroneous, however; P need not be integral. Indeed, the correct minimal period P is $1/\alpha$, which may be any real whatsoever, rational or irrational. This is proven in exactly the same way as for the Mathieu equation, of which our difference equation is, in some senses, a discrete counterpart.

The crucial fact in the proof is that the coefficients in the difference equation are themselves periodic in the variable m, with period $P=1/\alpha$. Therefore when m is replaced by $m+P$ in the difference equation, $g(m)$ becomes $g(m+P)$ but the coefficients are unaltered, which says that if $g(m)$ is a solution, then so is $g(m+P)$. Now there are two linearly independent solutions to a difference equation of second order (which ours is); let them be $g_1(m)$ and $g_2(m)$. Then $g_1(m+P)$ and $g_2(m+P)$ are also solutions; but since $g_1(m)$ and $g_2(m)$ form a basis, there must be numbers C_{ij} such that

$$\begin{pmatrix} g_1(m+P) \\ g_2(m+P) \end{pmatrix} = \begin{pmatrix} C_{11} & C_{12} \\ C_{21} & C_{22} \end{pmatrix} \begin{pmatrix} g_1(m) \\ g_2(m) \end{pmatrix}.$$

Now we can find a linear transformation which will diagonalize the 2×2 matrix; this transformation will mix g_1 and g_2 to produce new functions g_1' and g_2' with the property that

$$g_n'(m+P)=cg_n'(m)\quad (n=1,2),$$

where c is an eigenvalue of the C matrix. This proves the Bloch-Floquet theorem for the difference equation, with period $1/\alpha$. A corollary is that any solution $g(m)$ can be expressed as

$$g(m)=e^{i\mu m}G(m),$$

where

$$e^{i\mu P}=c,$$

and where $G(m)$ is a periodic function of period P. When c is 1 (which happens, as in the Mathieu equation, at one edge of each band), then $g(m)=G(m)$, so that the difference equation has a purely periodic solution of period P. In any case, the distance $1/\alpha$ plays the role of a fundamental period associated with the difference equation.

The difference equation per se allows us only to determine $G(m)$ when m is an integer. But the periodicity of $G(m)$ allows us to interpolate between integers, and to determine G there also. This comes about because the period $P=1/\alpha$ is (in general) not an integer. Suppose, for instance, that $\alpha=\frac{5}{17}$. Then the period is of length $\frac{17}{5}$. Now $G(0)$, $G(1)$, $G(2)$, and $G(3)$ all fall within one period, but $G(4)$ is beyond $G(\frac{17}{5})$, and hence equals $G(\frac{3}{5})$. Similarly, $G(5)=G(\frac{8}{5})$, $G(6)=G(\frac{13}{5})$, and so on. Finally, $G(17)=G(0)$ and the whole cycle starts over again. Therefore, we can plot the values $G(0)$ through $G(17)$ inside one period of length P; they will appear in some rearranged order. The two orders and their relation are shown below. Integer order:

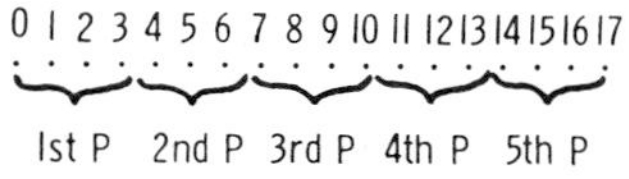

In the figure below, the five complete periods shown above are superimposed, to give the rearranged order (note that the scale of the two figures is different):

0 7 14 4 11 1 8 15 5 12 2 9 16 6 13 3 10 0

The sequence of integers in the rearranged order is the successive multiples of 7, taken modulo 17. This is because 7 occurs exactly $\frac{1}{17}$ beyond the period boundary in the upper figure, and $\frac{1}{17}$ is the minimum distance possible. The general rule for the rearranged order when $\alpha = p/q$ is to take the multiples of $\tilde{p}$ (modulo q), where $\tilde{p}$ is defined by the congruence

$$p\tilde{p} = 1 \pmod{q}.$$

So far we have concentrated on what happens when α is rational; but the same process of folding back all values of $G(m)$ into one period of length P can be carried out. In the irrational case, however, the reordering will create a dense distribution of points inside the whole period. This is one place where irrational fields seem to make more physical sense than rational fields, in that one can determine the values of their wave functions on a dense set, rather than at just a discrete set.

If one takes a sequence of rational values α_n which approach an irrational value (and whose denominators therefore must go to infinity), the various periods $1/\alpha_n$ are all approximately the same, and it is therefore possible to compare the reordered wave functions of these rationals, to see if some trend emerges, pointing the way to the reordered wave function at the irrational field. One must also be sure to choose eigenvalues which are very close to each other; that this can be done is a consequence of the continuity theorem stated above. Such a process of comparison was carried out numerically for the following sequence of fractions (shown with their continued-fraction expansions) and their largest eigenvalues:

$$\alpha = \tfrac{1}{5} \quad (\epsilon = 2.9664),$$

$$\alpha = \tfrac{2}{11} = \cfrac{1}{5 + \frac{1}{2}} \quad (\epsilon = 3.028\,50),$$

$$\alpha = \tfrac{17}{93} = \cfrac{1}{5 + \cfrac{1}{2 + \frac{1}{8}}} \quad (\epsilon = 3.023\,983\,268).$$

The wave functions in rearranged order are shown in Fig. 6. It appears that an overall shape is established by the fraction with a low denominator (in this case $\frac{1}{5}$), and details of the shape are determined by fractions with higher and higher denominators. Note how these fractions, which are close in value but which have very different denominators, have magnetic periods P of very nearly the same length, which allows the direct comparison of their wave functions. This figure is strong evidence for the idea that the limiting case — namely, the wave-function for an irrational α — is a continuous function which can be obtained from the discrete points supplied by the difference equation by translating them all into a single magnetic period of length $1/\alpha$.

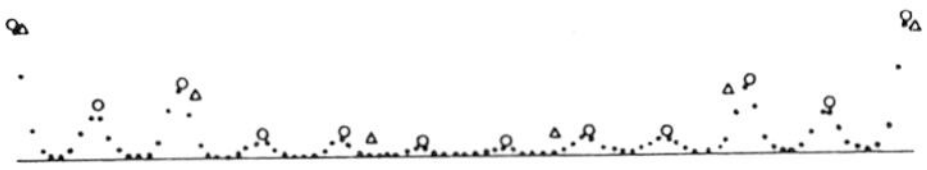

FIG. 6. Values of the wave function inside one magnetic period $P = 1/\alpha$, shown for three values of α (and their largest eigenvalues): triangles: $\alpha = \frac{1}{5}$ ($\epsilon = 2.9664$); circles: $\alpha = \frac{2}{11}$ ($\epsilon = 3.028\,50$); dots: $\alpha = \frac{17}{93}$ ($\epsilon = 3.023\,983\,268$). The x axis represents, in each case, a physical distance of $P(=1/\alpha)$ lattice spacings. The vertical scale is such that the highest dot represents the value 1.

In this connection, it is also interesting to point out that the one-dimensional "superlattice" of period $1/\alpha$ can be related to the magnetic translation group, in our model. It can be verified easily that all Landau-gauge magnetic translation operators[1]

$$T_{\text{mag}}(\vec{r}) = e^{i\vec{r}\cdot(\vec{p} + e\vec{A}'/c)/\hbar},$$

where

$$\vec{A}' = H(-y, 0, 0)$$

commutes with the effective Hamiltonian defined earlier. (This is not the case with the true Hamiltonian for a crystal electron in a magnetic field.) However, magnetic translation operators do not in general commute with other magnetic translation operators. The condition of commutation is that the parallelogram which they define should intercept an integral number of flux quanta. Now since the Landau gauge leads naturally to a one-dimensional mathematical treatment in which all the interesting phenomena happen along the x axis, it would seem natural to look for a commuting set of magnetic translation operators whose y spacing is "trivial" (i.e., is based on the lattice spacing), and whose x spacing contains information about the field. If we allow any magnetic translation in the y direction as long as it is through an integral number of lattice spacings, then the commutation condition quantizes the allowed magnetic translations along x; and the condition is precisely that they must be through $1/\alpha$ lattice spacings. The reason for this is that a rectangle of dimensions a/α along x, and a along y intercepts precisely one flux quantum.

This observation suggests that the best choice of unit cell for a "magnetic superlattice" may not be a square of q lattice spacings on a side (which only can be done for rational fields), but rather, a rectangle with one side equal to the lattice spacing, and the other side such that exactly one flux quantum is intercepted. And that is the superlattice defined by the period P for the difference equation.

Somewhat related to these ideas is the article by Chambers,[12] in which orbits, "hyperorbits" (and so on) are discussed. In particular there is a discussion of the correspondence between simple orbits and bands, "hyperorbits" and subbands, and so on.

X. POSSIBLE EXPERIMENTAL TEST

Finally, I would like to comment on the possibility of looking for the features predicted by this model experimentally. At first glance, the idea seems totally out of the range of possibility, since a value of $\alpha = 1$ in a crystal with the rather generous lattice spacing of $a = 2$ Å demands a magnetic field of roughly 10^9 G. It has been suggested, however (by Lowndes among others), that one could manufacture a synthetic two-dimensional lattice of considerably greater spacing than that which characterizes real crystals. The technique involves applying an electric field across a field-effect transistor (without leads). The effect of such a field is to drive electrons (or holes) to one side of the device, where they will crowd together in a thin layer, essentially creating a two-dimensional gas of charged particles. Now if the device is prepared in advance with a dielectric layer which is nonuniform, and which in fact is periodic in each of its two dimensions, then the two-dimensional gas will be moving in a periodic potential that can be manufactured to fit any specifications. In particular, one could make a tight-binding model so that the electronic energy bands are approximately given by our simple sum of two cosines. Moreover — and this is the crux of the idea — one can choose the lattice spacing; thus with a spacing of 200 Å instead of 2, a magnetic field of 100 kG gives a value of α equal to 1. All that remains to be done is to apply a uniform magnetic field perpendicular to the plane of the gas, and to measure the transitions when the sample is irradiated with electromagnetic radiation of various wavelengths. This is not to say that the idea is easy; but such an intriguing spectrum deserves a good experimental test.

ACKNOWLEDGMENTS

Most of the ideas set forth in this article were worked out during the author's stay at the Fachbereich Physik of the Universität Regensburg (W. Germany). In particular, the author enjoyed the hospitality of the Lehrstuhl Obermair, and worked in collaboration with the computer "Rumpelstilzchen," as well as with Professor G. Obermair, Professor A. Rauh, and Professor G. Wannier. The latter was the author's thesis advisor, and contributed many valuable comments and ideas. R. Boeninger and F. Claro provided much help in various ways. Professor R. Donnelly of the University of Oregon generously allowed me to use a small computer without which the work would never have gotten done. I would also like to thank the following individuals for conversations and ideas: Professor F. Bloch, Professor P. Csonka, Professor M. Demianski, Professor R. Feynman, Professor D. Lowndes, Professor J. McClure, Professor M. Moravcsik, Professor A. Nagel, and Professor R. Wallis. Finally, I would like to thank Professor P. Suppes, Director of the Institute for Mathematical Studies in the Social Sciences at Stanford University, for the hospitality afforded me while I was writing this article.

---

*Work supported by the NSF under Grant No. GH 39027.
†Present address: Institute for Mathematical Studies in the Social Sciences (Ventura Hall), Stanford University, Stanford, Calif. 94305.

[1]E. Brown, Solid State Phys. 22, 3313 (1968).
[2]F. A. Butler and E. Brown, Phys. Rev. 166, 630 (1968).
[3]R. E. Peierls, Z. Phys. 80, 763 (1933).
[4]J. M. Luttinger, Phys. Rev. 84, 814 (1951).
[5]W. Kohn, Phys. Rev. 115, 1460 (1959).
[6]G. H. Wannier, Rev. Mod. Phys. 34, 645 (1962).
[7]E. I. Blount, Phys. Rev. 126, 1636 (1962).
[8]W. Y. Hsu and L. M. Falicov, Phys. Rev. B 13, 1595 (1976).
[9]P. G. Harper, Proc. Phys. Soc. Lond. A 68, 874 (1955).
[10]M. Ya. Azbel', Zh. Eksp. Teor. Fiz. 46, 939 (1964) [Sov. Phys.-JETP 19, 634 (1964)].
[11]A. Rauh, Phys. Status Solidi B 65, K131 (1974).
[12]W. G. Chambers, Phys. Rev. 140, A135 (1965).
[13]A. Rauh, Phys. Status Solidi B 69, K9 (1975).
[14]D. R. Hofstadter, Ph.D. thesis (University of Oregon, 1975) (unpublished).
[15]J. G. Hocking and G. S. Young, *Topology* (Addison-Wesley, Reading, Mass., 1961).

PHYSICS REPORTS (Review Section of Physics Letters) 110, Nos. 5 & 6 (1984) 279–291. North-Holland, Amsterdam

Quantized Hall Effect in Two-Dimensional Periodic Potentials

David THOULESS
Physics Department, University of Washington, Seattle, WA 98195, U.S.A.

Unlike a number of the previous speakers, I have not had very much direct contact with Gregory Wannier, but I have been an admirer of his work, and it is an honor to be asked to talk at this symposium. As you will see, the work that he and his colleagues have done, particularly in recent years, has been one of my major sources of inspiration for the work that I shall talk about this afternoon.

To my mind, the discovery of the quantization of Hall conductance by von Klitzing, Dorda and Pepper [1] is one of the most exciting things that has happened in condensed matter physics in recent years. I regard it as particularly challenging because it is an experimental result, the quantization of something, which the theorists have completely missed, and there are no indications, as far as I know, in the literature that there should be accurate quantization of the Hall current. The preliminary experiments were done with sufficient precision that in the rather quickly produced paper that von Klitzing et al. published in 1980 they were able to quote the precision of 1 part in 10^5 for quantization. Rapidly, of course, people have pushed this several orders of magnitude better, so it is now competitive with the best determinations of the fine structure constant by other methods.

The work that I shall be talking about involves the use of strong magnetic fields on two-dimensional electron systems. Two systems have actually been used so far. One consists of the electrons on a crystalline silicon surface (in a MOSFET). In this nice system you can control the number of electrons by controlling the gate voltage. The other system that has been used is the junction between gallium arsenide and gallium-aluminum arsenide; here you cannot control the number of electrons so readily but you have higher mobilities and you can vary the field while keeping the electron number constant. What was seen in these structures is shown in fig. 1. The x-coordinate shows the gate voltage, which is proportional to the Fermi energy, so the number of electrons increases to the right. In the vertical direction one curve shows the longitudinal voltage, which is proportional to the resistance, while the other curve shows the transverse voltage, proportional to the Hall resistance. The two are drawn on different scales. In fact, the transverse voltage is always much greater than the longitudinal voltage, which means that the longitudinal component of resistance is proportional to the longitudinal component of conductance, a slightly confusing result arising from the fact that when you divide by the determinant of σ, the σ_{xy} term predominates. What you see is a series of steps in the Hall resistance which occur at points where the longitudinal resistance or the longitudinal conductance is vanishing. These are the energy gaps. These data are from a Si (100) surface, so there is a two-fold orbital degeneracy. There are two spin states and two orbital states. For lower values of the gate voltage the separate peaks corresponding to the four different states can be seen, but as one goes to higher Landau levels the structure gets washed out; first the two orbital states get smeared together, then at higher values of the gate voltage the two spin states are merged. Eventually the gaps between Landau levels

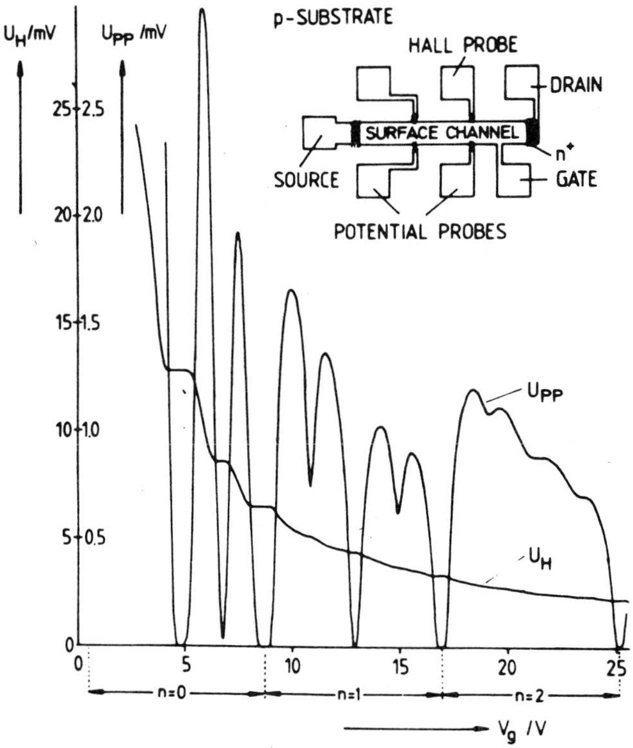

Fig. 1. Steps in the Hall voltage and minima of the longitudinal voltage corresponding to filled Landau levels of a Si MOSFET [1].

disappear. The lower spin component is not seen in this particular experiment, although I shall show some results later where more is known about those lower spin components.

The dramatic result was that these plateaus in the Hall resistance have the form

$$\rho_{xy} = h/ne^2 , \tag{1}$$

where n is an integer. The precision is remarkable for those of us who are happy to talk about 1 percent or 10 percent agreement in condensed matter physics. The fact that the QHE value is good to 1 part in 10^7 in absolute terms, and certainly in the order of 1 part in 10^8 if you do not have to worry about whether the standard resistance is properly calibrated or not, shows that it comes in the class of the very few quantization effects in condensed matter physics like the Josephson effect, the quantization of magnetic flux, and possibly the quantization of circulation in liquid helium, which demonstrate a macroscopic quantum effect. This was unsuspected by the theorists. Figure 2 shows in more detail similar results produced by Kawaji [2]. Here you see the steps in the Hall conductance which correspond to the filling of various components of the lowest Landau level. You can see the steps which occur when both components of the lowest spin state are full, when both components of the higher spin state are full, and the step between them when one orbital component of the upper spin state is full. At each of these steps the longitudinal conductance goes to zero. At this temperature the step correspond-

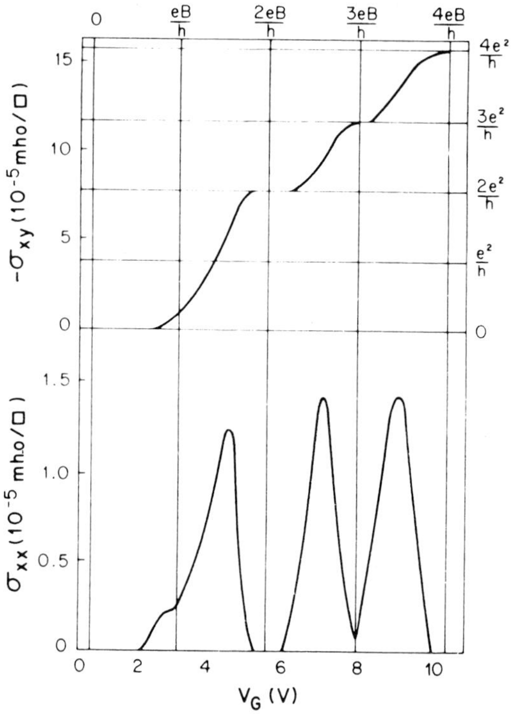

Fig. 2. Hall conductivity and longitudinal conductivity for a MOSFET in the lowest Landau level [2].

ing to filling of the lowest orbital state has disappeared, and the lowest orbital level seems to be entirely localized, and to give no contribution either to the longitudinal or to the Hall conductance.

Figure 3, also taken from Kawaji [2], shows one of the steps, at the complete filling of the lowest Landau level (all four components) in more detail. The plateau at 64530.2 ohms can be clearly seen. You can also see the longitudinal resistance going to a very low value, just where the plateau is flat.

I am not going to talk about the more recent experimental development where in higher mobility samples it has been found that there are also fractional values. That is a different story and one can deduce from the things that I am going to say today that these fractional effects must be due to many body effects. A non-interacting one-electron picture does not seem to be capable of giving fractional values except with the most implausible set of assumptions and special pleading built into it. When we started this work it did seem possible one might get fractions, but it is clear that that view was mistaken. I shall just run briefly through the elementary understanding of this but I am not going to go into details, but just remind you of the way the quantum theory of non-interacting electrons goes, first of all in the absence of a substrate. It is convenient to use the Landau gauge in which the vector potential is taken in the y direction. The solution of the Schroedinger equation is

$$\psi_{kn}(x, y) = e^{iky} f_n(x - \hbar k/eB) . \tag{2}$$

The solution is wave-like in the y direction and a harmonic oscillator in the x direction. When the

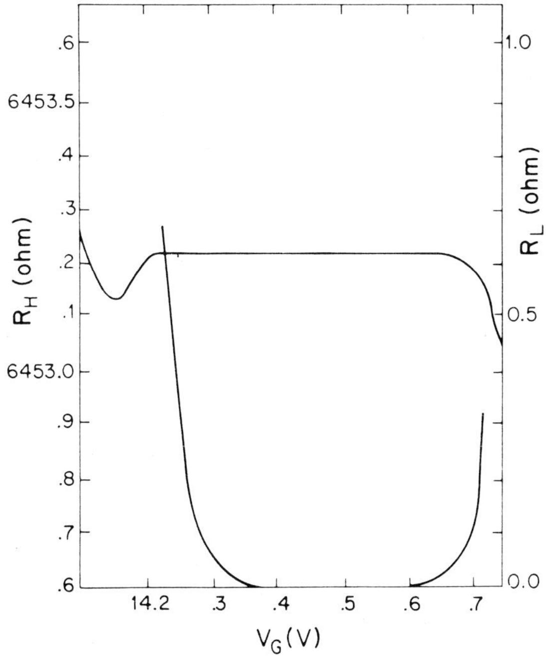

Fig. 3. High precision measurements of the components of resistance in the gap between the lowest and the second Landau level in a MOSFET [2].

electric field is put on there is a velocity which is just the ratio of electric to magnetic field and if you take full Landau levels, the current per electron, which is inversely proportional to B, cancels with the number of electrons per Landau level which is proportional to B, so the current is independent of the magnetic field but proportional to the electric field with the proportionality equal to the number of Landau levels times the famous factor 1/(25 813 ohms) which is the numerical value in eq. (1) of h/e^2. Well, that is very simple, of course, and there is nothing very surprising there. What has caught people by surprise is that this was completely insensitive to the substrate, impurities, etc. and, in fact, holds under circumstances in which this argument, as I presented it, gives you no clue that this will happen. The most important explanation of why it happened (a number of explanations were given simultaneously) was the one given by Laughlin [3] in which he used a gauge invariance argument. We will not go through this is detail, but will need the result of the argument. Consider a conducting annulus (this is Halperin's version [4] of Laughlin's argument) in a constant magnetic field with a solenoid running down the middle of it so that you can change the flux that threads the ring without changing the magnetic field. Then make a slow change in this flux which will, by gauge invariance, map the wave functions onto one another. With full Landau levels, there are no energy levels near the Fermi surface except those levels at the two edges of the system where the energy level is shifted from the bulk value and can therefore cut the chemical potential. If the flux is changed by one quantum unit, you end up with the same set of wave functions that you started with but they may be mapped differently. In particular, you may have shifted an integer number of electrons from one edge of the sample to the other.

Faraday's law provides us with an alternative way of thinking of this. Change of the magnetic flux by e/h produces an emf round the ring whose time integral is e/h. The time integral of the Hall current is the total electron transport, which is ne, according to the argument of the last paragraph. We put these two together, divide the current by the emf, and get ne^2/h whenever the Fermi energy is in a gap. Now, that says nothing about the levels being slightly distorted Landau levels or anything like that; it is a much more general argument, which is true whenever the Fermi energy lies in the gap.

Now we relate the QHE to what has been done by Wannier and his collaborators [5–8] and a number of others [9–12] on the effect of the combination of a uniform magnetic field and a periodic modulation. Consider raising the challenging question, "What happens if you split the Landau levels with a periodic potential?" This may be physically unrealistic, but you could certainly do it and there has been quite a lot of discussion recently about how you could take a MOSFET, modulate it with a periodic potential of a reasonably long wavelength, and look at what happens. My work has been carried out in collaboration with Kohmoto, Nightingale and den Nijs [13]. We knew from the earlier work that if the degeneracy of the Landau level was broken by a periodic potential, with q unit cells for every p flux quanta, then each Landau level split into p subbands. Laughlin's theorem [3] tells us that if the Fermi energy is sitting between one of these subbands into which the Landau level is split, the Hall conductance is still integer. So the result is no relation to a semiclassical result. If you did not know what Laughlin said, I think you would guess that with the Fermi energy between the subbands you would have some fractional value of the Hall conductance as you would have classically, but that is not the way that it must work out; Laughlin's result is quite clear on that issue. What we have found, of course, agrees with that. If we had disagreed with him I think it would have meant that we had done something wrong.

Figure 4 shows the intriguing figure for the square lattice that Hofstadter [6] produced for his thesis. You get the same diagram in two different ways, either by taking a tight binding model and putting a

Fig. 4. Hofstadter's diagram for the energy bands of a square lattice in the presence of a magnetic field [6]. The energy is plotted vertically and the number of flux quanta per unit cell (or its reciprocal) is plotted horizontally.

weak magnetic field on, in which case the s-band gets split into these components, or, by taking the case that I shall talk more about, where you take a Landau level and split it with a weak modulating potential. Energy is to be read on the vertical axis. In the tight binding, weak magnetic field, case the horizontal axis gives the number of flux quanta per unit cell, which goes from zero on the left side to unity on the right side. In the case of strong magnetic field, weak modulating potential, the horizontal axis gives the number of unit cells per flux quantum. You have, for example, in the weak modulating potential case one third of the way across the diagram three flux quanta per unit cell, and three subbands. One quarter of the way across you have four subbands but two of them touch in the middle, with a sort of accumulation point along where the subbands touch; for even denominators there are always two subbands touching at the center of the spectrum. Notice particularly the general structure of the diagram; there are well-defined energy gaps but the energy bands are terribly sensitive to the denominator, and are not well-behaved analytic functions of the number of flux quanta per unit cell. The gaps, however are well behaved, an important feature that I shall come back to.

Let me remind you of the formal side of this problem, in the weak modulation case. The degeneracy of a Landau level is broken by a perturbing potential $V_1' \cos(\alpha x) + V_2' \cos(\alpha y)$. The first term does not change the value of k in the wavefunction (2), and the second changes it by an amount α, so this perturbation leads to the matrix equation

$$V_2 c_{n-1} + 2V_1 \cos(n\alpha^2\hbar/eB + \nu)c_n + V_2 c_{n+1} = Ec_n , \tag{3}$$

which is known as Harper's equation [9]. The problem has mapped onto this linear chain type of equation with a modulated site energy, and then one integrates over all the phases ν of the modulation because those correspond to different values of k. You do not consider just a single chain but an ensemble of chains defined by different values of ν.

With a rational number of flux quanta per unit cell, $eB/2\pi h\alpha^2 = p/q$, the problem can be analyzed in terms of Bloch wave functions and, in fact, you get the precisely similar equation for the Bloch waves except that the diagonal and off-diagonal terms are interchanged. That leads to an important result known as Aubry duality. The dual equation has the form

$$V_1 e^{-i\nu} d_{n-1} + 2V_2 \cos(2\pi nq/p + K)\, d_n + V_1 e^{i\nu} d_{n+1} = Ed_n . \tag{4}$$

Aubry and André [14] have done some very important work based on this relation which has been exceedingly useful to us. Another way you can get it is to make a gauge transformation of the two-dimensional problem; the vector potential is taken in the x direction instead of the y direction. That, again, has the effect of interchanging V_1 and V_2, and interchanging this phase ν in the diagonal term with the Bloch wave number which I have called K. Also an important result is that if you make this duality transformation with different values for V_1 and V_2, Aubry has shown that you go over from wave functions that are localized in the x direction to wave functions that are extended in the x direction. The coordinates x and y are being interchanged, so when V_1 and V_2 are different, it is believed that the wave function will only be localized in one direction and will be extended in the perpendicular direction, in the case of an irrational number of flux quanta per unit cell. A second result of this work [14] is that the sum of the widths of the subbands seems to be equal to $4|V_1 - V_2|$. I have shown that this is a rigorous lower bound to the sum of the subband widths for rational values of the number of flux quanta per unit cell, and both numerical and theoretical arguments show that the deviation from this bound tends rapidly to zero as the numerator p tends to infinity [15]. The strange thing about this is that since V_1 and V_2 become equal as you go to the isotropic case in which the x

component and y component of the modulation are the same, the sum of the bandwidths shrinks to zero, and our numerical work shows that it shrinks to zero like the $1/p$; this result is actually implied by Hofstadter's observation of the self-similarity of fig. 4. This suggests that there is singular continuous spectrum in the case that V_1 and V_2 are equal.

To discuss the Hall effect in a periodic potential we need not assume that the modulation is weak; in general, you can define Bloch waves which, because of the magnetic field, have an extra gauge factor in them but otherwise they satisfy the standard boundary conditions. Using this generalization of Bloch waves [11], one can rewrite the Schroedinger equation for the wave function in terms of the Schroedinger equation for the coefficients u of the $\exp[i(k_1x + k_2y)]$ terms. This is the usual transformation in any solid state textbook except for the gauge. You have the effective Hamiltonian

$$H(k_1, k_2) = \frac{1}{2m}\left[\left(-i\hbar\frac{\partial}{\partial x} + k_1\right)^2 + \left(-i\hbar\frac{\partial}{\partial y} + k_2 - eBx\right)^2\right] + V(x, y), \tag{5}$$

with the velocity operators given by the partial derivatives of the Hamiltonian with respect to the k_1 and k_2, so one can use this to calculate the Kubo formula for the conductance, particularly for the transverse conductance, in the presence of a magnetic field. The result for the cross-correlation function of the two components of the current operator gives

$$\sigma_{xy} = \frac{ie^2}{A\hbar}\sum_{\varepsilon_\alpha<E_F}\sum_{\varepsilon_\beta>E_F}\frac{(\partial H/\partial k_1)_{\alpha\beta}(\partial H/\partial k_2)_{\beta\alpha} - (\partial H/\partial k_2)_{\alpha\beta}(\partial H/\partial k_1)_{\beta\alpha}}{(\varepsilon_\alpha - \varepsilon_\beta)^2}. \tag{6}$$

This is a product of dH/dk_1 times an energy denominator and dH/dk_2 times an energy denominator, and that looks suspiciously like perturbation theory for the change of the wave function with respect to k_1 and k_2, and that is indeed what it comes out as. It is the integral over the Brillouin zone

$$\int\int\left\{\frac{ie^2}{2\pi h}\left(\frac{\partial u^*}{\partial k_1}, \frac{\partial u}{\partial k_2}\right) - (\text{terms with } k_1 \text{ and } k_2 \text{ interchanged})\right\} dk_1\, dk_2, \tag{7}$$

which is a topological invariant which one of my mathematical colleagues tells me defines the first Chern class of a mapping of the Brillouin zone given by this wave function. But, from a physicist's point of view it is not so odd that this should be an integer, a topological invariant, because you can see that you can get rid of one of these integrals by using Stokes' theorem. Then you can use the periodicity in the Brillouin zone. The wave function at two opposite points of the Brillouin zone can differ only by a phase factor, so the integrand reduces to the derivative of this phase round the perimeter of the Brillouin zone, and this must give an integer multiple of 2π. An important result, which I return to later, is that if you could write the wave functions in such a way that they were analytic functions of $\boldsymbol{k}$ and at the same time single valued, the integral would be zero. You have two choices, actually, when the Hall current is non-vanishing. You can either write the wave functions as an analytic function of position in the Brillouin zone, in which case they do not match up on the boundaries (you have multiple valued functions in the extended zone scheme), or else you can define the phase uniquely only to find there must be at least one point in the Brillouin zone where your definition breaks down. That can be done by demanding that the value of the wave function at one particular point in space should be real-positive. Then, of course, you find somewhere in the Brillouin zone that particular point in real space has 0 as the value of the wave function and that is how you build in the singularity. This kind of topological behavior of the wave function is known as a texture in, for example, the work on singularities in superfluid helium-3. The singularities are put in artificially, but you cannot do things in such as way that you do not have singularities unless you put up with multiple-valued wave functions.

There is another way of writing this formula which is also useful. Instead of using the two variables k_1 and k_2 you can use the energy E and k_2, and integrate E round a contour which encloses the energy bands whose contribution to the Hall current one wants to evaluate. This gives the Hall conductance as

$$\frac{\mathrm{i}}{2\pi}\frac{e^2}{h}\oint \mathrm{d}E \int \mathrm{d}k_2 \left\{\left\langle\frac{\partial\psi}{\partial k_2}\middle|\frac{\partial\psi}{\partial E}\right\rangle - \left\langle\frac{\partial\psi}{\partial E}\middle|\frac{\partial\psi}{\partial k_2}\right\rangle\right\} . \tag{8}$$

Since the Hamiltonian is doubly periodic in these two variables the integral again gives an integer for the conductance. It is a different way of doing the same thing and may be convenient if you have overlapping bands at some point in energy space.

One question is, how to evaluate that formula. There are various tricks. I will talk about one of them which is actually particularly useful for the anisotropic model. One of the things that we did was to work out the case which the modulation in the x direction was much stronger than the modulation in the y direction and then try and continue to equal modulations. That turns out to work perfectly as far as I can tell with a square lattice. It does not work so nicely with a triangular lattice; in that case it breaks down at some points. Figure 5 shows a picture of representative points in the variable k of eq. (2), but, of course, that corresponds to the center of the wave function in x space. The figure shows the case $p = 5$ so we have five of these points in one wavelength of the modulating potential in the x direction. There is strong modulation in the x direction and weak scattering in the y direction. The modulation in the y direction gives mixing between the different points in k space. Now, what happens when we change the gauge, that is, when the flux through the solenoid is changed? All these points move along the curve and if there were no potential in the y direction they would just move smoothly along the curve and the Hall current would be associated with the fact that we have driven them a certain distance in the x direction. Now, the modulation in the y direction means that when two of these representative points which are separated by a Bragg vector of the modulating potential become almost degenerate

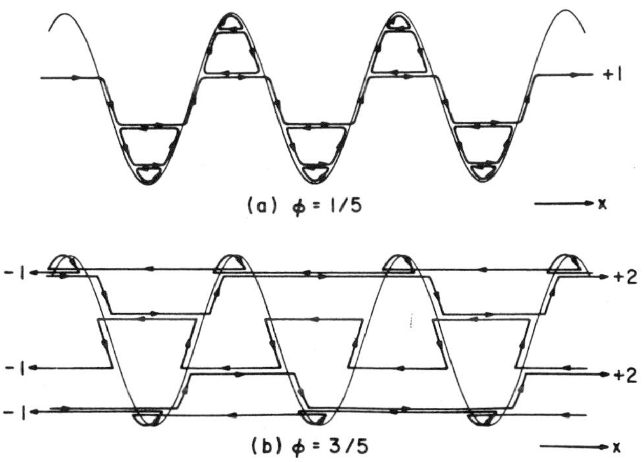

Fig. 5. Orbits in x or k space for a strongly anisotropic modulation. (a) shows $q/p = 1/5$, while (b) shows $q/p = 3/5$ [13].

they get mixed. So what happens is that Bragg scattering results in the open orbit which was represented by this sinusoidal curve being broken up into five different orbits. Four of these orbits are loops and so carry no Hall current, while the central subband has an open orbit and carries all the Hall current, so instead of having a fifth of the Hall current carried by each of the subbands, which is what you would get if there was no modulation in the y direction, you get no Hall current carried by four of the subbands, and five times as much current carried by the central subband. So one subband is doing all the work in this case. In fact, for a long time I was under the delusion that it would always be the central subband that carried all the Hall current, until we worked things out properly and realized that it was not the case. If you take five flux quanta per three unit cells then you get the situation shown in fig. 5b. The top, middle and bottom subbands each carry -1 units of current, and so are hole orbits, while the second and fourth subbands carry $+2$ units of current. The total carried by the five subbands still adds up to $+1$, so the total Hall current of the Landau level is conserved, but the subbands alternate between carrying Hall current $+2$ and -1. That is a simple version of what can happen in a more complicated form: you always find that there are two different classes of subbands which carry two different Hall currents, one positive and one negative or zero. These always add up to give a total Hall current of unity for the complete Landau level.

What I have discussed so far is the weak potential limit. In this case the electric field drives the representative point along the fixed sinusoidal curve. For the tight-binding model with a weak magnetic field you find that the electric field can be represented by keeping the points fixed in k space or x space, but moving the modulating potential at a steady rate in the x direction; any resultant Hall current is then produced by the Bragg scattering. One can then ask what would be the current if one moved both the modulating potential and the points at the same rate. The answer is very simple, since this gives a Galilean transformation of the stationary problem, and the total charge transported is equal to the total charge of the subbands one is considering. This gives rise to the formula

$$r = pt + qs\,, \tag{9}$$

where r is the number of filled subbands, t is the Hall conductance in the weak potential case, and s is the Hall conductance in the tight-binding case. In our published paper we produced an elaborate justification for eq. (9) and it gave us many headaches. I think we produced five different drafts of the argument for this equation, and the last one was not much more comprehensible than the first one. Fortunately I have recently published a paper in which this Galilean invariance argument is developed [16], and the derivation of this equation is then very simple. In this strongly anisotropic case which I had hoped one could continue to the anisotropic case, you have the condition that mod(s) is less than $p/2$, which determines the Hall current t uniquely. If the continuation is not valid eq. (9) is still rigorous, but the solution is not uniquely determined; you can always add a multiple of q to t and get a different solution. Figure 6 shows the plateaus of the Hall conductance obtained in this way for q/p equal to various ratios of the Fibonacci numbers: 3/5, 5/8, 8/13. The big gaps have nice small Hall currents 1 or 0. The small gaps, the ones you could only see if you have very little dirt around, very small measuring field, very low temperature, and so on, those are the exotic ones. You can see from the figure that the gaps with Hall currents +4 or −3 are very narrow gaps; you can hardly see the plateaus. It is generally true that the more exotic the Hall current the smaller the gap associated with it and the harder it would be to see it.

I have taken fig. 7 from Claro and Wannier [8]. It shows the triangular lattice, and I have marked what the Hall currents are for some of the gaps in the weak potential case. As Streda [17] has shown,

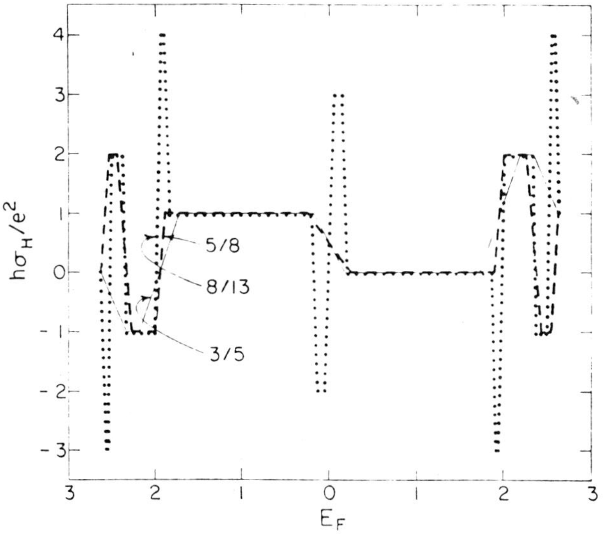

Fig. 6. Plateaus of the Hall conductance as a function of energy for values of q/p equal to ratios of Fibonacci numbers [13].

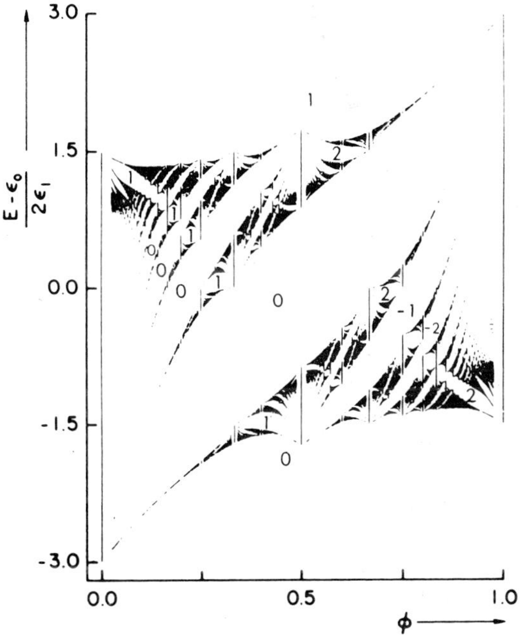

Fig. 7. Energy bands of a triangular lattice in a magnetic field [8]. The Hall conductance for the weak modulation limit is shown for a few of the gaps.

each gap has associated with it a Hall current equal to the intercept of that gap with the vertical axis. The gaps that run through the origin have Hall current zero. The ones that end up at the top left side of the diagram have Hall current 1. It is not always easy to trace the discontinuous gaps through the diagram. For the tight-binding model it is the slope rather than the intercept that gives the Hall current associated with a particular gap. This work has enabled us at least to classify these gaps in terms of the physically measurable quantity, namely the Hall conductance.

No fractional Hall effect comes out of this with one trivial (or, I think, physically unrealizable) exception. There are two cases which Claro and Wannier have commented on where the two subbands actually touch one another. These are the fourth gap of the $q/p = 1/6$ case and the second gap of the $q/p = 5/6$ case. If we took the Fermi energy in the pseudogap between these two touching subbands we could find half-integer values of the Hall conductance. In fact, as I learned from E. Davidson of the University of Washington Chemistry Department, there is a theorem by Herzberg and Longuet-Higgins [18] which says that the wave function changes its phase by π round a point at which energy levels cross. This is just what is needed to get a half-integer value for eqs. (7) or (8). The same happens at the band center for all even p in the square lattice.

There are a few isolated points that I would like to mention. Firstly, it is clearly possible, by altering the strength of the modulating potential, to go over continuously from the weak modulation limit to the tight-binding limit, keeping the same number of flux quanta per unit cell. For many of the gaps the Hall conductance is the same in the two limits, but in some cases the condition $|s| < p/2$ gives a different solution to eq. (9) than the condition $|t| < q/2$. The simplest case where that happens is for one flux quantum per unit cell, $p = q = 1$. The Hall conductance is then unity for the weak-modulation case when the lowest Landau level is filled, but it is zero in the tight-binding limit when the lowest band is filled. There must be some intermediate strength of the coupling where the two lowest Landau levels touch and the Hall conductance changes abruptly. It is in fact quite easy to see in detail how that happens. For weak modulation the highest orbital of the lowest Landau level has its maximum modulus at the maximum of the potential, and so it has odd parity about the minima of the potential, and corresponds to a p-state in the tight-binding limit. Similarly the lowest orbital of the second Landau level is symmetric about the minima of the potential and corresponds to an s-state in the tight-binding limit. These two states of opposite parity must cross at some definite value of the potential strength.

Figure 8 shows the results of a detailed calculation of the effect of increasing the strength of the modulation from a paper by Schellnhuber and Obermair [19]. In this case, with $q/p = 8$, you can see how the 8 lowest Landau levels on the left of the diagram form the s-band on the bottom right of the diagram. The eighth energy gap, which gives Hall current 8 on the left of the diagram, closes up at one point, and then gives Hall current 0 on the right of the diagram. Similarly, it can be seen where the seventh gap closes and changes over from 7 units of Hall current on the left to -1 units on the right.

There is an intriguing relation between the existence of open orbits in the semiclassical model, localization, and the width of the spectrum which I have studied but not fully understood [15]. We consider first of all a tight-binding model with a band energy

$$E(k_x, k_y) = 2V_1 \cos(k_x a) + 2V_2 \cos(k_y a). \tag{10}$$

In the isotropic case $V_1 = V_2$ the magnetic field drives the electrons round the contours of constant energy, which are all closed curves. When V_1 is greater than V_2 there is a range of energies near the center of the band for which there are open orbits, while there are still closed orbits near the band edges. One would not be very surprised if the open orbits gave states which were localized in the y

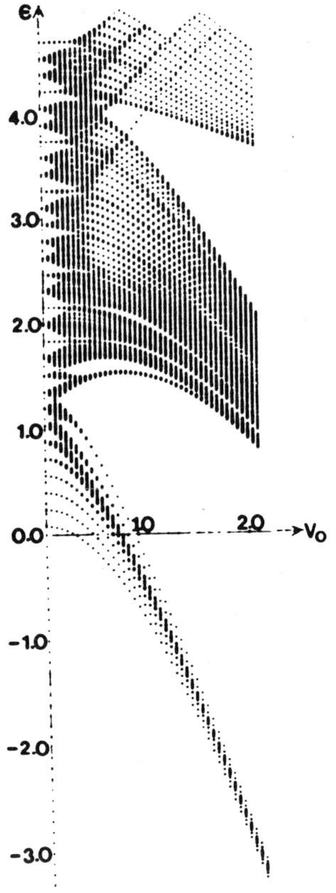

Fig. 8. Energy bands as a function of the strength of the potential in the case $q/p = 8$ [19].

direction and extended in the x direction, but which were broadened, while the Peierls–Onsager quantization condition should give sharp levels for the closed orbits. In fact the work of Aubry and André [14] and our own work [15] shows that, when the flux lattice is incommensurate with the crystal lattice (q/p irrational), in this anisotropic case all states are localized in the y direction and extended in the x direction, and the energy bands are all broadened. Only in the isotropic case there is no localization, and the sum of the band widths is zero. This analysis can be taken a little further by adding a next nearest neighbor hopping term to the tight-binding model, so that there is a term $4V_3\cos(k_xa)\cos(k_ya)$ added to eq. (10). If $2V_3$ is greater than V_1 or V_2, then the isotropic next nearest neighbor term dominates, there are no open orbits, no localized states, and the sum of the subband widths tends to zero in the incommensurate limit. If the anisotropic nearest neighbor terms dominate there is a range of open orbits, all states are localized in one direction, and the sum of the subband widths is non-zero.

Finally, it can be shown that the non-existence of proper Wannier functions for a periodic potential in the presence of a uniform magnetic field is connected to the quantized Hall current. If there were real Wannier functions, localized, complete, and normalizable, for a magnetic subband, it would be possible to combine them in the usual way to form a complete set of generalized Bloch waves, which would be analytic and single valued. These could then be substituted in eq. (7), and would give a Hall current of zero, for the reasons I have already given. It is therefore only for those subbands which carry no Hall current that proper Wannier functions exist [20].

References

[1] K. von Klitzing, G. Dorda and M. Pepper, Phys. Rev. Lett. 45 (1980) 494.
[2] S. Kawaji, preprint.
[3] R.B. Laughlin, Phys. Rev. B23 (1981) 5632.
[4] B.I. Halperin, Phys. Rev. B25 (1982) 2185.
[5] A. Rauh, G.H. Wannier and G. Obermair, Phys. Status Solidi(b) 63 (1974) 215.
[6] D. Hofstadter, Phys. Rev. B14 (1976) 2239.
[7] G.H. Wannier, Phys. Status Solidi(b) 88 (1978) 757.
[8] F. Claro and G.H. Wannier, Phys. Rev. B19 (1979) 6068.
[9] P.G. Harper, Proc. Phys. Soc. (London) A68 (1955) 874.
[10] M.Ya. Azbel, Soviet Phys. JETP 19 (1964) 634.
[11] J. Zak, Phys. Rev. 134 (1964) A1607.
[12] A.B. Pippard, Phil. Trans. Royal Soc. (London) A265 (1964) 317.
[13] D.J. Thouless, M. Kohmoto, M.P. Nightingale and M. den Nijs, Phys. Rev. Lett. 49 (1982) 405.
[14] S. Aubry and G. André, Ann. Israel Phys. Soc. 3 (1980) 133.
[15] D.J. Thouless, submitted to Phys. Rev. B.
[16] D.J. Thouless, Phys. Rev. B27 (1983) 6083.
[17] P. Streda, J. Phys. C15 (1982) L717.
[18] G.H. Herzberg and H.C. Longuet-Higgins, Discuss. Faraday Soc. 35 (1963) 77.
[19] H.J. Schellnhuber and G.M. Obermair, Phys. Rev. Lett. 45 (1980) 276.
[20] D.J. Thouless, J. Phys. C17 (1984) L325.

J. Phys. C: Solid State Phys., **15** (1982) L1299–L1303. Printed in Great Britain

LETTER TO THE EDITOR

Quantised Hall effect in a two-dimensional periodic potential

P Středa
Institute of Physics, Czechoslovak Academy of Sciences, 180 40 Praha 6, Na Slovance 2, Czechoslovakia

Received 6 October 1982

Abstract. The quantised Hall conductance σ_{xy} in strong magnetic fields due to electrons in a periodic two-dimensional potential has been studied using the linear response theory. This study reveals the connection of quantised values of σ_{xy} with the parameters of electron energy spectrum. The effect of the magnetic field strength and of the number of electrons has also been established.

The quantised Hall effect discovered by Klitzing *et al* (1980) is a remarkable quantum mechanical effect on a macroscopic scale. The Hall conductance σ_{xy} of a two-dimensional electron system in a strong magnetic field, at $T \sim 0$, is quantised in exact multiples of the unit e^2/h:

$$\sigma_{xy}^{Q} = -(e^2/h)\, i, \tag{1}$$

when the Fermi energy $E_{\rm F}$ is pinned in the gap between two adjacent Landau levels. The integer i is then usually interpreted as the number of fully occupied Landau levels.

A crystal potential acting on electrons in a magnetic field not only changes the energy width of a particular Landau level, but also forms new gaps (see e.g. Hofstadter 1976). According to the very general explanation of Laughlin (1981), a quantised Hall conductance should appear if the Fermi energy is located within any gap, i.e. also in a gap formed inside a Landau level. Several questions immediately arise: which values of i in equation (1) are admissable, how these values are connected with the position of gaps in the fairly complicated electron energy spectrum and how the electron Hall conductance is changed into the hole conductance, if the electron states are filled up. The first question was answered by Thouless *et al* (1982). To answer the remaining questions, quantised values of σ_{xy} will be established by making use of the Kubo formula in the form presented by the author (Středa 1982). The model of energy structure of independent electrons in magnetic field and periodic potential proposed by Hofstadter (1976) will be used. The dependence of quantised values of σ_{xy} on magnetic field strength as well as on the number of electrons will be presented, i.e. the effect of open orbits will be taken into account.

Let us suppose that the magnetic field $\boldsymbol{B} \equiv (0, 0, B)$ is perpendicular to the two-dimensional system and that only elastic scattering occurs. For simplicity the spin will not be taken into account. The one-electron Hamiltonian will have the following form:

$$H = (1/2m)(\boldsymbol{p} - e/c\,\boldsymbol{A})^2 + V(x, y) \tag{2}$$

where the vector potential $\boldsymbol{A}$ satisfies the equation $\boldsymbol{B} = \operatorname{curl} \boldsymbol{A}$, $\boldsymbol{p}$ is the momentum operator, $V(x, y)$ is an arbitrary potential, and e and m are electron charge and mass respectively. Under these assumptions the Kubo formula yields the expression for the Hall conductance (Středa 1982)

$$\sigma_{xy} = \frac{e^2}{2} \mathrm{i}\hbar \operatorname{Tr}[v_x G^+(E_\mathrm{F}) v_y \delta(E_\mathrm{F} - H) - v_x \delta(E_\mathrm{F} - H) v_y G^-(E_\mathrm{F})] + ec \left. \frac{\partial N(E)}{\partial B} \right|_{E = E_\mathrm{F}}. \tag{3}$$

The Green function G and the ith component of velocity operator v_i are defined as follows:

$$G^{\pm}(E) = (E - H \pm \mathrm{i}0)^{-1}, \qquad \delta(E - H) = -\frac{1}{2\pi \mathrm{i}}(G^+ - G^-), \qquad v_i = \frac{1}{\mathrm{i}\hbar}[r_i, H] = \frac{1}{m}(p_i - (e/c) A_i). \tag{4}$$

$N(E)$ is the number of states per unit area having energy lower than E:

$$N(E) = \int_{-\infty}^{E} \operatorname{Tr} \delta(\eta - H)\, \mathrm{d}\eta. \tag{5}$$

If the Fermi energy lies within any gap (i.e. $\delta(E_\mathrm{F} - H) \to 0$), the first term on the right-hand side of equation (3) is just equal to zero and the quantised values of the Hall conductance is obtained:

$$\sigma_{xy}^{Q} = ec \left. \frac{\partial N(E)}{\partial B} \right|_{E = E_\mathrm{F}}. \tag{6}$$

At the same time, the transverse conductivity is zero (Středa 1982).

The energy structure of an electron in a magnetic field and in a periodic potential has been intensively studied by Wannier and his collaborators (Hofstadter 1976, Wannier 1978, Claro and Wannier 1979) for the case when one electron band is formed at zero field ($B = 0$). It strongly depends on the magnetic flux φ through a lattice unit cell area Ω:

$$\varphi = \frac{|e| B}{hc} \Omega. \tag{7}$$

The energy spectrum is periodic in φ with period $\Delta\varphi = 1$. Since $\varphi = 1$ corresponds to a very large magnetic field, unattainable in today's laboratories, we limit ourselves to the interval $\varphi \in (0, 1)$. The calculation of the electron energy spectrum can only be performed for rational values $\varphi = p/q$ (p and q are integers with no common factor). The number of allowed energy bands equals q.

Following Wannier (1978), the position of gaps can be described by straight lines

$$\Omega N = m + n\varphi \tag{8}$$

where m and n are integers and ΩN is the number of states per unit cell below the gap determined by m and n (figure 1). In the paper of Wannier (1978) the quantity ΩN is denoted W and is called the statistical weight. At cross-points the nearest bands touch and some of the gaps can disappear. The energy width of gaps decreases with an increase of m and n.

Energy bands can be broadened, e.g. by crystal lattice imperfections and phonons. Thus in real two-dimensional crystals only the gaps corresponding to $|m|$ and $|n|$ less than some parameter M remain conserved.

Combining equations (6), (7) and (8), expressions for the Hall conductance σ_{xy}^Q and resistance ρ_{xy}^Q are immediately obtained:

$$\sigma_{xy}^Q = -\frac{e^2}{h} n, \qquad \rho_{xy}^Q = \frac{h}{e^2}\frac{1}{n}. \tag{9}$$

Let us suppose for simplicity that only gaps corresponding to integers m and n satisfying the condition $|n|$, $|m| \leq 4$ did not disappear due to crystal imperfections (figure 1). All

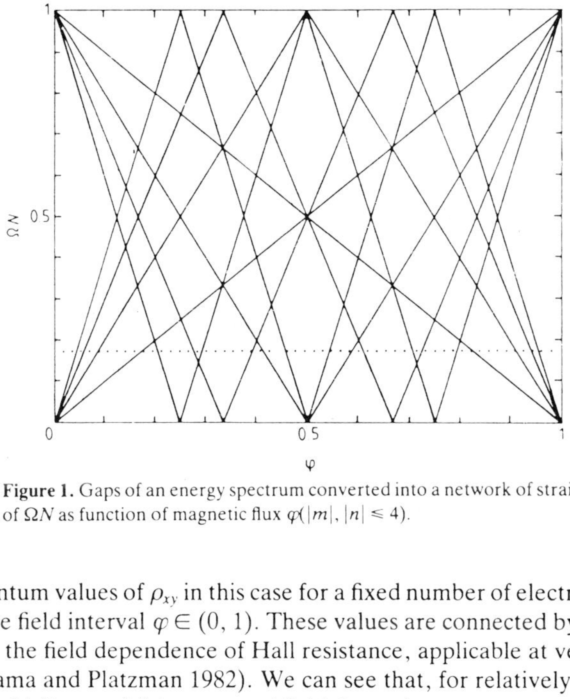

Figure 1. Gaps of an energy spectrum converted into a network of straight lines in the graph of ΩN as function of magnetic flux φ($|m|$, $|n| \leq 4$).

available quantum values of ρ_{xy} in this case for a fixed number of electrons are drawn in figure 2 for the field interval $\varphi \in (0, 1)$. These values are connected by a step function, which images the field dependence of Hall resistance, applicable at very low temperatures (Fukuyama and Platzman 1982). We can see that, for relatively small intensities of magnetic field B, ρ_{xy} exhibits a normal field dependence, corresponding to the usual model based on free-electron-like Landau levels. For the example presented ($\Omega N = 0.173$) this picture breaks down, if the Fermi energy immerses into the lowest Landau level. Going to higher fields, the gap structure of the Landau level is manifested.

Changing the number of electrons at a given magnetic field, a similar picture is obtained (figure 3). For a small number of electrons, ρ_{xy} falls by steps as Landau levels are filling up. In the middle of the band ($\Omega N \sim 0.5$), where open orbits appear, ρ_{xy} often changes sign and has a very complicated form. At the top of the band the electron spectrum changes into the hole spectrum and the hole Hall resistance appears (ρ_{xy} looks like resistance at the bottom of the band, but reversed in sign).

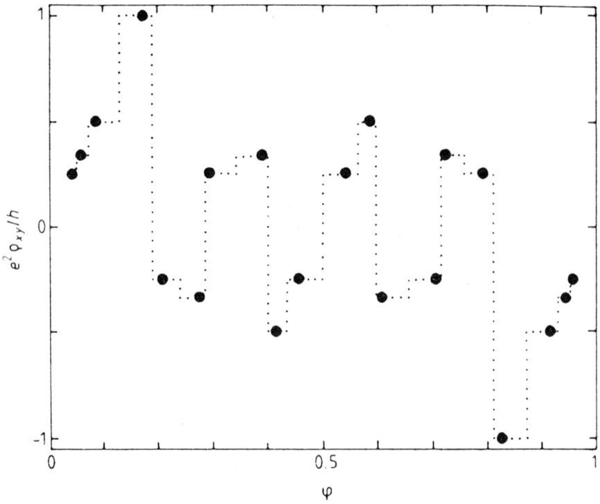

Figure 2. Quantised values of the resistance ρ_{xy} (denoted ●) as a function of the magnetic flux φ for a given number of electrons $\Omega N = 0.173$ (dotted line in figure 1) and $|n|, |m| \leq 4$.

The construction of the N and B dependences is very simple except for the points, where for the given ΩN and B equation (8) is satisfied for several values of n. In those cases one must be very careful to decide, using more precise calculations of bandstructure (Hofstadter 1976), which gap disappears and which does not.

At first glance, an experimental verification of the theory presented here is almost impossible. It can hardly be believed that magnetic fields stronger by several orders than those available now will be reached in the near future. But there might be another way of checking the proposed dependences. If a superstructure in the two-dimensional layer is formed, the magnetic flux through a unit cell becomes larger at a given B. In such a case the quantised conductance could be obtained, when the Fermi energy lies in an inner gap of the lowest Landau level, at reasonably high fields.

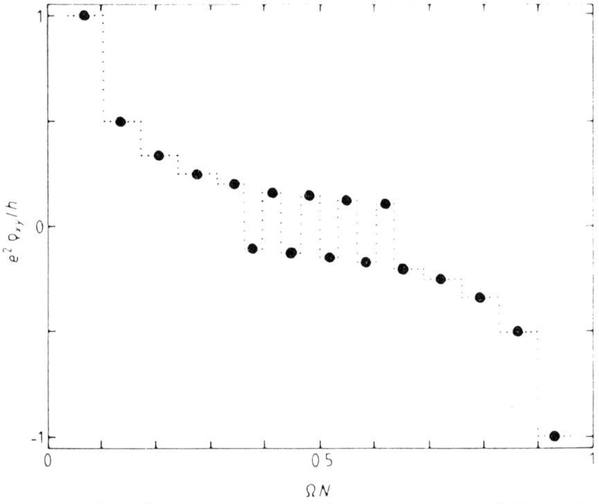

Figure 3. Quantised values of the resistance ρ_{xy} (denoted ●) as a function of the number of electrons ΩN for a given magnetic flux through the unit cell $\varphi = 0.069$. Only gaps corresponding to the condition $|n|, |m| \leq 9$ were taken into account.

I wish to thank D J Thouless for sending me a copy of his work prior to publication and for the valuable comment on my preprint.

References

Claro F H and Wannier G H 1979 *Phys. Rev.* B**19** 6068–74
Fukuyama H and Platzman P M 1982 *Phys. Rev.* B**25** 2934–6
Hofstadter D R 1976 *Phys. Rev.* B**14** 2239–49
Klitzing K, Dorda G and Pepper M 1980 *Phys. Rev. Lett.* **45** 494–7
Laughlin R B 1981 *Phys. Rev.* B**23** 5632–3
Středa P 1982 *J. Phys. C: Solid State Phys.* **15** L717–21
Thouless D J, Kohmoto M, Nightingale M P and den Nijs M 1982 preprint
Wannier G H 1978 *Phys. Status Solidi* b **88** 757–65

Two-Dimensional Magnetotransport in the Extreme Quantum Limit

D. C. Tsui,[a),(b)] H. L. Stormer,[(a)] and A. C. Gossard
Bell Laboratories, Murray Hill, New Jersey 07974
(Received 5 March 1982)

A quantized Hall plateau of $\rho_{xy} = 3h/e^2$, accompanied by a minimum in ρ_{xx}, was observed at $T < 5$ K in magnetotransport of high-mobility, two-dimensional electrons, when the lowest-energy, spin-polarized Landau level is $\frac{1}{3}$ filled. The formation of a Wigner solid or charge-density-wave state with triangular symmetry is suggested as a possible explanation.

PACS numbers: 72.20.My, 71.45.-d, 73.40.Lq, 73.60.Fw

In the presence of an intense perpendicular magnetic field B, a system of two-dimensional (2D) electrons is expected to form a Wigner solid[1,2] at low temperatures (T). In the infinite-B limit, an analogy can be drawn to the classical electron gas on the surface of liquid helium, which crystallizes into a solid[3] when the ratio of the electron's average potential energy to thermal energy $\Gamma \equiv e^2\pi^{1/2}n/\epsilon kT = 137$ (n is the electron areal density). At finite B, quantum effects become important and it has been suggested that a charge-density-wave (CDW) state[4] may be possible at considerably higher T as a precursor to Wigner crystallization. Early experiments were carried out on the Si inversion layer at the Si-SiO_2 interface. Kawaji and Wakabayashi[5] and Tsui[6] made high-B magnetoconductivity measurements and observed structures and electric field dependences which cannot be explained by the independent-electron theory of Ando and Uemura.[7] Subsequently, Kennedy *et al.*[8] observed a shift in cyclotron resonance, concomitant with a drastic line narrowing, in the high-B limit, when the average electron separation exceeds the cyclotron diameter. Wilson, Allen, and Tsui[9] studied the dependence of this phenomenon on the Landau-level filling factor ($\nu = nh/eB$), and found that a pinned-CDW model[10] gave the most satisfactory account of the cyclotron resonance data. However, in the range of n at which these experiments were performed, localization due to disorder at the Si-SiO_2 interface is known to be important even in the absence of B, and consequently, it has not been possible to discern true Coulomb effects from those due to disorder.

In this Letter, we report some striking, new results on the transport of high-mobility, 2D electrons, in GaAs-AlGaAs heterojunctions, in the extreme quantum limit ($\nu < 1$), when the lowest-energy, spin-polarized Landau level is partially filled. We found that at temperatures $T < 5$ K, the diagonal part ρ_{xx} of the resistivity tensor shows a dip at $\nu = \frac{1}{3}$, which becomes stronger at lower T. For $\nu < \frac{1}{3}$, ρ_{xx} follows an approximately exponential increase with inverse T. The Hall resistivity ρ_{xy}, on the other hand, approaches a step of $3h/e^2$ at $\nu = \frac{1}{3}$ as T decreases, but remains essentially independent of T away from this Hall plateau. These features of the data resemble those of the quantized Hall resistance and the zero-resistance state expected exclusively for integral values of ν. We suggest that these striking results are evidence for a new electronic state at $\nu = \frac{1}{3}$. They are consistent with the notion that a Wigner solid, or a CDW state with triangular crystal symmetry, is favored at $\nu = \frac{1}{3}$ when the unit cell area of the lattice is a multiple of the area of a magnetic flux quantum.

The samples, consisting of 1-μm undoped GaAs, 500-Å undoped $Al_{0.3}Ga_{0.7}As$, 600-Å Si-doped $Al_{0.3}Ga_{0.7}As$, and 200-Å Si-doped GaAs single crystals, were sequentially grown on insulating GaAs substrates using molecular-beam-epitaxy techniques.[11] The 2D electron gas, resulting from ionized donors placed 500 Å inside AlGaAs,[12] is established at the undoped GaAs side of the GaAs-AlGaAs heterojunction. Samples were cut into standard Hall bridges and Ohmic contacts to the electron layer were made with In at 400°C. Low-field transport measurements were used to determine n and μ. Our samples have n from 1.1×10^{11} to 1.4×10^{11} cm^{-2} and μ from 80 000 to 100 000 cm^2/V sec. The high-B measurements were performed at the Francis Bitter National Magnet Laboratory, Cambridge, Mass.

Figure 1 shows ρ_{xy} and ρ_{xx} of one specimen as a function of B at four different temperatures. The scale at the top of the figure shows the Landau level filling factor ν, which gives the number of occupied levels. At integral values of ν, the data show the characteristic features of the quantized Hall plateaus and the vanishing of ρ_{xx},[13-15] when the Fermi energy E_F is pinned in the gap between two adjacent levels. Removal of spin de-

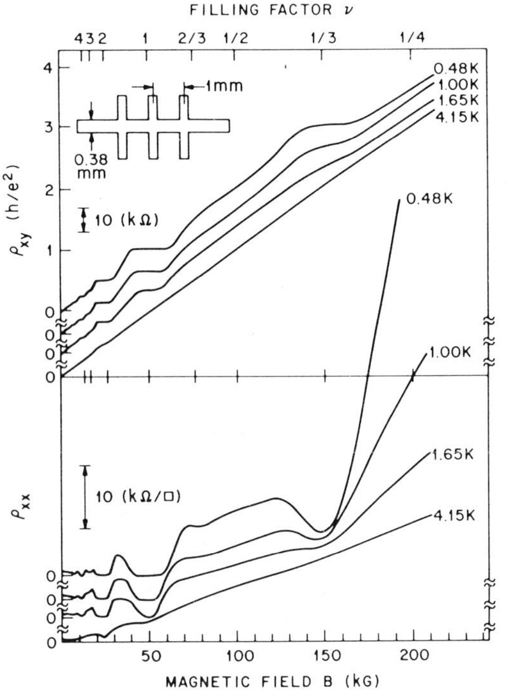

FIG. 1. ρ_{xy} and ρ_{xx} vs B, taken from a GaAs-$Al_{0.3}$-$Ga_{0.7}As$ sample with $n = 1.23\times10^{11}/cm^2$, $\mu = 90\,000$ cm^2/V sec, using $I = 1$ μA. The Landau level filling factor is defined by $\nu = nh/eB$.

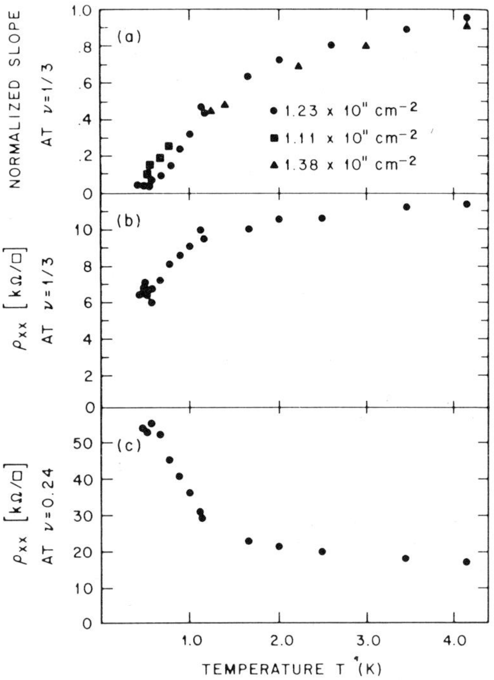

FIG. 2. T dependence of (a) the slope of ρ_{xy} at $\nu = \frac{1}{3}$, normalized to the slope at ~ 30 K, (b) ρ_{xx} at $\nu = \frac{1}{3}$, and (c) ρ_{xx} at $\nu = 0.24$.

generacy[16] is seen in the appearance of these features at odd-integer values of ν. As observed earlier,[17] the plateaus in ρ_{xy} as well as the vanishing of ρ_{xx} become increasingly pronounced as T is decreased.

In the extreme quantum limit, $\nu < 1$, only the lower spin state of the lowest Landau level, i.e., the (0, ↑) level, remains partially occupied. In this regime (i.e., $B > 50$ kG in Fig. 1), the system is completely spin polarized. For $T > 4.2$ K, $\rho_{xy} = B/ne$, and ρ_{xx} shows also nearly linear dependence on B, as expected from the free-electron theory of Ando and Uemura.[7,16] At lower T, ρ_{xy} deviates from $\rho_{xy} = B/ne$ at $\nu \sim \frac{1}{3}$. This deviation becomes more pronounced as T decreases and approaches a plateau of $\rho_{xy} = 3h/e^2$, within an accuracy better than 1% at 0.42 K. The appearance of this plateau is accompanied by a minimum in ρ_{xx}, as apparent in the lower panel of Fig. 1. The development of these features is similar to that of the quantized Hall resistance and the concomitant vanishing of ρ_{xx}, observed at integral values of ν at higher T. Moreover, for $\nu < \frac{1}{3}$ and away from the plateau region, ρ_{xx} shows strong increase with decreasing T, while ρ_{xy} shows very weak decrease or essentially independence of T. This behavior has been seen to $\nu = 0.21$, the smallest ν attained in this experiment.

Figure 2 illustrates the development of ρ_{xx} and ρ_{xy} at fixed B as a function of T. Figure 2(a) shows the slope of ρ_{xy} at $\nu = \frac{1}{3}$, normalized to the slope at high T (~ 30 K), for three samples with slightly different n. Figure 2(b) shows the accompanying ρ_{xx} minimum (at $\nu = \frac{1}{3}$), and Fig. 2(c) shows ρ_{xx} at $\nu = 0.24$ to illustrate its T dependence for $\nu < \frac{1}{3}$, away from the Hall plateau. Several points should be noted. First, the slope of ρ_{xy} at $\nu = \frac{1}{3}$ approaches zero at $T \sim 0.4$ K, indicative of a true quantized Hall plateau. Second, replotting the data in Fig. 2(a) on logarithmic slope versus inverse T scale shows a linear portion for data taken at $T \geqslant 1.1$ K. This fact allows us to extrapolate the normalized slope to 1 at $T_0 = 5$ K, which we identify as the temperature for the onset of this phenomenon. Third, ρ_{xx} at $\nu = \frac{1}{3}$ is ~ 6 kΩ/□

at our lowest T of ~ 0.42 K, and much lower T is needed to determine if ρ_{xx} at $\nu = \frac{1}{3}$ indeed vanishes with vanishing T. Finally, if the data for ρ_{xx} at $\nu = 0.24$ [Fig. 2(c)] are replotted on logarithmic ρ_{xx} versus inverse T scale, an exponential dependence on $1/T$ is seen for $T \gtrsim 0.6$ K, with a preexponential factor of $\sim 13 \text{k}\Omega/\square$. This result may be interpreted as due to thermally activated transport with an activation energy of 0.94 K. The preexponential factor is considerably lower than the maximum metallic sheet resistance of ~ 40 k$\Omega/\square$, predicted for Anderson localization in the tails of Landau levels,[18] but comparable to that signifying a metal-insulator transition for 2D systems in the absence of B.[19] Moreover, the data for $\nu < \frac{1}{3}$ suggest a state of localization, in which the electron mobility is thermally activated,[17] as seen in ρ_{xx}. The electron density, as seen in the slope of ρ_{xy} vs B, remains essentially independent of T.

The existence of quantized Hall resistance accompanied by the vanishing of ρ_{xx} at integral values of ν is now well known. Their observation is attributed to the existence of an energy gap between the extended states in two adjacent Landau levels and the presence of localized states, which pin E_F in the gap, to keep all the extended states in finite numbers of Landau levels completely occupied for finite regions of B or n. Laughlin's argument[20] based on gauge invariance demonstrated that the quantized Hall resistance, given by $\rho_{xy} = h/ie^2$ $(i = 1, 2, \ldots)$, results from complete occupation of all the extended states in the Landau level, regardless of the presence of localization. Our observation of a quantized Hall resistance of $3h/e^2$ at $\nu = \frac{1}{3}$ is a case where Laughlin's argument breaks down. If we attribute it to the presence of a gap at E_F when $\frac{1}{3}$ of the lowest Landau level is occupied, his argument will lead to quasiparticles with fractional electronic charge of $\frac{1}{3}$, as has been suggested for $\frac{1}{3}$-filled quasi one-dimensional systems.[21]

At the present, there is no satisfactory explanation for all of our observations. The fact that this phenomenon always occurs at $\nu = \frac{1}{3}$ and that it is most striking in samples with the highest electron mobility suggest the formation of a new spin-polarized electronic state, such as Wigner solid or CDW, with a triangular symmetry,[22] which is favored at $\nu = \frac{1}{3}$. In this picture, the observed features of ρ_{xx} and ρ_{xy} may be attributed to transport of the collective ground state. At $T = 0$, the transport is free of dissipation and ρ_{xx} is expected to vanish. Since the number of electrons in this ground state is $n = eB/3h$, the Hall resistivity is $\rho_{xy} = B/ne = 3h/e^2$. As discussed by Baraff and Tsui,[23] observation of the quantized Hall plateau may be attributed to the presence of donor states inside $Al_xGa_{1-x}As$. The thermal activation of ρ_{xx} at $\nu = \frac{1}{3}$ may result from activation of defects in the condensate, which give rise to dissipation.

Finally, our data also show weaker, but similar, structures in ρ_{xx} near $\nu = \frac{2}{3}$ and near $\nu = \frac{3}{2}$, accompanied by slight changes in the slope of ρ_{xy}. These structures, though discernible in Fig. 1, are well resolved in the data taken from the sample with $n = 1.4 \times 10^{11}/\text{cm}^2$ and $\mu = 100\,000$ cm^2/V sec at 1.2 K. In our picture, the structure at $\nu = \frac{2}{3}$ may be identified as due to the formation of a Wigner solid or CDW of holes, expected from electron-hole symmetry. At $\nu = \frac{3}{2}$, the $(0, \uparrow)$ Landau level is completely filled, but the $(0, \downarrow)$ level is half filled. Consequently, $\frac{2}{3}$ of all the electrons occupy the low-lying $(0, \uparrow)$ level, and only the remaining $\frac{1}{3}$ with spin $\downarrow$ may participate in the formation of a collective ground state. Our data appear to suggest that this condition is also favored for a Wigner solid or CDW ground state.

In summary, we observed striking structures in the magnetotransport coefficients of high-μ, 2D electrons in GaAs-$Al_xGa_{1-x}As$ heterojunctions at $\nu = \frac{1}{3}$, and similar, but much weaker, structures at $\nu = \frac{2}{3}$ and $\frac{3}{2}$. Their development as a function of T is reminiscent of the quantized ρ_{xy} and the concomitant vanishing of ρ_{xx}, expected only for integral values of ν. We suggest as a possible explanation the formation of a new electronic state, such as a Wigner solid or CDW state with a triangular symmetry.

We thank P. M. Tedrow for the He^3 refrigerator; R. B. Laughlin, P. A. Lee, V. Narayanamurti, and P. M. Platzman for discussions; and K. Baldwin, G. Kaminsky, and W. Wiegmann for technical assistance. This work was supported in part by the National Science Foundation.

---

[a]Visiting scientist at the Francis Bitter National Magnet Laboratory, Cambridge, Mass. 02139.

[b]Present address: Department of Electrical Engineering and Computer Science, Princeton University, Princeton, N.J. 08544.

[1]E. P. Wigner, Phys. Rev. 46, 1002 (1934).

[2]Y. E. Lozovik and V. I. Yudson, Pis'ma Zh. Eksp. Teor. Fiz. 22, 26 (1975) [JETP Lett. 22, 11 (1975)].

[3]C. C. Grimes and G. Adams, Phys. Rev. Lett. 42, 795 (1979).

[4]H. Fukuyama, P. M. Platzman, and P. W. Anderson, Phys. Rev. B 19, 5211 (1979).

[5]S. Kawaji and J. Wakabayashi, Solid State Commun. 22, 87 (1977).
[6]D. C. Tsui, Solid State Commun. 21, 675 (1977).
[7]T. Ando and Y. Uemura, J. Phys. Soc. Jpn. 36, 959 (1974).
[8]T. A. Kennedy, R. J. Wagner, B. D. McCombe, and D. C. Tsui, Solid State Commun. 22, 459 (1977).
[9]B. A. Wilson, S. J. Allen, and D. C. Tsui, Phys. Rev. Lett. 44, 479 (1980).
[10]H. Fukuyama and P. A. Lee, Phys. Rev. B 18, 6245 (1978).
[11]A. Y. Cho and J. R. Arthur, Prog. Solid State Chem. 10, 157 (1975).
[12]H. L. Stormer, A. Pinczuk, A. C. Gossard, and W. Wiegmann, Appl. Phys. Lett. 38, 691 (1981); T. J. Drummond, H. Morkoc, and A. Y. Cho, J. Appl. Phys. 52, 1380 (1981).
[13]K. von Klitzing, G. Dorda, and M. Pepper, Phys. Rev. Lett. 45, 494 (1980).
[14]D. C. Tsui and A. C. Gossard, Appl. Phys. Lett. 37, 550 (1981).
[15]D. C. Tsui, H. L. Stormer, and A. C. Gossard, Phys. Rev. B 25, 1405 (1982).
[16]D. C. Tsui, H. L. Stormer, A. C. Gossard, and W. Wiegmann, Phys. Rev. 21, 1589 (1980); Th. Englert, D. C. Tsui, and A. C. Gossard, Surface Sci. 113, 295 (1982).
[17]M. A. Paalanen, D. C. Tsui, and A. C. Gossard, Phys. Rev. B 25, 5566 (1982).
[18]H. Aoki and H. Kamimura, Solid State Commun. 21, 45 (1977).
[19]D. J. Bishop, D. C. Tsui, and R. C. Dynes, Phys. Rev. Lett. 44, 1153 (1980).
[20]R. B. Laughlin, Phys. Rev. B 23, 5632 (1981).
[21]W. P. Su and J. R. Schrieffer, Phys. Rev. Lett. 46, 738 (1981).
[22]D. Yoshioka and H. Fukuyama, J. Phys. Soc. Jpn. 47, 394 (1979), and 50, 1560 (1981).
[23]G. A. Baraff and D. C. Tsui, Phys. Rev. B 24, 2274 (1981).

Surface Science 170 (1986) 141–147
North-Holland, Amsterdam

ODD AND EVEN FRACTIONALLY QUANTIZED STATES IN GaAs–GaAlAs HETEROJUNCTIONS

R.G. CLARK, R.J. NICHOLAS and A. USHER
Clarendon Laboratory, Parks Road, Oxford OX1 3PU, UK

and

C.T. FOXON and J.J. HARRIS
Philips Research Laboratories, Redhill, Surrey RH1 5HA, UK

Received 31 July 1985; accepted for publication 13 September 1985

Fractional quantization at values $\nu = p/q$ has been observed in two ultra high mobility heterojunctions. For the $N = 0$ Landau level the set of fractions corresponding to $q = 3$, 5 and 7 has been observed for both spin states, while for the $N = 1$ Landau level even fractions are observed at 1/4, 1/2 and 3/4. Both the temperature and electron concentration dependence of the resistivity minima at fractional occupancies have been studied.

The fractional quantum Hall effect (FQHE) has been observed to give rise to a hierarchy of minima in the resistivity ρ_{xx} of a 2D electron gas, with corresponding plateaus in ρ_{xy} at fractional occupancies ν $(= nh/eB)$ of the form p/q, with p = integers and q = odd integers. This began with the observation of 1/3 and 2/3 states [1], and has now progressed to the observation of $q = 5$, 7, and possibly 9 fractions in the lowest energy spin split level [2,3], and $q = 3$ fractions in the higher energy spin state of the first Landau level [3–5] $(1 < \nu < 2)$. The theories of this effect (refs. [6–8] and references therein) invoke the existence of an interacting many body electron fluid, with a new ground state and an electron–hole symmetry about occupancies of 1/2 in each spin split Landau level. In this work we demonstrate the existence of a complete set of fractional states with $q = 3$, 5, 7 for both spin states of the $N = 0$ Landau level, and investigate their dependence upon temperature and electron concentration, while for the $N = 1$ Landau level we find evidence preliminary for the existence of even fractions.

The two samples studied were very high mobility GaAs–$GaAl_{0.32}As$ heterojunctions grown by MBE on a Varian GenII machine [9]. The first, G63, had mobilities varying from 0.6×10^6 to 2.15×10^6 $cm^2/V \cdot s$ as the carrier concentration was increased from 0.7×10^{11} cm^{-2} to 2.1×10^{11} cm^{-2}, and the second, G29, had mobilities of 0.1 to 1.9×10^6 $cm^2/V \cdot s$ as the carrier concentration went from 1.15×10^{11} to 3.5×10^{11} cm^{-2}. The carrier concentrations were increased progressively by illumination from short bursts of light from a red LED. The samples were mounted on a laminated copper cold finger in a 6mK dilution refrigerator, and were cooled to 20 mK. The temperatures were measured by a carbon resistor, mounted next to the sample, and were verified at lowest temperatures by a ^{54}MnNi nuclear orientation thermometer mounted at the end of the cold finger well beyond the sample. Magnetic fields were provided by a superconducting solenoid.

Fig. 1 shows recordings of ρ_{xx} and ρ_{xy} for sample G63, after illumination. In the high field region $\nu < 1$, well defined minima can be seen in the series 2/3, 3/5, 4/7 and a pronounced feature is seen at 5/9. A sharp 4/5 minimum is also seen, together with a broader feature close to 3/4, as reported by Ebert et al. [4]. Plateaus can be seen in the Hall voltage even for the 4/7 occupancy. The detailed structure obtained in the lower field region for $\nu > 1$ is amplified in fig. 2, after a few percent more carriers have been introduced. The resistivity

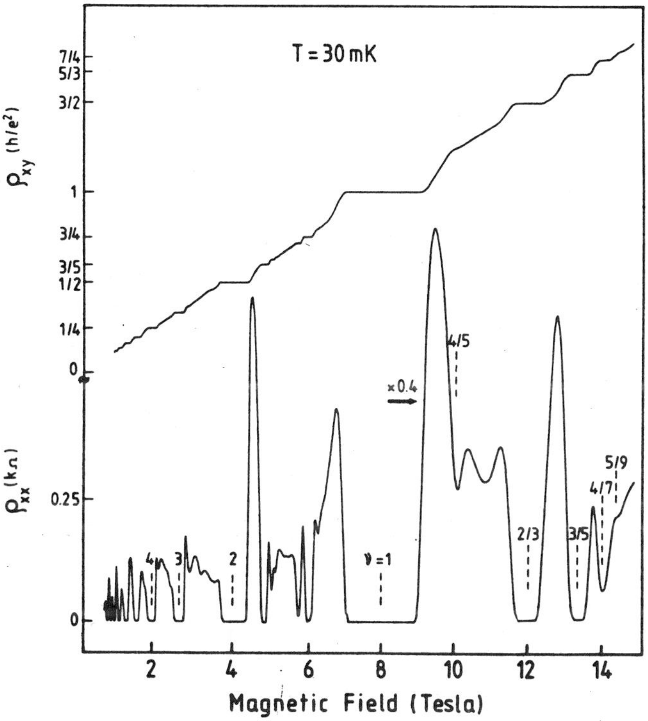

Fig. 1. The resistivity and Hall voltage (upper trace) for sample G63 at 30 mK. The sample is in the shape of a Hall bar, with a width of 150 μm and a separation between contacts of 1.5 mm. The fractional occupancies for $\nu < 1$ are indicated on the trace. The Hall resistivity is given in units of (h/e^2). The magnetic field scale is determined from an NMR calibrated magnetoresistance coil.

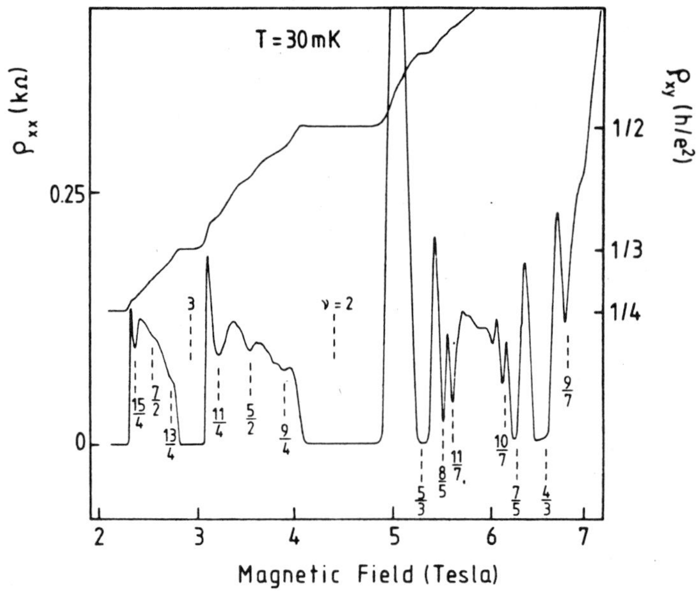

Fig. 2. Low field resistivity and Hall traces in sample G63, conditions as in fig. 1.

again shows very pronounced minima in positions corresponding accurately to the fractions 1 + . . . 1/3, 2/3, 2/5, 3/5, 2/7, 3/7, 4/7, with a further weak minimum close to the 4/9 occupancy, and a broader minimum close to 1 + 1/2, as observed near $\nu = 1/2$ [2,4]. The minima are again observed to correspond well to plateaus in the Hall voltage.

For the $N = 1$ Landau level rather different behaviour is observed ($2 < \nu < 4$). The resistivity shows three minima developing at occupancies of 1/4, 1/2 and 3/4, as indicated in fig. 2, for both spin orientations. Plateaus are also beginning to form in the Hall voltage at magnetic field positions, and hence occupancies, corresponding to the even fractions. The developing plateaus do not lie on the straight line connecting the centres of all of the integer and odd fractional Hall plateaus, and hence the values of the quantized Hall resistance do not at this stage correspond with any fractional assignments.

The observation of even fractions would seem to be a consequence of the occupation of the $N = 1$ Landau level, rather than an artefact of the rather low field at which they are observed, since it is possible to observe odd fractions in the $N = 0$ Landau level at the same magnetic field, before illumination, when the electron concentration and the mobility are lower. This is illustrated in fig. 3, where the upper trace shows minima for the fractions 1 + . . . 1/3, 2/5 and 2/3 for magnetic fields between 3 and 4 T, and the $N = 1$ Landau level is beginning to develop some structure even at around 2T! The most likely cause of ambiguity in the assignment of the even fractional values is the weak

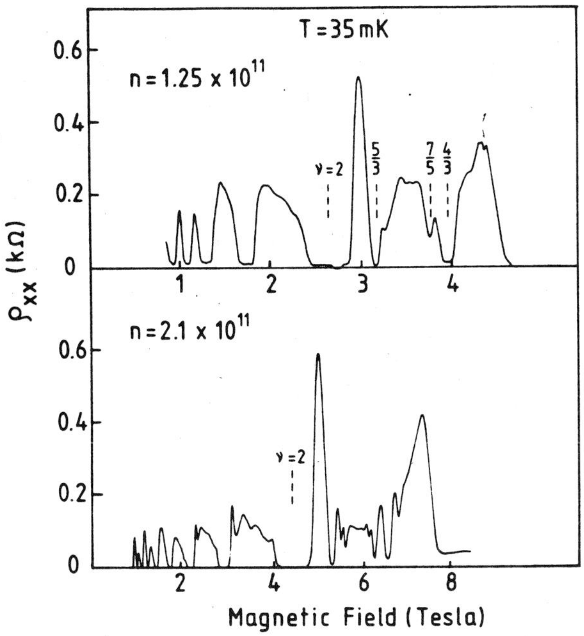

Fig. 3. Low field resistivity in sample G63 at 35 mK for two different electron concentrations.

resolution of the minima. An example of how this may affect the fractional value is shown in fig. 4, in which the development of the fractional minima for $1 < \nu < 2$ is studied as a function of electron concentration and mobility in sample G29. The mobility increases very rapidly as carriers are photoexcited, and the strength of the minima is dramatically enhanced by only a 20% increase in electron concentration. The positions for 4/3 and 5/3 occupancies are marked in the upper and lower traces, and show how the minima occur closer to the centre of the level when they are not fully resolved. Similar behaviour was observed in our earlier work [5]. If a similar effect were occurring for the $N = 1$ Landau level in G63, however, we would expect that the minima would move outward as they became sharpened; therefore the 1/4 and 3/4 minima would move further away from the 1/3 and 2/3 occupancies, which are predicted always to be the strongest features by the majority of current theories [7,10]. This will be discussed in detail in a future work [11].

The temperature evolution of the fractional features is shown in fig. 5. The first structures to appear are at the highest fields, for the < 1 region. For the region $\nu > 1$ the 7/5 and 5/3 minima rather surprisingly appear before the 4/3, although the 4/3 then overtakes the 7/5 at lower temperatures. Approximate activation energies Δ deduced from this study, defined by $\rho_{xx} \sim e^{-\Delta/kT}$, are shown in table 1. There is considerable uncertainty, however,

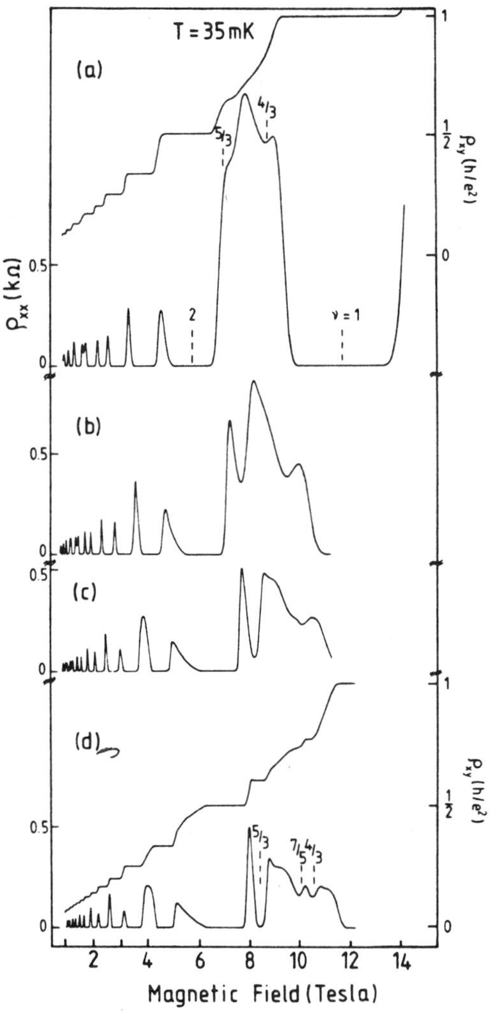

Fig. 4. The electron concentration dependence for ρ_{xx} and ρ_{xy} in sample G29 at 35 mK. The electron concentration has been increased by a series of short bursts of light from a red LED. At lower concentrations only integer quantization can be seen.

particularly for the higher order fractions ($q = 5, 7$), due to a slight hysteresis which seems to be present between up and down sweeps of the magnetic field. This may be associated with the change in the ground state as the quantum fluid changes its occupancy. The activation energy for 2/3 is in good agreement with values reported by other authors at similar fields for either 2/3 or 1/3 [4,12,13]. The values for 4/3 and 5/3, 0.5 K and 0.7 K, are substantially

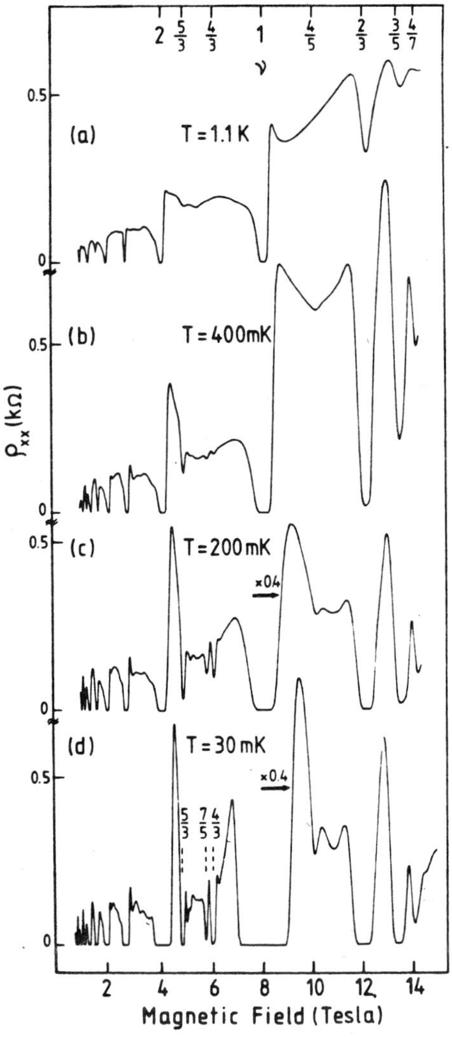

Fig. 5. The temperature dependence of ρ_{xx} for sample G63, with an electron concentration of 1.9×10^{11} cm^{-2} and a mobility of 2.1×10^{6} cm^2/V·s.

Table 1
Approximate activation energies of the resistivity minima

Fraction	Field (T)	Energy (K)
2/3	12.0	2.2
4/3	6.0	0.5
5/3	4.8	0.7
3/5	13.3	0.7
7/5	5.7	0.2
4/7	14.0	0.2

larger and closer together than those reported by Ebert et al. [4], who found a value of only 0.09 K for 5/3 and 0.3 K for 4/3. Our value for 3/5 is also in reasonable agreement with that reported by Chang et al. [2], and the theoretical estimate of 2 K made by Laughlin [10]. A final interesting feature of the curves is the very rapid increase in the conductivity at the low field side of each level, as it is approaching complete occupancy. For example the peak at $\nu = 0.85$ has increased in conductivity by a factor of 5 on cooling from 1 K down to 30 mK, while the sharp peak at $\nu = 1.8$ also shows a rapid increase.

References

[1] D.C. Tsui, H.L. Störmer and A.C. Gossard, Phys. Rev. Letters 48 (1982) 1559.

[2] A.M. Chang, P. Berglund, D.C. Tsui, H.L. Störmer and J.C.M. Hwang, Phys. Rev. Letters 53 (1984) 997.

[3] E.E. Mendez, L.L. Chang, M. Heiblum, L. Esaki, M. Naughton, K. Martin and J. Brooks, Phys. Rev. B30 (1984) 7310.

[4] G. Ebert, K. von Klitzing, J.C. Maan, G. Remenyi, C. Probst and G. Weimann, J. Phys. C (Solid State Phys.) 17 (1984) L775.

[5] R.G. Clark, R.J. Nicholas, M.A. Brummell, A. Usher, S. Collocott, J.C. Portal, F. Alexandre and J.M. Masson, Solid State Commun. 56 (1986) 173.

[6] R.B. Laughlin, Phys. Rev. Letters 50 (1983) 1395.

[7] R. Tao, Phys. Rev. B29 (1984) 636.

[8] A.H. MacDonald, Phys. Rev. B30 (1984) 3550.

[9] C.T. Foxon, J.J. Harris, R.G. Wheeler and D.E. Lacklison, to be published.

[10] R.B. Laughlin, Surface Sci. 142 (1984) 163.

[11] R.J. Nicholas, R.G. Clark, A. Usher, C.T. Foxon and J.J. Harris to be published.

[12] A.M. Chang, M.A. Paalanen, D.C. Tsui, H.L. Störmer and J.C.M. Hwang, Phys. Rev. B28 (1983) 6133.

[13] J. Wakabayashi, S. Kawaji, J. Yoshino and H. Sakaki, in: Proc. 17th Intern. Conf. on Physics of Semiconductors, San Francisco, 1984, Eds. D.J. Chadi and W.A. Harrison (Springer, Berlin, 1985) p. 283.

Magnetic Field Dependence of Activation Energies in the Fractional Quantum Hall Effect

G. S. Boebinger
Department of Physics and Francis Bitter National Magnet Laboratory, Massachusetts Institute of Technology, Cambridge, Massachusetts 02139

A. M. Chang[(a)]
AT&T Bell Laboratories, Holmdel, New Jersey 07733

H. L. Stormer[(a)]
AT&T Bell Laboratories, Murray Hill, New Jersey 07974

and

D. C. Tsui[(a)]
Department of Electrical Engineering and Computer Science, Princeton University, Princeton, New Jersey 08544
(Received 19 July 1985)

We have studied the temperature dependence of the fractional quantum Hall effect at Landau-level filling factors $\nu=\frac{1}{3}, \frac{2}{3}, \frac{4}{3}, \frac{5}{3}, \frac{2}{5}$, and $\frac{3}{5}$ in magnetic fields up to 28 T to determine the magnitude of the associated energy gaps. The data suggest a single activation energy for $\nu=\frac{1}{3}, \frac{2}{3}, \frac{4}{3}$, and $\frac{5}{3}$. Its magnitude, much smaller than predicted by current theories, vanishes for $B \lesssim 6$ T and saturates at $B \gtrsim 18$ T. The data also suggest a single activation energy for $\nu=\frac{2}{5}$ and $\frac{3}{5}$ which is smaller than predicted.

PACS numbers: 72.20.My, 73.40.Lq, 73.60.Fw

The fractional quantum Hall effect, FQHE, is observed in high-mobility ($\mu \gtrsim 100\,000$ cm^2/V-s) two-dimensional electron systems at low temperatures ($T \lesssim 4$ K) and high magnetic fields ($B \gtrsim 5$ T).[1-7] The FQHE is phenomenologically similar to the integral quantum Hall effect, IQHE: Plateaus are observed in the Hall resistivity, ρ_{xy}, concomitant with minima in the diagonal resistivity, ρ_{xx}. While the IQHE exists at magnetic fields corresponding to integral Landau-level filling, ν, the FQHE is observed at fractional Landau-level filling $\nu=p/q$ where q is always odd ($\nu=nh/eB$, $n=$area density, and eB/h =Landau-level degeneracy). In both cases, observation of a zero-resistance state implies the existence of a gap in the excitation spectrum of the system. Measurements of the activation energy in the IQHE reproduce closely the Landau-level splitting representing the gaps in the single-particle density of states of the system.[8] The FQHE is of many-particle origin and, hence, activation-energy data on ρ_{xx} in a given fractional state is expected to provide a measure of the size of the gap in the excitation spectrum of the correlated electronic ground state.

Theories have been developed assessing the nature of the underlying electronic state. In particular, a model due to Laughlin interprets the FQHE as the signature of an incompressible quantum fluid at fractional filling factors $\nu=p/q$ with q odd.[9-12] At finite temperatures, it predicts thermal excitation of fractionally charged quasielectrons and quasiholes across an energy gap above the ground state. The magnitude of the gap depends on ν and is expected to scale with magnetic field as e^2/l_0, where $l_0=(\hbar c/eB)^{1/2}$ is the magnetic length, the only relevant length scale in the system. Although many interesting properties of the electron liquid can be deduced from the existing model, presently the only experimentally accessible quantity is the size of the energy gap associated with a given fractional state.

In this Letter we report the determination of the magnetic field dependence and the relative magnitude of the activation energies of the FQHE at $\nu=\frac{1}{3}, \frac{2}{3}, \frac{4}{3}, \frac{5}{3}$, and $\nu=\frac{2}{5}, \frac{3}{5}$. The four specimens are modulation-doped GaAs/AlGaAs heterostructures with typical mobilities $\mu=5000\,000$–$850\,000$ cm^2/V-s and electron densities $n=(1.5\text{–}2.3)\times 10^{11}$ cm^{-2}. The electron densities are tunable by a backside gate bias which also affects the sample mobility.[13] Standard Hall bridge specimens were used to measure ρ_{xx} in samples A and B, while quasi Corbino-geometry specimens were used to measure the diagonal conductance σ_{xx} in samples C and D. Near the ρ_{xx} minima, $\rho_{xx} \ll \rho_{xy}$ and thus σ_{xx} and ρ_{xx} differ only by a constant factor: $\sigma_{xx}=\rho_{xx}/(\rho_{xx}^2+\rho_{xy}^2) \sim \rho_{xx}/\rho_{xy}^2$. A specifically designed dilution refrigerator was used which reached 64 mK in a hybrid magnet at 28.6 T. The specimens were immersed in the dilute ^3He-^4He liquid near a carbon resistor thermometer. No significant temperature hysteresis was observed during temperature sweeps and the slight magnetoresistance of the carbon resistor was determined as in Naughton *et al.*[14]

The temperature dependences of ρ_{xx} and σ_{xx} have been previously interpreted as activation energies in the FQHE.[7,13,15] The value of ρ_{xx} or σ_{xx} at the minimum corresponding to a given fractional factor is determined as a function of temperature from 120 mK to 1.4 K. Figure 1 shows such graphs for $\nu = \frac{2}{3}$ at two different magnetic fields.

The data of Fig. 1(a) follow a straight line, indicating activated conduction. The activation energy, Δ, is determined from $\rho_{xx} = \rho_0 \exp(-\Delta/2T)$. As defined here, Δ represents the quasiparticle pair-creation energy. (Note that this definition of Δ differs from the activation energies defined in Refs. 7, 13, and 15 by a factor of 2.) At higher T, ρ_{xx} deviates from a simple activated dependence as a result of the weak minimum riding on a slightly temperature-dependent background. All data taken at magnetic fields between 6 and ~ 10 T indicate simple activated behavior. Data taken at $B \gtrsim 10$ T deviate from simple activated dependence also at the lowest temperatures.

Figures 1(b) and 1(c) show the two qualitatively different types of low-temperature behavior. The majority (eleven out of fourteen) of our high-field data resemble Fig. 1(b) in which the deviation is smooth and curved. Data like those in Fig. 1(b) fit very well over the entire temperature range to a sum of activated conduction at higher T and hopping conduction at lower T [solid curve in Fig. 1(b)]. This dependence suggests that the quasiparticles in the FQHE become localized at low temperatures, in analogy to the localization of electrons in the IQHE.[16,17]

The formula used to model the hopping conduction in a magnetic field is from Ono,[18] $\sigma = \sigma_0(T) \times \exp[-(T_0/T)^{1/2}]$, although the two-dimensional Mott variable-range hopping formula,[19] $\sigma = \sigma_0(T) \times \exp[-(T_0/T)^{1/3}]$, fits the data as well. The data could not be fitted if we assumed only hopping conduction over the entire temperature range. Ihm and Phillips[20] have suggested the existence of a second activation energy in the FQHE, due to excitations of electrons which, in the presence of potential fluctuations, have not condensed into the Laughlin quantum liquid. Attempts to fit the data by use of two activation energies are equally successful. In any case, the activation energy at higher T is only slightly dependent on the formula chosen to fit the lower-T data. This dependence is within the error bars given in Fig. 2.

Three sets of data, all from sample D, resemble Fig. 1(c) in which the deviation from simple activated behavior is a sharp break to a second linear region. This behavior is similar to that reported by Kawaji *et al.*[22] The existence of two linear regions does suggest two separate activation energies, Δ_1 and Δ_2. The linear regions of the data are best described by $\Delta_1 = 6.4$ K and $\Delta_2 = 1.4$ K. However, a plot of their sum,

$$\sigma_{xx} = \sigma_1 \exp(-\Delta_1/2T) + \sigma_2 \exp(-\Delta_2/2T),$$

does not adequately fit the sharp break in the data. A

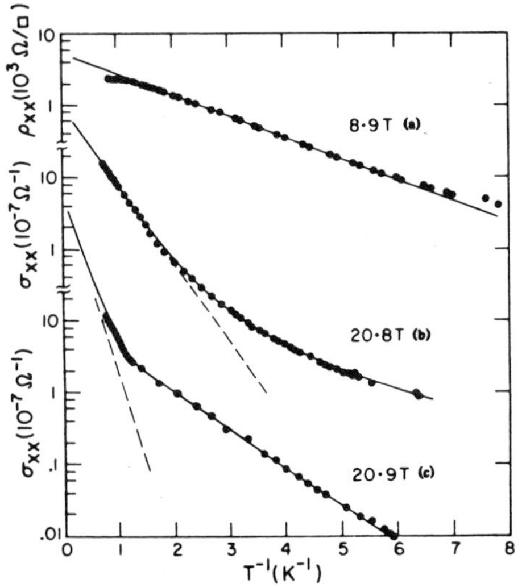

FIG. 1. Temperature dependence of the minimum at $\nu = \frac{2}{3}$. (a) at $B = 8.9$ T, showing simple activated behavior; (b) at $B = 20.8$ T, showing the smooth, curved deviation from activated behavior at lower temperatures; (c) at $B = 20.9$ T, showing the sharp, linear deviation.

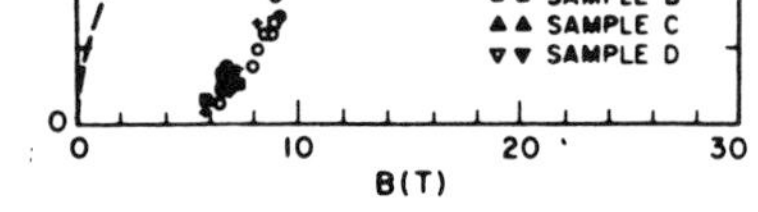

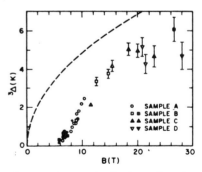

FIG. 2. Activation energies for the minima of $\nu = \frac{1}{3}, \frac{2}{3}, \frac{4}{3}, \frac{5}{3}$ vs magnetic field. Open symbols indicate data from $\nu = \frac{2}{3}$. Filled symbols at $B = 5.9$ and 7.3 T are from $\nu = \frac{5}{3}$ and $\frac{4}{3}$, respectively. All other filled symbols are from $\nu = \frac{1}{3}$. The dashed line is given by $0.03e^2/\epsilon l_0$. Data from sample A are from Ref. 13. The four data points shown by plusses are from Refs. 7 and 22.

successful least-squares fit using the above equation yields $\Delta_1 = 11.4$ K and $\Delta_2 = 2.4$ K [solid curve in Fig. 1(c)]. However, Δ_1 and Δ_2 exhibit wide variations apparently uncorrelated with B. (At $B = 19.3$ T, $\Delta_1 = 4.2$ K and $\Delta_2 = 1.4$ K. At $B = 26.8$ T, $\Delta_1 = 6.0$ K and $\Delta_2 = 1.8$ K.) In view of the fact that data sets containing a sharp break were the exception and are uncorrelated with magnetic field, we regard them as an artifact probably caused by a nonequilibrium configuration of the electronic state within the sample. An additional observation is consistent with this speculation: A second set of data taken immediately after the data in Fig. 1(c) but with a slower cooling rate follows a smooth curve similar to Fig. 1(b).

Figure 2 presents the activation energies from the data on $\nu = \frac{1}{3}, \frac{2}{3}, \frac{4}{3}$, and $\frac{5}{3}$. The three atypical sets of data resembling Fig. 1(c) have not been included. The data from Kawaji *et al.*[22] are not included because they also resemble Fig. 1(c). Four features of Fig. 2 should be stressed:

(1) There is no apparent sample dependence among these samples of similar mobility.

(2) The data for $\nu = \frac{1}{3}$ and $\frac{2}{3}$ overlap at $B \sim 20$ T. Also, the data for $\nu = \frac{4}{3}$ and $\frac{5}{3}$ are consistent with the data for $\nu = \frac{2}{3}$ at similar magnetic fields. This suggests a single activation energy, ${}^3\Delta$, for all of the filling factors: $\nu = \frac{1}{3}, \frac{2}{3}, \frac{4}{3}$, and $\frac{5}{3}$.

(3) The observed activation energies are much smaller than theoretically predicted. These theories all yield quasiparticle pair-creation energies for $\nu = \frac{1}{3}$ and $\frac{2}{3}$ of the form $\Delta = Ce^2/\epsilon l_0$, where $\epsilon \sim 12.8$ is the dielectric constant of GaAs. The constant of proportionality, C, is model dependent. From hypernetted-chain calculations, Laughlin determines[12] $C = 0.056$ and Chakraborty determines $C = 0.053$.[23] Monte Carlo calculations by Morf and Halperin give $C = 0.094$.[24] From calculations on finite numbers of electrons, Haldane has extrapolated to $N \rightarrow \infty$ to yield $C = 0.105$.[25] A single-mode approximation, in analogy with Feynman's theory for ^{4}He by Girvin, MacDonald, and Platzman yields $C = 0.106$.[26] To compare these results with our experimental data, Fig. 2 contains a curve of ${}^3\Delta$ vs B for $C = 0.030$, almost a factor of 2 smaller than the lowest theoretical value.

It has been suggested that the quasiparticle-quasihole interactions can result in bound states. Laughlin[27] discusses a quasiexciton state with a minimum energy equivalent to $C = 0.014$ at $kl_0 = 0$, where k is the wave vector of the quasiexciton. However, this calculation is unreliable for $k \rightarrow 0$. Haldan and Rezayi[25] and Girvin, MacDonald, and Platzman[26] find a roton minimum, analogous to the roton minimum in superfluid ^{4}He, at $kl_0 \sim 1.4$, where $C \sim 0.075$ for both. These activation energies, with the exception of Laughlin's quasiexciton, lie well above the observed values. It is not clear, however, that the electrically neutral bound states would be observed in magnetotransport measurements.

(4) ${}^3\Delta$ vs B does not follow the predicted $B^{1/2}$ magnetic field dependence. Rather, the phenomenon has a finite threshold at $B \sim 5.5$ T. For higher magnetic fields, there is a roughly linear increase in ${}^3\Delta$ up to $B \sim 18$ T, followed by an apparent saturation of ${}^3\Delta \sim 5.2$ K for $B \gtrsim 18$ T.

For completeness, we have also studied the temperature dependence of ρ_{xx} at $\nu = \frac{2}{5}$ and $\frac{3}{5}$ for 14 T $\lesssim B \lesssim 28$ T. Within our temperature range, ρ_{xx} changes by less than an order of magnitude. The data deviate from simple activated behavior but can be fitted with activated conduction at higher T and any of the discussed models at lower T. Attempts to fit the data over the entire temperature range with a hopping conduction formula were unsuccessful. The fifteen sets of experimental data again suggest a single activation energy, ${}^5\Delta$, for $\nu = \frac{2}{5}$ and $\frac{3}{5}$, which varies monotonically from ${}^5\Delta \sim 1.4$ K at $B = 14$ T to ${}^5\Delta \sim 2.5$ K at $B = 28$ T. Halperin[11] estimates that the pair-creation energies at $\nu = p/q$ should scale as $q^{-2.5}$, which yields ${}^5\Delta \sim 0.28\,{}^3\Delta$, and $C \sim 0.015$–0.030 for $\nu = \frac{2}{5}$ and $\frac{3}{5}$. This corresponds to ${}^5\Delta \gtrsim 2.9$ K and $\gtrsim 4.0$ K at $B = 14$ and 28 T, respectively. A more extensive description of these observations at $\nu = \frac{2}{5}$ and $\frac{3}{5}$ will be given in a forthcoming article.

There exists a startling discrepancy between the experimental results and the theoretical calculations of C and the magnetic field dependence of ${}^3\Delta$ which remains to be explained. A reduction of the many-particle gap due to disorder and subsequent thermal excitation to a mobility edge provides a qualitative explanation for the reduced values of C as well as for the finite threshold field.[13] Recent theoretical work is attempting to assess quantitatively the effects of disorder and finite thickness of the two-dimensional electron system on the energy gaps in the FQHE.[28] Initial results from these calculations arrive at an average Δ which approaches the experimental results; however, the observed B dependence is not well reproduced.

In conclusion, we find a single activation energy, ${}^3\Delta$, for $\nu = \frac{1}{3}, \frac{2}{3}, \frac{4}{3}$, and $\frac{5}{3}$ in magnetic fields up to 28 T. ${}^3\Delta$ is much smaller than expected and does not exhibit the expected $B^{1/2}$ magnetic field dependence. Instead, ${}^3\Delta$ has a finite magnetic field threshold above which it has a roughly linear increase with magnetic field. ${}^3\Delta$ apparently saturates at the highest magnetic fields. We also find a single activation energy, ${}^5\Delta$, for $\nu = \frac{2}{5}$ and $\frac{3}{5}$. ${}^5\Delta$ is also much smaller than predicted. The discrepancies between these experimental results and the theoretical values remain to be explained.

We thank P. A. Wolff and L. Rubin for support at the National Magnet Laboratory; J. C. M. Hwang for growth of the molecular-beam-epitaxy crystals; and R. B. Laughlin, S. M. Girvin, and J. C. Phillips for discussions. One of us (G.S.B.) is supported by the Hertz Foundation, and one of us (D.C.T.) is supported by the National Science Foundation through Grant No. DMR-8212167. The Francis Bitter National Magnet Laboratory is supported by the National Science Foundation through its Division of Materials Research.

---

[a]Guest Scientist at the Francis Bitter National Magnet Laboratory, Cambridge, Mass. 02139.

[1]D. C. Tsui, H. L. Stormer, and A. C. Gossard, Phys. Rev. Lett. **48**, 1559 (1982).

[2]H. L. Stormer, A. Chang, D. C. Tsui, J. C. M. Hwang, A.C. Gossard and W. Wiegmann, Phys. Rev. Lett. **50**, 1953 (1983).

[3]A. M. Chang, P. Berglund, D. C. Tsui, H. L. Stormer, and J. C. M. Hwang, Phys. Rev. Lett. **53**, 997 (1984).

[4]G. S. Boebinger, A. M. Chang, H. L. Stormer, and D. C. Tsui, to be published.

[5]E. E. Mendez, M. Heiblum, L. L. Chang, and L. Esaki, Phys. Rev. B **28**, 4886 (1983).

[6]E. E. Mendez, L. L. Chang, M. Heiblum, L. Esaki, M. Naughton, K. Martin, and J. Brooks, Phys. Rev. B **30**, 7310 (1984).

[7]G. Ebert, K. von Klitzing, J. C. Maan, G. Remenyi, C. Probst, G. Weimann, and W. Schlapp, J. Phys. C **17**, L775 (1984).

[8]For example, B. Tausenfreund and K. von Klitzing, Surf. Sci. **142**, 220 (1984).

[9]R. B. Laughlin, Phys. Rev. Lett. **50**, 1395 (1983).

[10]F. D. M. Haldane, Phys. Rev. Lett. **51**, 605 (1983).

[11]B. I. Halperin, Phys. Rev. Lett. **52**, 1583 (1984).

[12]R. B. Laughlin, Surf. Sci. **142**, 163 (1984).

[13]A. M. Chang, M. A. Paalanen, D. C. Tsui, H. L. Stormer, and J. C. M. Hwang, Phys. Rev. B **28**, 6133 (1983), and unpublished data.

[14]M. J. Naughton, S. Dickinson, R. C. Samaratunga, J. S. Brooks, and K. P. Martin, Rev. Sci. Instrum. **54**, 1529 (1983).

[15]D. C. Tsui, H. L. Stormer, J. C. M. Hwang, J. S. Brooks, and M. J. Naughton, Phys. Rev. B **28**, 2274 (1983).

[16]D. C. Tsui, H. L. Stormer, and A. C. Gossard, Phys. Rev. B **25**, 1405 (1982).

[17]G. Ebert, K. von Klitzing, C. Probst, E. Schuberth, K. Ploog, and G. Weimann, Solid State Commun. **45**, 625 (1983).

[18]Y. Ono, J. Phys. Soc. Jpn. **51**, 237 (1982).

[19]N. F. Mott and E. A. Davis, *Electronic Processes in Non-Crystalline Materials* (Clarendon, Oxford, 1979), 2nd ed.

[20]J. Ihm and J. C. Phillips, J. Phys. Soc. Jpn. **54**, 1506 (1985).

[21]V. M. Pudalov and S. G. Semenchinsky, Solid State Commun. **52**, 567 (1984).

[22]S. Kawaji, J. Wakabayashi, J. Yoshino, and H. Sakaki, J. Phys. Soc. Jpn. **53**, 1915 (1984).

[23]T. Chakraborty, Phys. Rev. B **31**, 4026 (1985).

[24]R. Morf and B. I. Halperin, to be published.

[25]F. D. M. Haldane and E. H. Rezayi, Phys. Rev. Lett. **54**, 237 (1985).

[26]S. M. Girvin, A. H. MacDonald, and P. M. Platzman, Phys. Rev. Lett. **54**, 581 (1985).

[27]R. B. Laughlin, Physica (Amsterdam) **126B**, 254 (1984).

[28]A. H. MacDonald, S. M. Girvin, and P. M. Platzman, Bull. Am. Phys. Soc. **30**, 301 (1985), and to be published.

Journal of the Physical Society of Japan
Vol. 56, No. 9, September, 1987, pp. 3005-3008

Second Activation Energy in the Fractional Quantum Hall Effect

Junichi WAKABAYASHI, Satoru SUDOU, Shinji KAWAJI,
Kazuhiko HIRAKAWA† and Hiroyuki SAKAKI†

Department of Physics, Gakushuin University, Mejiro, Toshima-ku, Tokyo 171
†Institute of Industrial Science, University of Tokyo, Roppongi, Minato-ku, Tokyo 106

(Received July 14, 1987)

The temperature dependence of the resistivity minima of the 2/3 effect in the fractional quantum Hall effect has been measured for a GaAs/AlGaAs heterostructure with a backside gate. The results at the highest negative gate bias have shown a single activated conduction. The systematic change of the temperature dependence controlled by the gate bias indicates that this activation energy does not correspond to the excitation energy of the quasi-particles, that is, there is a second activation process in the fractional quantum Hall effect.

The temperature dependence of the resistivities ρ_{xx} and ρ_{xy} in the fractional quantum Hall effect (FQHE) suggests that there is a finite energy gap in the excitation spectrum. The activated temperature dependence of ρ_{xx} minima at each fractional filling ν is considered to originate from the excitation of the fractionally charged quasi-particles which carry the current.[1] The filling factor ν is defined by $\nu=N_s h/eB$ where N_s is the areal electron density and eB/h is the degeneracy of a Landau level.

Different types of temperature dependence of ρ_{xx} have been observed by several groups.[2-6] Chang *et al.*[2] observed a single activated temperature dependence at $\nu=2/3$ in a sample with a backside gate. The two activated conduction processes in the FQHE were first introduced by Kawaji *et al.*[3] for the temperature dependence at $\nu=1/3$ and 2/3 in a very high mobility sample. Boebinger *et al.*[4] analyzed their data considering three types of temperature dependence. Wakabayashi *et al.*[5,6] have performed systematic experiments on the temperature dependence of ρ_{xx} in several samples having a wide range of mobility. Their results at $\nu=1/3$ in three samples were all expressed well by a sum of two activated conduction processes as

$$\rho_{xx}(T)=\rho_{01}\,\mathrm{e}^{-W_1/T}+\rho_{02}\,\mathrm{e}^{-W_2/T}, \qquad (1)$$

where W_1 is the activation energy measured in kelvin in the high-temperature region and W_2 is that in the low-temperature region. The results at $\nu=2/3$ were also expressed as a sum of two activated conduction processes although the separation of the two processes was not so clear as at $\nu=1/3$. The activation energy W_1 in the high-temperature region was considered to correspond to the theoretical excitation energy of the quasi-particles, based on the magnitude of the activation energy. The results at low temperatures by Wakabayashi *et al.*[6] have error bars corresponding to one standard deviation and, in particular, the temperature dependence at low temperatures of the 2/3 effect indicated a weak deviation from the activated temperature dependence at high temperatures. Therefore, it was difficult to confirm that the temperature dependence at low temperatures was an activation type and not the hopping conduction type which gives rise to the temperature dependence of the form of $A/T\cdot\exp[-(T_0/T)^{1/2}]$[7] or $A/T^{2/3}\cdot\exp[-(T_0/T)^{1/3}]$,[8] where A and T_0 are constants relating to the electron-phonon coupling and the density of states at the Fermi level, respectively.

In the present paper, we report on the temperature dependence of resistivity ρ_{xx} at $\nu=2/3$. We have extended the temperature range well below the temperature where the deviation from the activated dependence at high temperatures was previously observed. The calibration procedure of the carbon thermometer is described elsewhere.[6)] The measurement has been done during the sweeping down of the temperature. The temperature sweep was stopped at a couple of intermediate temperatures to check the thermal equilibrium condition. The sample used was a modulation-doped GaAs/$Al_{0.3}Ga_{0.7}As$ heterostructure with a backside gate made by molecular beam epitaxy. The concentration of silicon donors and the thickness of the doped AlGaAs layer were 1.7×10^{23} m^{-2} and 110 nm, respectively. The spacer, the undoped AlGaAs layer, was 30 nm thick. The backside gate electrode was made by gold deposition after polishing the Cr-doped semi-insulating GaAs substrate to the thickness of about 100 μm. The width and the length of the sample were 50 and 600 μm, respectively, and potential probes were located 1/3 and 2/3 along the sample. The electron density N_s and the electron mobility μ at zero gate bias were 1.9×10^{15} m^{-2} and 49 m^2/V·s, respectively. The electron density was varied down to 1.5×10^{15} m^{-2} by applying a negative bias of -58 V to the gate and varied up to 2.5×10^{15} m^{-2} by applying a positive bias of 86 V. Concurrently, the electron mobility was varied between 27 m^2/V·s and 88 m^2/V·s. The electron mobility has shown a power-law dependence on the electron density as $\mu\propto N_s^{2.4}$. The usual lock-in technique at 31 ~ 37 Hz was used to measure ρ_{xx} and ρ_{xy}.

Figure 1 shows the temperature dependence of ρ_{xx} at each gate bias between -58 V and 86 V. A current of 0.2 nA was used for the data between 9.2 T and 11.9 T and 2.0 nA for the data between 12.9 T and 15.2 T. Each solid line except the top curve in Fig. 1 represents a fitted curve obtained by the eq. (1). The top curve was fitted by a single activation. It has been shown that the activation energy W_1 depends systematically on μB rather than magnetic field B.[5,6)] Figure 2 shows the μB dependence of the activation energies and the pre-exponential factors obtained from the curve fitting shown in Fig. 1. The activation energy W_1 and the pre-exponential factor ρ_{01} increases as μB is increased by biassing the backside gate positively. On the contrary, ρ_{02} decreases rapidly as μB increases. These results show that the W_1 process is the dominant process at $\mu B=\omega_c\tau=\infty$, that is, it is the process in an ideal sample. This demonstrates clearly that W_1 corresponds to the excitation energy of quasi-particles in the FQHE. Similar μB dependence of the activation energies and the pre-exponential factors has been observed for the 1/3 effect.[9)]

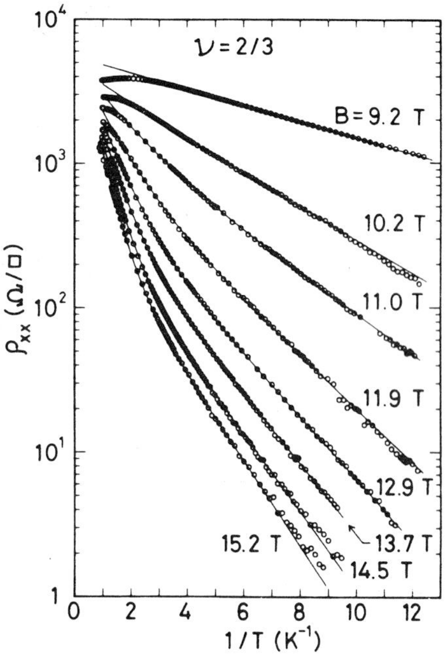

Fig. 1. Temperature dependence of ρ_{xx} minima at $\nu=2/3$ in a single sample with a backside gate. The current used is 0.2 nA for the field between 9.2 T and 11.9 T, and 2.0 nA between 12.9 T and 15.2 T. Each solid line represents a fitted curve for $\rho_{xx}(T)=\rho_{01}\exp(-W_1/T)+\rho_{02}\exp(-W_2/T)$ except the top curve.

When a large negative bias was applied to the backside gate, the temperature dependence of ρ_{xx} showed a single activated conduction as shown by the top curve in Fig. 1. This is clearly shown in Fig. 3 by plotting the deviation of $\rho_{xx}(T)$ from the single activated conduction against $1/T$. The deviation below 90 mK is presumably due to the breakdown of the thermal contact between the sample and the copper heatsink. It is important that the activated process in this case is the W_2 process, which dominates at low temperatures, and not the W_1 process, which corresponds to the excitation of quasi-particles. This can be seen from the μB dependence of the pre-exponential fac-

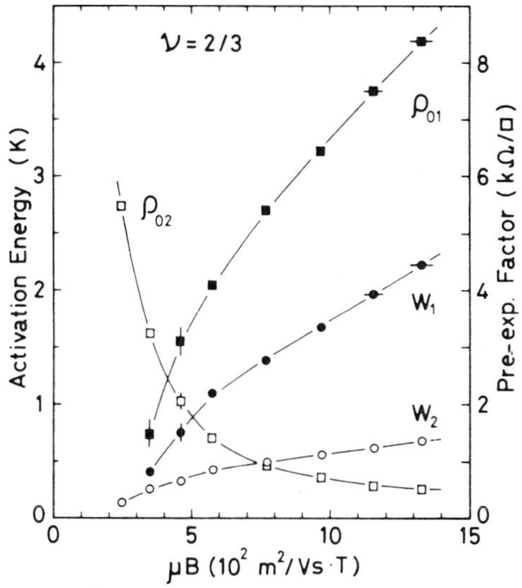

Fig. 2. Pre-exponential factor ρ_{01}, ρ_{02} and the activation energy W_1 and W_2 against μB obtained from the data in Fig. 1.

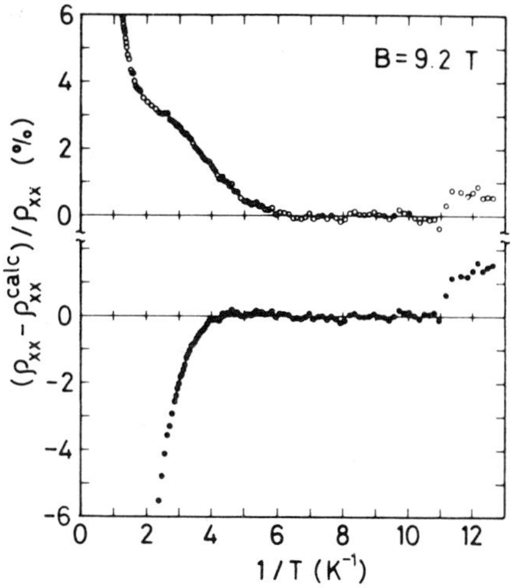

Fig. 3. Deviation of $\rho_{xx}(T)$ at $B=9.2$ T in Fig. 1 from the calculated resistivity of a single activation type (closed circle) and 2D hopping conduction type (open circle).

tors ρ_{01} and ρ_{02} shown in Fig. 2, that is, $\rho_{01} \ll \rho_{02}$ when a large negative bias is applied.

It is worthwhile to see whether hopping conduction is a possible process which can explain the temperature dependence of ρ_{xx} at low temperatures. It was generally difficult to fit the formula of Mott's variable range hopping[8] to the data in Fig. 1. The deviation of $\rho_{xx}(T)$ at $B=9.2$ T from the 2D hopping conduction formula derived by Ono[7] is shown in Fig. 3. It may appear to have a good fit at low temperatures. However, we consider the hopping conduction to be unlikely for the following reasons. First, it is difficult to explain reasonably the deviation in high temperatures shown in Fig. 3. Secondly, the hopping conduction is not suitable to explain such a high conductive state ($\sigma_{xx} \sim \rho_{xx}/\rho_{xy}^2$ in the present system, where ρ_{xy} is a constant). Finally, the density of states at the Fermi level $D(E_f)$ obtained from the parameter T_0[7] is two orders of magnitude larger than the density of states in a zero magnetic field. The density of states $D(E_f)$ obtained from the parameter T_0 should give the quasi-particle density of states. It is unreasonable that $D(E_f)$ is greater than the density of states in a zero magnetic field because the Fermi level lies in the energy gap at $\nu=2/3$ in the FQHE.

On the other hand, it is also possible to fit the data in high magnetic fields using the 2D hopping conduction formula[7] for the second term in the eq. (1). However, it should be noted that these data in high magnetic fields at low temperatures, that is, low values of ρ_{xx}, are easily affected by the residual resistance due to, for example, inhomogeneities in the sample. We consider that the second activation process does not change continuously to the hopping conduction process as the magnetic field is increased. However, there is a possibility that the hopping process may appear as a third conduction process when the temperature and the resistivity are further lowered.

From the above discussions, we conclude that the conduction process at low temperatures represented by W_2 is also activated. In other words, there is a second activation process in the FQHE.

Finally, we discuss the possible origin of the W_2 process. Results in Fig. 2 show that the W_2 process becomes dominant when a negative bias is applied to the backside gate and the mobility is decreased. This fact suggests that W_2 process is highly correlated with the disorder. Ihm and Phillips[10] discussed the two activation processes in terms of the coexistence of the condensed quantum liquid region and the uncondensed region in the sample. According to their results, the activation energies Δ_1 and Δ_2 should satisfy the inequality

$$\Delta_1/2<\Delta_2<\Delta_1. \quad (2)$$

Our early data[3,11] have indicated such relations between Δ_1 and Δ_2 which were determined straightforwardly from a log ρ_{xx} versus $1/T$ plot. However, it has been generally found that the recent data[5,6,9] of activation energy W_1 and W_2 do not satisfy the inequality (2), that is, $W_1>2W_2$.

Another possible origin of the second activation energy W_2 is the energy at the "roton minimum" appearing in the collective excitation spectrum of the quasi-particles.[12-14] Because a "roton" is easily created around impurities, the excitation energy will become smaller as impurities increase. This explains qualitatively the results of W_2 and ρ_{02} in Fig. 2. However, collective excitation at the "roton minimum" is a bound quasi-electron-quasi-hole pair which is neutral and cannot carry the current. Hence, there must be another process to decompose the bound quasi-electron-quasi-hole pair. If the decomposing process is faster than the excitation process, an activated conduction process corresponding to the excitation to the "roton minimum" will appear in the transport measurements. On the other hand, if the decomposing process is the rate-determining process, the decomposing energy, that is, the difference between the zone boundary energy and the roton minimum energy, will appear in the transport measurements.[14] However, the origin of such a decomposing process is not clear at present.

In summary, we have reported on the systematic measurements of the temperature dependence of the resistivity minima of the 2/3 fractional quantum Hall effect using a backside gate. The temperature dependence at each gate bias was expressed by a sum of two activated conduction processes. The μB dependence of the pre-exponential factors and the activation energies has shown that the activation energy in a high-temperature region corresponds to the excitation energy of the quasi-particles. A single activated conduction process observed at the highest negative gate bias indicates that there is a second activation process in the fractional quantum Hall effect.

Acknowledgments

The authors thank Professor D. Yoshioka for valuable discussions. This work is supported by a Grant-in-Aid for Special Distinguished Research from the Ministry of Education, Culture and Science.

References

1) *The Quantum Hall Effect*, ed. R. E. Prange and S. M. Girvin (Springer-Verlag, 1987).
2) A. M. Chang, M. A. Paalanen, D. C. Tsui, H. L. Stormer and J. C. M. Hwang: Phys. Rev. **B28** (1983) 6133.
3) S. Kawaji, J. Wakabayashi, J. Yoshino and H. Sakaki: J. Phys. Soc. Jpn. **53** (1984) 1915.
4) G. S. Boebinger, A. M. Chang, H. L. Stormer and D. C. Tsui: Phys. Rev. Lett. **55** (1985) 1606.
5) J. Wakabayashi, S. Kawaji, J. Yoshino and H. Sakaki: Surf. Sci. **170** (1986) 136.
6) J. Wakabayashi, S. Kawaji, J. Yoshino and H. Sakaki: J. Phys. Soc. Jpn. **55** (1986) 1319.
7) Y. Ono: J. Phys. Soc. Jpn. **51** (1982) 237.
8) N. F. Mott and E. A. Davis: *Electronic Properties in Non-Crystalline Materials* (Clarendon, Oxford, 2nd ed.)
9) J. Wakabayashi, S. Sudou, S. Kawaji, K. Hirakawa and H. Sakaki: *Proc. 7th Int. Conf. on the Electronic Properties of Two-Dimensional Systems, Santa Fe, 1987*, to be published.
10) J. Ihm and J. C. Phillips: J. Phys. Soc. Jpn. **54** (1985) 1506.
11) J. Wakabayashi, S. Kawaji, J. Yoshino and H. Sakaki: *Proc. 17th Int. Conf. Physics of Semicond., San Francisco, 1984* (Springer-Verlag, New York, 1985) p. 283.
12) F. D. M. Haldane and E. H. Rezayi: Phys. Rev. Lett. **54** (1985) 237.
13) S. M. Girvin, A. H. MacDonald and P. M. Platzman: Phys. Rev. **B33** (1986) 2481.
14) D. Yoshioka: J. Phys. Soc. Jpn. **55** (1986) 885.

Anomalous Quantum Hall Effect: An Incompressible Quantum Fluid with Fractionally Charged Excitations

R. B. Laughlin
Lawrence Livermore National Laboratory, University of California, Livermore, California 94550
(Received 22 February 1983)

This Letter presents variational ground-state and excited-state wave functions which describe the condensation of a two-dimensional electron gas into a new state of matter.

PACS numbers: 71.45.Nt, 72.20.My, 73.40.Lq

The "$\frac{1}{3}$" effect, recently discovered by Tsui, Störmer, and Gossard,[1] results from the condensation of the two-dimensional electron gas in a GaAs-Ga_xAl_{1-x}As heterostructure into a new type of collective ground state. Important experimental facts are the following: (1) The electrons condense at a particular density, $\frac{1}{3}$ of a full Landau level. (2) They are capable of carrying electric current with little or no resistive loss and have a Hall conductance of $\frac{1}{3}e^2/h$. (3) Small deviations of the electron density do not affect either conductivity, but large ones do. (4) Condensation occurs at a temperature of ~1.0 K in a magnetic field of 150 kG. (5) The effect occurs in some samples but not in others. The purpose of this Letter is to report variational ground-state and excited-state wave functions that I feel are consistent with all the experimental facts and explain the effect. The ground state is a new state of matter, a quantum fluid the elementary excitations of which, the quasielectrons and quasiholes, are fractionally charged. I have verified the correctness of these wave functions for the case of small numbers of electrons, where direct numerical diagonalization of the many-body Hamiltonian is possible. I predict the existence of a sequence of these ground states, decreasing in density and terminating in a Wigner crystal.

Let us consider a two-dimensional electron gas in the x-y plane subjected to a magnetic field H_0 in the z direction. I adopt a symmetric gauge vector potential $\vec{A} = \frac{1}{2}H_0[x\hat{y} - y\hat{x}]$ and write the eigenstates of the ideal single-body Hamiltonian $H_{sp} = |(\hbar/i)\nabla - (e/c)\vec{A}|^2$ in the manner

$$|m,n\rangle = (2^{m+n+1}\pi m!\,n!)^{-1/2}\exp[\tfrac{1}{4}(x^2+y^2)]\left(\frac{\partial}{\partial x} + i\frac{\partial}{\partial y}\right)^m\left(\frac{\partial}{\partial x} - i\frac{\partial}{\partial y}\right)^n \exp[-\tfrac{1}{2}(x^2+y^2)], \quad (1)$$

with the cyclotron energy $\hbar\omega_c = \hbar(eH_0/mc)$ and the magnetic length $a_0 = (\hbar/m\omega_c)^{1/2} = (\hbar c/eH_0)^{1/2}$ set to 1. We have

$$H_{sp}|m,n\rangle = (n+\tfrac{1}{2})|m,n\rangle. \quad (2)$$

The manifold of states with energy $n+\frac{1}{2}$ constitutes the nth Landau level. I abbreviate the states of the lowest Landau level as

$$|m\rangle = (2^{m+1}\pi m!)^{-1/2} z^m \exp(-\tfrac{1}{4}|z|^2), \quad (3)$$

where $z = x + iy$. $|m\rangle$ is an eigenstate of angular momentum with eigenvalue m. The many-body Hamiltonian is

$$H = \sum_j \{|(\hbar/i)\nabla_j - (e/c)\vec{A}_j|^2 + V(z_j)\} + \sum_{j>k} e^2/|z_j - z_k|, \quad (4)$$

where j and k run over the N particles and V is a potential generated by a uniform neutralizing background.

I showed in a previous paper[2] that the $\frac{1}{3}$ effect could be understood in terms of the states in the lowest Landau level solely. With $e^2/a_0 \lesssim \hbar\omega_c$, the situation in the experiment, quantization of interelectronic spacing follows from quantization of angular momentum: The only wave functions composed of states in the lowest Landau level which describe orbiting with angular momentum m about the center of mass are of the form

$$\psi = (z_1 - z_2)^m (z_1 + z_2)^n \exp[-\tfrac{1}{4}(|z_1|^2 + |z_2|^2)]. \quad (5)$$

My present theory generalizes this observation to N particles.

I write the ground state as a product of Jastrow functions in the manner

$$\psi = \{\prod_{j<k} f(z_j - z_k)\}\exp(-\tfrac{1}{4}\textstyle\sum_l |z_l|^2), \quad (6)$$

and minimize the energy with respect to f. We

observe that the condition that the electrons lie in the lowest Landau level is that $f(z)$ be polynomial in z. The antisymmetry of ψ requires that f be odd. Conservation of angular momentum requires that $\prod_{j<k} f(z_j - z_k)$ be a homogeneous polynomial of degree M, where M is the total angular momentum. We have, therefore, $f(z) = z^m$, with m odd. To determine which m minimizes the energy, I write

$$|\psi_m|^2 = |\{\prod_{j<k}(z_j - z_k)^m\}\exp(-\tfrac{1}{4}\sum_l |z_l|^2)|^2 = e^{-\beta\Phi}, \tag{7}$$

where $\beta = 1/m$ and Φ is a classical potential energy given by

$$\Phi = -\sum_{j<k} 2m^2 \ln|z_j - z_k| + \tfrac{1}{2} m \sum_l |z_l|^2. \tag{8}$$

Φ describes a system of N identical particles of charge $Q = m$, interacting via logarithmic potentials and embedded in a uniform neutralizing background of charge density $\sigma = (2\pi a_0^2)^{-1}$. This is the classical one-component plasma (OCP), a system which has been studied in great detail. Monte Carlo calculations[3] have indicated that the OCP is a hexagonal crystal when the dimensionless plasma parameter $\Gamma = 2\beta Q^2 = 2m$ is greater than 140 and a fluid otherwise. $|\psi_m|^2$ describes a system uniformly expanded to a density of $\sigma_m = m^{-1}(2\pi a_0^2)^{-1}$. It minimizes the energy when σ_m equals the charge density generating V.

In Table I, I list the projection of ψ_m for three particles onto the lowest-energy eigenstate of angular momentum $3m$ calculated numerically. These are all nearly 1. This supports my assertion that a wave function of the form of Eq. (6) has adequate variational freedom. I have done a similar calculation for four particles with Coulombic repulsions and find projections of 0.979 and 0.947 for the $m = 3$ and $m = 5$ states.

ψ_m has a total energy per particle which for small m is more negative than that of a charge-density wave (CDW).[4] It is given in terms of the radial distribution function $g(r)$ of the OCP by

$$U_{\text{tot}} = \pi \int_0^\infty \frac{e^2}{r}[g(r) - 1] r\,dr. \tag{9}$$

In the limit of large Γ, U_{tot} is approximated within a few percent by the ion disk energy:

$$U_{\text{tot}} \simeq -\sigma_m \int \frac{e^2}{|r|} d^2r + \frac{\sigma_m^2}{2} \iint \frac{e^2}{|r_{12}|} dr_1^2\, dr_2^2 = (4/3\pi - 1)2e^2/R, \tag{10}$$

where the integration domain is a disk of radius $R = (\pi\sigma_m)^{-1/2}$. At $\Gamma = 2$ we have the exact result[5] that $g(r) = 1 - \exp[-(r/R)^2]$, giving $U_{\text{tot}} = -\frac{1}{2}\pi^{1/2}e^2/R$. At $m = 3$ and $m = 5$ I have reproduced the Monte Carlo $g(r)$ of Caillol *et al.*[3] using the modified hypernetted chain technique described by them. I obtain $U_{\text{tot}} = (-0.4156 \pm 0.0012)e^2/a_0$ and $U_{\text{tot}}(5) = (-0.3340 \pm 0.0028)e^2/a_0$. The corresponding values for the charge-density wave[4] are $-0.389e^2/a_0$ and $-0.322e^2/a_0$. U_{tot} is a smooth function of Γ. I interpolate it crudely in the manner

$$U_{\text{tot}}(m) \simeq \frac{0.814}{\sqrt{m}}\left(\frac{0.230}{m^{0.64}} - 1\right)\frac{e^2}{a_0}. \tag{11}$$

This interpolation converges to the CDW energy near $m = 10$. The actual crystallization point cannot be determined from that of the OCP since the CDW has a lower energy than the crystal described by ψ_m for $m > 71$.

I generate the elementary excitations of ψ_m by piercing the fluid at z_0 with an infinitely thin solenoid and passing through it a flux quantum $\Delta\varphi = hc/e$ adiabatically. The effect of this operation on the single-body wave functions is

$$(z - z_0)^m \exp(-\tfrac{1}{4}|z|^2) \to (z - z_0)^{m+1}\exp(-\tfrac{1}{4}|z|^2). \tag{12}$$

Let us take as approximate representations of these excited states

$$\psi_m^{+z_0} = A_{z_0}\psi_m = \exp(-\tfrac{1}{4}\sum_l |z_l|^2)\{\prod_i (z_i - z_0)\}\{\prod_{j<k}(z_j - z_k)^m\}, \tag{13}$$

and

$$\psi_m^{-z_0} = A_{z_0}^{+}\psi_m = \exp(-\tfrac{1}{4}\sum_l |z_l|^2)\left\{\prod_i \left(\frac{\partial}{\partial z_i} - \frac{z_0}{a_0^2}\right)\right\}\{\prod_{j<k}(z_j - z_k)^m\}, \tag{14}$$

TABLE I. Projection of variational three-body wave functions ψ_m in the manner $\langle\psi_m|\Phi_m\rangle/(\langle\psi_m|\psi_m\rangle\langle\Phi_m|\Phi_m\rangle)^{1/2}$. Φ_m is the lowest-energy eigenstate of angular momentum $3m$ calculated with $V = 0$ and an interelectronic potential of either $1/r$, $-\ln(r)$, or $\exp(-r^2/2)$.

m	$1/r$	$-\ln(r)$	$\exp(-r^2/2)$
1	1	1	1
3	0.999 46	0.996 73	0.999 66
5	0.994 68	0.991 95	0.999 39
7	0.994 76	0.992 95	0.999 81
9	0.995 73	0.994 37	0.999 99
11	0.996 52	0.995 42	0.999 96
13	0.997 08	0.996 15	0.999 85

for the quasihole and quasielectron, respectively. For four particles, I have projected these wave functions onto the analogous ones computed numerically. I obtain 0.998 for ψ_3^{-0} and 0.994 for ψ_5^{-0}. I obtain 0.982 for $\bar{\psi}_3^{+\bar{0}} = \{\prod_i (z_i - \bar{z})\}\psi_3$, which is ψ_3^{+0} with the center-of-mass motion removed.

These excitations are particles of charge $1/m$. To see this let us write $|\psi^{+z_0}|^2$ as $e^{-\beta\Phi'}$, with $\beta = 1/m$ and

$$\Phi' = \Phi - 2\sum_l \ln|z_l - z_0|. \tag{15}$$

Φ' describes an OCP interacting with a phantom point charge at z_0. The plasma will completely screen this phantom by accumulating an equal and opposite charge near z_0. However, since the plasma in reality consists of particles of charge 1 rather than charge m, the real accumulated charge is $1/m$. Similar reasoning applies to ψ^{-z_0} if we approximate it as $\prod_j (z_j - z_0)^{-1} P_{z_0}\psi_3$, where P_{z_0} is a projection operator removing all configurations in which any electron is in the single-body state $(z - z_0)^0 \exp(-\frac{1}{4}|z|^2)$. The projection of this approximate wave function onto $\psi_3^{-z_0}$ for four particles is 0.922. More generally, one observes that far away from the solenoid, adiabatic addition of $\Delta\varphi$ moves the fluid rigidly by exactly one state, per Eq. (12). The charge of the particles is thus $1/m$ by the Schrieffer counting argument.[6]

The size of these particles is the distance over which the OCP screens. Were the plasma weakly coupled ($\Gamma \leq 2$) this would be the Debye length $\lambda_D = a_0/\sqrt{2}$. For the strongly coupled plasma, a better estimate is the ion-disk radius associated with a charge of $1/m$: $R = \sqrt{2}\, a_0$. From the size we can estimate the energy required to make a particle. The charge accumulated around the phantom in the Debye-Hückel approximation is

$$\delta\rho = \frac{e/m}{2\pi\lambda_D^2} K_0(r/\lambda_D),$$

where K_0 is a modified Bessel function of the second kind. The energy required to accumulate it is

$$\Delta_{\text{Debye}} = \frac{1}{2}\iint \frac{\delta\rho\,\delta\rho}{|r_{12}|} = \frac{\pi}{4\sqrt{2}} \frac{1}{m^2} \frac{e^2}{a_0}. \tag{16}$$

This estimate is an upper bound, since the plasma is strongly coupled. To make a better estimate let $\delta\rho = \sigma_m$ inside the ion disk and zero outside, to obtain

$$\Delta_{\text{disk}} = \frac{3}{2\sqrt{2}\pi} \frac{1}{m^2} \frac{e^2}{a_0}. \tag{17}$$

For $m = 3$, these estimates are $0.062e^2/a_0$ and $0.038e^2/a_0$. This compares well with the value $0.033e^2/a_0$ estimated from the numerical four-particle solution in the manner

$$\Delta \simeq \frac{1}{2}\{E(\psi_3^{-0}) + E(\bar{\psi}_3^{+\bar{0}}) - 2E(\psi_3)\}, \tag{18}$$

where $E(\psi_3)$ denotes the eigenvalue of the numerical analog of ψ_3. This expression averages the electron and hole creation energies while subtracting off the error due to the absence of V. I have performed two-component hypernetted chain calculations for the energies of $\psi_3^{+z_0}$ and $\psi_3^{-z_0}$. I obtain $(0.022 \pm 0.002)e^2/a_0$ and $(0.025 \pm 0.005)e^2/a_0$. If we assume a value $\epsilon = 13$ for the dielectric constant of GaAs, we obtain $0.02e^2/\epsilon a_0 \simeq 4$ K when $H_0 = 150$ kG.

The energy to make a particle does not depend on z_0, so long as its distance from the boundary is greater than its size. Thus, as in the single-particle problem, the states are degenerate and there is no kinetic energy. We can expand the creation operator as a power series in z_0:

$$A_{z_0} = \sum_{j=0}^{N} A_j(z_1 \cdots, z_N) z_0^{N-j}. \tag{19}$$

These A_j are the elementary symmetric polynomials,[7] the algebra of which is known to span the set of symmetric functions. Since every antisymmetric function can be written as a symmetric function times ψ_1, these operators and their adjoints generate the entire state space. It is thus appropriate to consider them N linearly independent particle creation operators.

The state described by ψ_m is incompressible because compressing or expanding it is tantamount to injecting particles. If the area of the system is reduced or increased by δA the energy rises by $\delta U = \sigma_m \Delta |\delta A|$. Were this an elastic solid characterized by a bulk modulus B, we would have $\delta U = \frac{1}{2}B(\delta A)^2/A$. Incompressibility causes the longitudinal collective excitation roughly equivalent to a compressional sound wave to be absent, or more precisely, to have an energy $\sim\Delta$ in the long-wavelength limit. This facilitates current conduction with no resistive loss at zero temperature. Our prototype for this behavior is full Landau level ($m = 1$) for which this collective excitation occurs at $\hbar\omega_c$. The response of this system to compressive stresses is analogous to the response of a type-II superconductor to the application of a magnetic field. The system first generates Hall currents without compressing, and then at a critical stress collapses by an area quantum $m2\pi a_0^2$

and nucleates a particle. This, like a flux line, is surrounded by a vortex of Hall current rotating in a sense opposite to that induced by the stress.

The role of sample impurities and inhomogeneities in this theory is the same as that in my theory of the ordinary quantum Hall effect.[8] The electron and hole bands, separated in the impurity-free case by a gap 2Δ, are broadened into a continuum consisting of two bands of extended states separated by a band of localized ones. Small variations of the electron density move the Fermi level within this localized state band as the extra quasiparticles become trapped at impurity sites. The Hall conductance is $(1/m)\times(e^2/h)$ because it is related by gauge invariance to the charge of the quasiparticles e^* by $\sigma_{\rm Hall}=e^*e/h$, whenever the Fermi level lies in a localized state band. As in the ordinary quantum Hall effect, disorder sufficient to localize all the states destroys the effect. This occurs when the collision time τ in the sample in the absence of a magnetic field becomes smaller than $\tau < \hbar/\Delta$.

I wish to thank H. DeWitt for calling my attention to the Monte Carlo work and D. Boercker for helpful discussions. I also wish to thank P. A. Lee, D. Yoshioka, and B. I. Halperin for helpful criticism. This work was performed under the auspices of the U. S. Department of Energy by Lawrence Livermore National Laboratory under Contract No. W-7405-Eng-48.

[1]D. C. Tsui, H. L. Störmer, and A. C. Gossard, Phys. Rev. Lett. 48, 1559 (1982).

[2]R. B. Laughlin, Phys. Rev. B 27, 3383 (1983).

[3]J. M. Caillol, D. Levesque, J. J. Weis, and J. P. Hansen, J. Stat. Phys. 28, 325 (1982).

[4]D. Yoshioka and H. Fukuyama, J. Phys. Soc. Jpn. 47, 394 (1979); D. Yoshioka and P. A. Lee, Phys. Rev. B 27, 4986 (1983), and private communication.

[5]B. Jancovici, Phys. Rev. Lett. 46, 386 (1981). $\Gamma=2$ corresponds to a full Landau level, for which the total energy equals the Hartree-Fock energy $-\sqrt{\pi/8}\,e^2/a_0$. This correspondence may be viewed as the underlying reason an exact solution at $\Gamma = 2$ exists.

[6]W. P. Su and J. R. Schrieffer, Phys. Rev. Lett. 46, 738 (1981).

[7]S. Lang, *Algebra* (Addison-Wesley, Reading, Mass., 1965), p. 132.

[8]R. B. Laughlin, Phys. Rev. B 23, 5632 (1981).

Fractional Quantization of the Hall Effect: A Hierarchy of Incompressible Quantum Fluid States

F. D. M. Haldane
Department of Physics, University of Southern California, Los Angeles, California 90089
(Received 28 June 1983)

With use of spherical geometry, a translationally invariant version of Laughlin's proposed "incompressible quantum fluid" state of the two-dimensional electron gas is formulated, and extended to a hierarchy of continued-fraction Landau-level filling factors ν. Observed anomalies at $\nu=\frac{2}{5},\frac{2}{7}$ are explained by fluids deriving from a $\nu=\frac{1}{3}$ parent.

PACS numbers: 71.45.Nt, 72.20.Nt, 73.40.Lq

The quantum Hall effect (quantization of the Hall resistance $\rho_{xy}=h/\nu e^2$ at simple rational values of ν at low temperatures, together with a dramatic fall in the sheet resistance ρ_{xx}) observed in GaAs-$Ga_xAl_{1-x}As$ heterostructures[1,2] may be explained (naively) if the ground state of the two-dimensional (2D) electron gas in high perpendicular magnetic fields has *no gapless excitations* (and hence no dissipation at low temperatures) when the Landau-level occupation factor takes one of the quantized values ν. This is trivially the case for free electrons when ν is integer, as seen in the earlier experiments,[1] but the effect (or its precursor anomalies) has recently been observed[2] with *fractional* quantization, to date at $\nu=\frac{2}{7},\frac{1}{3},\frac{2}{5},\frac{3}{5},\frac{2}{3},\frac{4}{5},\frac{4}{3}$, and $\frac{5}{3}$, all with *odd* denominators (when $\nu>1$, the electrons are not fully spin-polarized; $\nu=\frac{4}{3},\frac{5}{3}$ values may be understood as the $\nu=1$ effect for majority spins, plus the $\nu=\frac{1}{3},\frac{2}{3}$ effects for minority spins). A "Wigner solid" charge-density-wave ground state is expected[3] at low occupations, but such a state has a gapless Goldstone mode because translational and rotational symmetry (described by the Euclidean group) is broken. A state without gapless excitations may instead be characterized[4,5] as an "incompressible quantum fluid," and variational wave functions of Jastrow form that describe such states have recently been proposed by Laughlin[4] at occupations $\nu=1/m$, m an odd integer.

The Laughlin wave functions are not translationally invariant, but describe a circular droplet of fluid, which must be confined in an external potential. Laughlin circumvented this problem by formally relating the properties of the fluid to those of the classical 2D one-component plasma, which has a thermodynamic limit, and calculating plasma properties. In this Letter, I describe a variant of Laughlin's scheme with fully translationally invariant wave functions, and extend it to describe a hierarchy of fluid states with occupation factors given by the continued fractions

$$\cfrac{1}{m+\cfrac{\alpha_1}{p_1+\cfrac{\alpha_2}{\ddots+\cfrac{\alpha_n}{p_n}}}}$$

where $m=1,3,5,\ldots$, $\alpha_i=\pm1$, and $p_i=2,4,6\ldots$; this number will be denoted by $[m,\alpha_1p_1,\alpha_2p_2,\ldots,\alpha_np_n]$, and is a rational with an *odd denominator*. The fluid state at $\nu=[m,p_1,\ldots,p_n]$ cannot occur unless its "parent" state at $\nu=[m,p_1,\ldots,p_{n-1}]$ also occurs; whether or not a given fluid state occurs will depend on the details of the interactions. The experimentally observed anomalies with $\nu<1$ correspond to $[3,2]$, $[3]$, $[3,-2]$, $[1,2,-2]$, $[1,2]$, and $[1,4]$; they all derive from the $m=1$ and $m=3$ hierarchies.

The technical innovation that I make is to place a 2D electron gas of N particles on a *spherical* surface of radius R, in a radial (monopole) magnetic field $B=\hbar S/eR^2$ (>0) where $2S$, the total magnetic flux through the surface in units of the flux quantum $\Phi_0=h/e$, is integral as required by Dirac's monopole quantization condition.[6] This device allows the construction of homogeneous states with finite N; in the limit R, N, and $S\to\infty$, the Euclidean group of the plane is recovered from the rotation group $O^+(3)$ of the sphere.

Single-particle states.—The single-particle Hamiltonian is

$$H=|\vec{\Lambda}|^2/2M=\tfrac{1}{2}\omega_c|\vec{\Lambda}|^2/\hbar S,$$

where M is the effective mass, and $\omega_c=eB/M$ is the cyclotron frequency. $\vec{\Lambda}=\vec{r}\times[-i\hbar\nabla+e\vec{A}(\vec{r})]$ is the dynamical angular momentum; $\nabla\times\vec{A}=B\hat{\Omega}$, $\hat{\Omega}=\vec{r}/R$. $\vec{\Lambda}$ has no component normal to the surface: $\vec{\Lambda}\cdot\vec{\Omega}=\vec{\Omega}\cdot\vec{\Lambda}=0$; its commutation relations are $[\Lambda^\alpha,\Lambda^\beta]=i\hbar\epsilon^{\alpha\beta\gamma}(\Lambda^\gamma-\hbar S\Omega^\gamma)$. The generator of rotations is instead given by $\vec{L}=\vec{\Lambda}+\hbar S\hat{\Omega}$:

$[L^\alpha, X^\beta] = i\hbar\epsilon^{\alpha\beta\gamma}X^\gamma$, $\vec{X} = \vec{L}, \hat{\Omega}$, or $\vec{\Lambda}$; this *has* a normal component* $\vec{L}\cdot\hat{\Omega} = \hat{\Omega}\cdot\vec{L} = \hbar S$. This algebra implies the spectrum $|\vec{L}|^2 = \hbar^2 l(l+1)$, $l = S + n$, $n = 0, 1, 2, \ldots$, and that $2S$ is integral (the Dirac condition[6]); $|\vec{\Lambda}|^2 = |\vec{L}|^2 - \hbar^2 S^2 = \hbar^2\{n(n+1) + (2n+1)S\}$. $\hat{\Omega}$ can be specified by spinor coordinates $u = \cos(\frac{1}{2}\theta)\exp(\frac{1}{2}i\varphi)$, $v = \sin(\frac{1}{2}\theta)\exp(-\frac{1}{2}i\varphi)$: $\hat{\Omega}(u,v) = (\sin\theta\cos\varphi, \sin\theta\sin\varphi, \cos\theta)$. To describe the wave functions, I choose the gauge $\vec{A} = (\hbar S/eR) \times \hat{\varphi}\cot\theta$; the singularities at the two poles (each admitting flux $S\Phi_0$) have no physical consequence. The Hilbert space of the lowest Landau level ($l = S$, with energy $\frac{1}{2}\hbar\omega_c$) is spanned by the coherent states $\psi_{(\alpha,\beta)}{}^{(S)}$ defined by $\{\hat{\Omega}(\alpha,\beta)\cdot\vec{L}\}\psi_{(\alpha,\beta)}{}^{(S)} = \hbar S\psi_{(\alpha,\beta)}{}^{(S)}$; these are polynomials in u and v of total degree $2S$:

$$\psi_{(\alpha,\beta)}{}^{(S)}(u,v) = (\alpha^* u + \beta^* v)^{2S}, \quad |\alpha|^2 + |\beta|^2 = 1.$$

Within this subspace, the electron may be represented as a spin S, the orientation of which indicates the point on the sphere about which the state is localized.[7] The operator $\vec{L}$ can be written as $L^+ = \hbar u\,\partial/\partial v$, $L^- = \hbar v\,\partial/\partial u$, $L^z = \frac{1}{2}\hbar(u\,\partial/\partial u - v\,\partial/\partial v)$, and $S = \frac{1}{2}(u\,\partial/\partial u + v\,\partial/\partial v)$; u and v may also be represented as independent boson creation operators, and $\partial/\partial u$ and $\partial/\partial v$ as their conjugate destruction operators.

Two-particle states.—The operator $|\vec{L}_1 + \vec{L}_2|^2$ has eigenvalues $\hbar^2 J_{12}(J_{12}+1)$, $J_{12} = 0, 1, \ldots, 2S$; the coherent states with $J_{12} = J$, $\{\hat{\Omega}(\alpha,\beta)\cdot(\vec{L}_1 + \vec{L}_2)\}\psi_{(\alpha,\beta)}{}^{(S,J)} = \hbar J\psi_{(\alpha,\beta)}{}^{(S,J)}$, have wave functions

$$\psi_{(\alpha,\beta)}{}^{(S,J)} = (u_1 v_2 - u_2 v_1)^{2S-J} \prod_{i=1,2} (\alpha^* u_i + \beta^* v_i)^J.$$

Fermi statistics requires that $2S - J_{12}$ be odd, and Bose statistics, that it be even. Note that the factor $u_1 v_2 - u_2 v_1$ commutes with $\vec{L}_1 + \vec{L}_2$. If Π_S is the projection operator on states of the lowest Landau level, the projection on rotationally invariant operators $V(\hat{\Omega}_1\cdot\hat{\Omega}_2)$ (such as the interparticle interaction) can be expanded as

$$\Pi_S V(\hat{\Omega}_1\cdot\hat{\Omega}_2)\Pi_S = \sum_{J=0}^{2S} V_J{}^{(S)} P_J(\vec{L}_1 + \vec{L}_2),$$

where $P_J(L)$ is the projection operator on states with $|\vec{L}|^2 = \hbar^2 J(J+1)$. In particular, $\Pi_S(\hat{\Omega}_1\cdot\hat{\Omega}_2)\Pi_S = \vec{L}_1\cdot\vec{L}_2/\{\hbar(S+1)\}^2$; the smaller the value of $2S - J_{12}$, the smaller the mean separation between the particles, which are precessing about their common center of mass at $\hat{\Omega}(\alpha,\beta)$.

N-particle states.—In the spirit of Laughlin,[4] I discuss the N-particle wave function,

$$\Psi_N{}^{(m)} = \prod_{i<j} (u_i v_j - u_j v_i)^m, \quad S = \tfrac{1}{2}m(N-1).$$

The case $m = 1$ can be alternatively expressed as the antisymmetric Slater determinant describing complete filling of the lowest Landau level, with $N = 2S + 1$. Because $\vec{L}_{tot} = \sum_i \vec{L}_i$ commutes with $u_i v_j - u_j v_i$, $\Psi_N{}^{(m)}$ is explicitly translationally and rotationally invariant on the surface of the sphere: $\vec{L}_{tot}\Psi_N{}^{(m)} = 0$. It is totally antisymmetric (Fermi statistics) for odd m, and symmetric (Bose statistics) for even m. The Laughlin droplet wave functions,[4] centered at $\hat{\Omega}(\alpha,\beta)$, can be recovered by multiplying $\Psi_N{}^{(m)}$ by a factor $\prod_i(\alpha^* u_i + \beta^* v_i)^n$, and taking the limit $n\to\infty$, $R\to\infty$, $R^2/2n = a_0{}^2$, where $a_0 = (\hbar/eB)^2$ is the Larmor radius of the lowest Landau level.

Remarks.—(1) $\Psi_{n=3}{}^{(m)}$ is an exact eigenstate of *any* pair interaction $\sum_{i<j}\{\Pi_S V(\hat{\Omega}_i\cdot\hat{\Omega}_j)\Pi_S\}$, because $J_{12} = J_{23} = J_{31} = S = m$; in the planar geometry, Laughlin's $N = 3$ droplet states are reportedly not exact: Overlaps with numerically calculated exact eigenstates[4] (e.g., 0.994 68 for the Coulomb interaction, $m = 5$) are close to, but *not* exactly, unity. (2) for $N \geq 4$, $m > 1$, $\Psi_N{}^{(m)}$ is *not* an exact eigenstate of a general interaction potential: This would require that it is an exact eigenstate with $J_{ij} = J$ of the angular momentum of any pair of particles. The spectrum of values of J_{ij} contained in $\Psi_N{}^{(m)}$ is easily determined by writing it as the product of three factors (i) involving coordinates i,j only, (ii) involving coordinates $k \neq i,j$ only, and (iii) the cross term $\prod_k (v_k u_i - u_k v_i)^m (v_k u_j - u_k v_j)^m$ which determines J_{ij}: $J_{ij} \leq m(N-2) = 2S - m$. The special character of the states $\Psi_N{}^{(m)}$ is thus that they have no components with $J_{ij} = 2S - m + 2, 2S - m + 4, \ldots \leq 2S$ that would be present in a more general wave function of the appropriate symmetry: The states of closest approach of the pair of particles are suppressed. In particular, when $S = S(N; m) \equiv \frac{1}{2}m(N-1)$, $\Psi_N{}^{(m)}$ may be characterized as the *exact nondegenerate ground state* of the projection-operator interaction potential

$$\Pi_S H_{m,S}{}^{int}\Pi_S = \sum_{i<j}\{\sum_{J>2S-m} P_J(\vec{L}_i + \vec{L}_j)\}.$$

This is essentially a kind of hard-core interaction; $\Psi_N{}^{(m)}$ will thus be a particularly good variational approximation for the ground state of systems with strong repulsion at close separations.

Excited states.—In this geometry, the natural excitation operators, analogous to those suggested by Laughlin,[4] are

$$A_N^\dagger(\alpha,\beta)=\prod_{i=1}^{N}(\beta u_i-\alpha v_i)\quad(\text{"holes"}),$$

$$A_N(\alpha,\beta)=\prod_{i=1}^{N}\left(\beta^*\frac{\partial}{\partial u_i}-\alpha^*\frac{\partial}{\partial v_i}\right)\quad(\text{"particles"}),$$

which, respectively, increase or decrease the flux quantum number S by $\frac{1}{2}$, and decrease or increase $\hat{\Omega}(\alpha,\beta)\cdot\vec{L}_{tot}$ by $\frac{1}{2}N\hbar$. The single-excitation states $A_N^\dagger(\alpha,\beta)\Psi_N^{(m)}$ and $A_N(\alpha,\beta)\Psi_N^{(m)}$ have $J_{tot}=\frac{1}{2}N$, and describe defects in the fluid localized around[7] $\hat{\Omega}(\alpha,\beta)$. Since

$$[A_N^\dagger(\alpha,\beta),A_N^\dagger(\alpha',\beta')]=[A_N(\alpha,\beta),A_N(\alpha',\beta')]=0,$$

the two-hole and two-particle states are symmetric in the excitation coordinates, and the excitations thus obey Bose statistics {note, however, that $[A_N(\alpha,\beta),A_N^\dagger(\alpha',\beta')]\neq 0$}. A state with N_p^{ex} particle and N_h^{ex} hole excitations has $S=S(N;m)+\frac{1}{2}(N_h^{ex}-N_p^{ex})$; on the other hand, if the system is excited by addition or removal of an electron at fixed magnetic field, the final state has $S=S(N+1;m)\mp\frac{1}{2}m$. The comparison indicates that the hole excitations carry a fractional charge $e^*=+e/m$, and the particles a fractional charge $-e^*$, as proposed by Laughlin.[4] The degeneracy $N+1$ of the single-excitation states supports the same conclusion: In the thermodynamic limit there is one state for each unit $\Phi_m\equiv m\Phi_0=h/e^*$ of magnetic flux through the surface.

Hierarchy of fluid states.—I will assume, following Laughlin,[4] that for some m, the ground state of the 2D electron gas with $S=S(N;m)$ is well represented by the approximate wave function $\Psi_N^{(m)}$, and that there is a gap in the excitation spectrum, the lowest-energy excitations being (bound) particle-hole pairs. Consider now a slightly different field strength so that $S=S(N;m)+\frac{1}{2}N^{ex}$; the low-energy states at this field strength can be considered as deriving from the fluid state $\Psi_N^{(m)}$ with an imbalance of particle and hole excitations, $N^{ex}=N_h^{ex}-N_p^{ex}$. Since there is, by assumption, a gap for making particle-plus-hole excitations, the lowest-energy states will belong to a manifold of purely hole states ($N^{ex}>0$) or purely particle states ($N^{ex}<0$), separated by a gap from higher-energy states. If the interaction energy of the *excitations* is small compared to this energy gap, the problem of constructing the collective ground state of the *excitation* fluid is precisely analogous to the original problem of constructing the ground state of the *electron* fluid, but with S replaced by $\frac{1}{2}N$, N replaced by $|N^{ex}|$, and Fermi statistics replaced by Bose statistics. A Laughlin fluid state of the excitations[8] can be constructed if

$$\tfrac{1}{2}N=S(|N^{ex}|;p),$$

where p is now *even* (Bose statistics): $p=2,4,6,\ldots$. This leads to $|N^{ex}|=(N/p)+1$; this second family of fluid states thus can occur at field strengths $S=S(N;m,\pm p)\equiv\frac{1}{2}m(N-1)\pm\frac{1}{2}[(N/p)+1]$, and requires that N be divisible by p. If this fluid state exists, with a sufficiently strong gap, the argument can be iterated by constructing a type-$[|p_1|,p_2]$ fluid state of the excitations of the primary type-$[m]$ electron fluid, and so on; the hierarchical set of equations is

$$S(N;m,p_1,\ldots,p_n)=S(N;m)+\tfrac{1}{2}|N^{ex}|\operatorname{sgn}(p_1);$$

$$\tfrac{1}{2}N=S(|N^{ex}|;|p_1|,p_2,\ldots,p_n).$$

The filling factor ν is given by $N/2S$ in the thermodynamic limit; the hierarchical equations become

$$\{\nu(m,p_1,\ldots,p_n)\}^{-1}=m+\operatorname{sgn}(p_1)\,\nu(|p_1|,p_2,\ldots p_n),$$

with the solution

$$\nu(m,p_1,\ldots,p_n)=[m,p_1,\ldots,p_n].$$

The charge of the excitations is easily found by determining how many are produced by adding an electron at fixed magnetic field: The result is that if ν is expressed as the rational P/Q, Q is odd, $e^*=e/Q$, and the Hall resistance can be written $\rho_{xy}=\Phi_0/Pe^*$, consistent with Laughlin's "gauge invariance" argument.[9]

The above analysis indicates how "incompressible fluid" states may derive from parent "incompressible fluid" states at simpler rational filling factors ν; the most stable fluid states will correspond to the simplest rationals with small values of m and p_i, where the fluid densities are highest, and hence short-range repulsion effects strongest. What is so far missing is a calculational scheme for the direct determination of *whether* a given fluid state exists for a given interaction potential, e.g., the Coulomb interaction. The new formalism based on a spherical geometry may simplify this task. From a variational viewpoint,[4] the correlation energy of $\Psi_N^{(m)}$ and its excitations $A_N^\dagger(\alpha,\beta)\Psi_N^{(m)}$ must be determined; this reduces to (i) the determination of the expansion coefficients $V_J^{(S)}$ of the interaction potential, and (ii) analysis of the wave functions to determine the relative weights of the components with a given pair angular momentum J_{ij}. Progress may be possible in this formalism. Beyond the variational approach, the problem has been reduced to a generalized version of an infinite-coordination Heisenberg problem involving N spin-S objects, and direct numerical calculation of the low-lying energy levels at an increasing sequence of values of N with $S = S(N;m,p_1,\ldots,p_n)$, coupled with a *finite-size scaling* analysis of how the gap behaves as $N\rightarrow\infty$, may prove possible at simple rationals. It may be remarked that, in this geometry, the gapless Wigner lattice would also derive from an isotropic state $L_{tot}=0$ as $N\rightarrow\infty$: The sphere cannot be tiled with a triangular lattice without introducing *disclination defects*; these will be mobile, and will restore translational and rotational invariance.

This work was brought to conclusion during a stay at the Laboratoire de Physique des Solides, Université Paris-Sud, Orsay, and the hospitality and support of R. Jullien and Professor J. Friedel is gratefully acknowledged. The assistance of M. Kolb in numerical study of the case $N=4$ was invaluable in the development of the above ideas.

[1]K. v. Klitzing, G. Dorda, and M. Pepper, Phys. Rev. Lett. 45, 494 (1980); D. C. Tsui and A. C. Gossard, Appl. Phys. Lett. 37, 550 (1981).

[2]D. C. Tsui, H. L. Stormer, and A. C. Gossard, Phys. Rev. Lett. 48, 1559 (1982); H. L. Stormer *et al.*, Phys. Rev. Lett. 50, 1953 (1983).

[3]L. Bonsall and A. A. Maradudin, Phys. Rev. B 15, 1959 (1977); D. Yoshioka and H. Fukuyama, J. Phys. Soc. Jpn. 47, 394 (1979); D. Yoshioka and P. A. Lee, Phys. Rev. B 27, 4986 (1983); K. Maki and X. Zotos, to be published.

[4]R. B. Laughlin, Phys. Rev. Lett. 50, 1395 (1983).

[5]D. Yoshioka, B. I. Halperin, and P. A. Lee, Phys. Rev. Lett. 50, 1219 (1983).

[6]P. A. M. Dirac, Proc. Roy. Soc. London, Ser. A 133, 60 (1931).

[7]The convention used here is that negative-charge coherent single-particle states (electrons, "particles") transform *contragrediently* under rotations, while positive-charge states ("holes") transform *cogrediently* ($\vec{L}$ points *away* from their position).

[8]The possibility of a hole fluid is raised, along with other suggestions such as a fluid of electron pairs, in a discussion of how Laughlin's wave functions might be generalized to other rational values of ν by B. I. Halperin, to be published.

[9]R. B. Laughlin, Phys. Rev. B 23, 5632 (1981).

Ground State of Two-Dimensional Electrons in Strong Magnetic Fields and $\frac{1}{3}$ Quantized Hall Effect

D. Yoshioka[(a)]
Bell Laboratories, Murray Hill, New Jersey 07974

and

B. I. Halperin
Laboratoire de Physique de l'Ecole Normale Supérieure, F-75231 Paris Cédex 5, France, and Department of Physics,[(b)] Harvard University, Cambridge, Massachusetts 02138

and

P. A. Lee
Department of Physics, Massachusetts Institute of Technology, Cambridge, Massachusetts 02139
(Received 28 February 1983)

The authors have diagonalized numerically the Hamiltonian of a two-dimensional system of up to six interacting electrons, in the lowest Landau level, in a rectangular box with "periodic" boundary conditions. They find that the ground state has a pair correlation function quite different from that of a Wigner crystal, and its energy is significantly lower. They also find some indications of a downward cusp in the energy at $\frac{1}{3}$ filling.

PACS numbers: 72.20.My, 71.45.-d, 73.40.Qv

The origin of the quantized Hall effect of a two-dimensional electron system in a strong magnetic field[1] is now well understood.[2-5] However, there has been no satisfactory explanation for the anomalous quantized Hall effect, which is observed in AlGaAs-GaAs heterojunctions.[6,7] In this experiment the Hall conductivity σ_{xy} shows plateaus at $\sigma_{xy}=\frac{1}{3}e^2/h$ and $\frac{2}{3}e^2/h$. Since the AlGaAs-GaAs heterojunction has a very high electron mobility, it is natural to suspect that the effect comes from the Coulomb interaction between electrons. However, an attempt to explain the effect by the formation of a Wigner-crystal-like charge-density-wave (CDW) state was not successful.[8] It was found that $\frac{1}{3}$ or $\frac{2}{3}$ filling of the lowest Landau level did not lead to any observable singularity in the energy, and it was also shown that in the crystalline state, σ_{xy} takes as values only *integer* multiples of the quantum e^2/h, if the crystal is pinned by impurities.[9,10] We need a new state which has lower energy than the crystal to explain the anomalous quantized Hall effect.

In the present paper, we investigate numerically the eigenstates of an electron system, in the first Landau level, in a rectangular cell with "periodic" boundary conditions, and up to six particles. We find that the ground state has significantly lower energy than that of a Hartree-Fock Wigner crystal, and that the pair correlation function $g(\vec{r})$ looks quite different from that of a crystal. States resembling the Wigner crystal with regards to energy and $g(\vec{r})$ appear at higher energy.

We take the coordinate system such that the boundary of the cell is given by $x=0$, $x=a$, $y=0$, $y=b$, with the vector potential $\vec{A}=(0,xB)$. Our boundary condition requires that $ab/2\pi l^2$ *be an integer* m, where $2\pi l^2=hc/eB$. Then there are m different single-electron states in the cell, whose wave functions are given by

$$\varphi_j(\vec{r})=\left(\frac{1}{b\pi^{1/2}l}\right)^{1/2}\sum_{k=-\infty}^{\infty}\exp\left[i\frac{(X_j+ka)y}{l^2}-\frac{(X_j+ka-x)^2}{2l^2}\right]. \quad (1)$$

Here integer j, $1\leq j\leq m$, specifies the state, and $X_j=2\pi l^2 j/b$ is the center coordinate of the cyclotron motion.

The electrons in the cell interact with each other and with the uniform positive background charge by the Coulomb interaction. Because of the boundary condition, the Coulomb potential in real space is given by

$$V(\vec{r})=\sum_s\sum_t e^2/\epsilon|\vec{r}+sa\hat{x}+tb\hat{y}|, \quad (2)$$

where ϵ is the dielectric constant, and $\hat{x}$ and $\hat{y}$ are the unit vectors along the x and y axes, respectively. Since we consider only the lowest Landau level, the Hamiltonian consists entirely of the Coulomb interaction term:

$$H=\sum_j S a_j{}^\dagger a_j+\sum_{j_1}\sum_{j_2}\sum_{j_3}\sum_{j_4}A_{j_1j_2j_3j_4}a_{j_1}{}^\dagger a_{j_2}{}^\dagger a_{j_3}a_{j_4}, \tag{3}$$

where a_j is the destruction operator for the jth state. The single-electron part comes from the interaction between an electron and its image, so that S is a known constant related to the Coulomb energy of the classical rectangular Wigner crystal.[11]

The two-electron part is given by

$$A_{j_1j_2j_3j_4}=\tfrac{1}{2}\int d^2r_1\int d^2r_2\,\varphi_{j_1}^*(\vec{r}_1)\varphi_{j_2}^*(\vec{r}_2)V(\vec{r}_1-\vec{r}_2)\varphi_{j_3}(\vec{r}_2)\varphi_{j_4}(\vec{r}_1)$$

$$=\frac{1}{2ab}\sum_q{}'\sum_s\sum_t \delta_{q_x,2\pi s/a}\,\delta_{q_y,2\pi t/b}\,\delta'_{j_1-j_4,t}\,\frac{2\pi e^2}{\epsilon q}\exp\left[-\frac{l^2q^2}{2}-2\pi i s\frac{j_1-j_3}{m}\right]\delta'_{j_1+j_2,j_3+j_4}. \tag{4}$$

The Kronecker δ with prime means that the equation is defined modulo m, and the summation over q excludes $q=0$.

We specify the number n of electrons in our cell, so the filling factor ν is given by n/m. The Hamiltonian of the system has particle-hole symmetry, after a constant term, which equals $-(\pi/8)^{1/2}\nu^2(e^2/\epsilon l)$ in the infinite-n limit, has been removed. Our calculations are done for $\nu\leq 0.5$ and extended to $\nu>0.5$ using this symmetry.

The basis n-electron wave function is specified by the occupation of the single-electron state: $(j_1, j_2, \ldots, j_n)$. The total number of the basis is given by $\binom{m}{n}$. However, the Hamiltonian H conserves the total momentum in the y direction, $J=j_1+j_2+\ldots j_n(\mathrm{mod}\,m)$. So the number of the basis for fixed m, n, and J is approximately $m^{-1}\binom{m}{n}$, which gives the dimension of the Hamiltonian matrix.

Two values of J which differ by a multiple of n are equivalent, since $(j_1, j_2, \ldots, j_n)$ and $(j_1+1, j_2+1, \ldots, j_n+1)$ differ only by translation in the x direction. Hence when m and n have no common factor, the energy spectrum of the Hamiltonian is independent of J and every eigenenergy is at least m-fold degenerate. On the other hand, when m and n have a common factor, the states are less degenerate and the ground state is realized only at certain choices of J. For example, at $\nu=\frac{1}{3}$ and $n=4$, the threefold-degenerate ground state is found at $J=2$, 6, 10.

Since we are interested in the ground states near $\nu=\frac{1}{3}$, the diagonalization was done numerically for $n=4$, 5, and 6, and $0.25\leq n/m=\nu\leq 0.5$ except for $n=6$ where we calculate only up to $m=20$ or $\nu=0.3$. Figure 1 shows the ground-state energy of n-electron systems as a function of the filling factor $\nu=n/m$ for the choice of aspect ratio $a/b=n/4$. (This choice of a/b seems to give an approximate local minimum in the energy.) To investigate the nature of the eigenstates of the Hamiltonian we also calculated the pair correlation function $g(\vec{r})$, which is the same for all states in a degenerate multiplet. For $n=4$ the $g(\vec{r})$ for the ground state has fourfold rotational symmetry and has peaks at $\vec{r}=(\pm a/2, 0)$ and $(0, \pm b/2)$, but *not* at $\vec{r}=(\pm a/2, \pm b/2)$, where we would expect to have peaks if the state were a square crystal. States which correspond to a square crystal and a triangular crystal are found at higher energy for even m. The energy of the triangular crystal is lower than that of the square crystal and it has a minimum at $a/b=2/\sqrt{3}$. Here $g(\vec{r})$ has the form

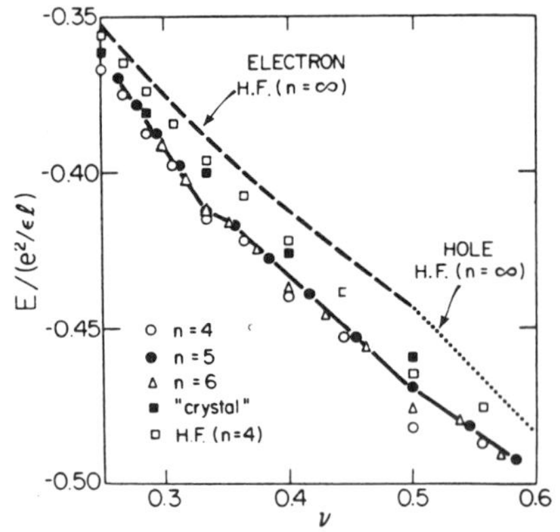

FIG. 1. The energies per particle of two-dimensional electron systems vs the fractional filling of the first Landau level. The dashed and dotted lines show energy of the electron and hole crystals resulting from the Hartree-Fock approximation for the infinite system. Open circles, closed circles, and triangles show the ground-state energies for $n=4$, 5, and 6 electrons for $\nu\leq\frac{1}{2}$ and $n=4$, 5, and 6 holes for $\nu>\frac{1}{2}$. Closed squares show the crystal state for the $n=4$ system. Open squares show the energy of the crystal state for the $n=4$ system obtained by the Hartree-Fock approximation. The solid line drawn through the $n=5$ ground-state energies is a guide to the eye only.

similar to that of the triangular crystal obtained by the Hartree-Fock (HF) approximation in the infinite system. The energy of this crystalline state is also shown in Fig. 1.

In order to clarify further the nature of the ground state and to clarify the boundary effect, we also apply the Hartree-Fock approximation to the Hamiltonian of the four-electron system. We assume order parameters $\Delta_{j_1 j_2} \equiv \langle a_{j_1}{}^\dagger a_{j_2} \rangle$, to be finite, decouple the Hamiltonian, and obtain self-consistent solutions for $\Delta_{j_1 j_2}$.[12] The state we obtain is always a triangular CDW state except for $\nu = \frac{1}{2}$, where we get a unidirectional CDW state. The energy becomes minimum at $a/b = 2/\sqrt{3}$. This energy is also shown in Fig. 1.

The HF energy of the four-electron system is slightly lower than the HF energy of the infinite crystal, shown in Fig. 1 by the dashed and dotted curves for the electron and hole crystals, respectively. This difference comes from the boundary condition. In the finite system the Coulomb potential between an electron and its image separated by $R = (ma, nb)$ is always $e^2/\epsilon R$, whereas an average over the Gaussian charge distribution is required in the infinite crystal. The energy difference between the finite and infinite systems is completely explained by this effect.

Next, we note that the $n = 4$ HF energies are slightly higher than the energies of the "crystal states" obtained in the exact diagonalization. The difference, which we would like to attribute to the correlation between nearest electrons, is about 2 times larger than the correlation energy estimated by Yoshioka and Lee[8] by second-order perturbation theory, which seems reasonable. However, as mentioned above, the crystal states are not the lowest states of our systems.

As seen in Fig. 1, the *ground-state energies* of our small systems tend to have downward cusps at simple rational values of ν. Clearly it is not possible to extrapolate our data to $n = \infty$. Nevertheless, it is interesting that the downward cusp at $\nu = \frac{1}{3}$ remains roughly constant for the three systems calculated ($n = 4, 5, 6$). A cusp is also visible at $\nu = \frac{2}{5}$, but unfortunately we have data points only for two values of n. By contrast, the ground-state energy at $\nu = \frac{1}{2}$ shows a large, nonmonotonic variation with n. As a guide to the eye we have connected the points in Fig. 1 for $n = 5$, for which there are no low-order rationals except for $\nu = \frac{1}{5}$ and $\frac{1}{2}$. It is interesting that this curve shows almost no cusp at $\nu = \frac{1}{2}$.

At this point, we can make a number of speculations regarding the infinite system and the relation to the experiment. We regard our data as supportive of the idea that the ground state is not crystalline, but a translationally invariant "liquid." We speculate that this liquid has commensurate energy at $\nu = \frac{1}{3}$ (and possibly other simple rational values), and that for a large but finite system, the ground state at $\nu = \frac{1}{3}$ is threefold degenerate and separated by an energy gap from a variety of excited states. By going to a moving frame, it is then clear that at $\nu = \frac{1}{3}$ a Hall current will flow without dissipation, even in the presence of impurities. At ν *close* to $\frac{1}{3}$, we further suppose that the ground state, which is now highly degenerate, can be described as the $\nu = \frac{1}{3}$ ground state plus an additional small density of quasi "particles" or "holes." This leads naturally to a downward cusp in the energy as function of ν. The Hall plateau at $\sigma_{xy} = \frac{1}{3}e^2/h$ can then be explained if the quasiparticles are localized by impurities and thus do not contribute to the Hall current, which is simply carried by the underlying $\nu = \frac{1}{3}$ state. Very recently, we have learned of a very original proposal by Laughlin of a wave function for a liquid state at $\nu = 1/p$, for p odd, which appears to have the requisite commensurate energy.[13]

An alternative explanation of the Hall conductance plateau is also possible, assuming the existence of a commensurate energy E_c at $\nu = \frac{1}{3}$, if one believes that the electron system in the GaAs accumulation layer is in equilibrium with the donor states in GaAlAs by tunnelling.[14] For ν near $\frac{1}{3}$, it is energetically favorable to pin the density at $\frac{1}{3}$, provided that the energy gain $|\Delta n| E_c$ exceeds the charging energy $2\pi L_d (e\Delta n)^2/\epsilon$ to transfer $\Delta n \equiv (\nu - \frac{1}{3})/2\pi l^2$ electrons across a depletion layer of thickness L_d. If we extract a rough estimate of $E_c \approx 0.008(e^2/\epsilon l)$ from the depth of the cusp in Fig. 1 near $\nu = \frac{1}{3}$, and we use $L_d = 240$ Å and $l = 66$ Å ($B = 15$ T), this would lead to a full width of the Hall step at $\nu = \frac{1}{3}$ of $\delta\nu/\nu \approx 20\%$, which is consistent with current experiments. Clearly, this alternative explanation depends on the details of the layer structure and we find it less appealing than the first explanation we offered.

We thank M. A. Paalmen, H. L. Stormer, and D. C. Tsui for discussions of the experimental data, and we thank the Aspen Physics Center, where part of this work was performed, for its hospitality. This work was supported in part by National Science Foundation Grant No. DMR 82-07431, and by U. S. Joint Services Electronics Program Grant No. DAAG29-83-K-0003.

[a]Present and permanent address: Institute for Solid State Physics, University of Tokyo, Roppongi, Minato-ku, Tokyo, Japan.
[b] Permanent address.

[1]K. V. Klitzing, G. Dorda, and M. Pepper, Phys. Rev. Lett. 45, 494 (1980).
[2]R. B. Laughlin, Phys. Rev. B 23, 5632 (1981).
[3]D. J. Thouless, J. Phys. C 14, 3475 (1981).
[4]H. Aoki and T. Ando, Solid State Commun. 38, 1079 (1981).
[5]B. I. Halperin, Phys. Rev. B 25, 2185 (1982).
[6]D. C. Tsui, H. L. Stormer, and A. C. Gossard, Phys. Rev. Lett. 48, 1559 (1982).
[7]H. L. Stormer, D. C. Tsui, A. C. Gossard, and J. C. M. Hwang, in Proceedings of the Sixteenth International Conference on the Physics of Semiconductors (to be published).
[8]D. Yoshioka and P. A. Lee, Phys. Rev. B 27, 4986 (1983).
[9]H. Fukuyama and P. M. Platzman, Phys. Rev. B 25, 2934 (1982).
[10]D. Yoshioka, Phys. Rev. B 27, 3637 (1983).
[11]L. Bonsall and A. Maradudin, Phys. Rev. B 15, 1959 (1977).
[12]H. Aoki, J. Phys. C 12, 633 (1979).
[13]R. B. Laughlin, private communications.
[14]G. A. Baraff and D. C. Tsui, Phys. Rev. B 24, 2274 (1981).

Statistics of Quasiparticles and the Hierarchy of Fractional Quantized Hall States

B. I. Halperin
Physics Department, Harvard University, Cambridge, Massachusetts 02138
(Received 9 November 1983)

Quasiparticles at the fractional quantized Hall states obey quantization rules appropriate to particles of fractional statistics. Stable states at various rational filling factors may be constructed iteratively by adding quasiparticles or holes to lower-order states, and the corresponding energies have been estimated.

PACS numbers: 05.30.–d, 03.65.Ca, 71.45.Nt, 73.40.Lq

Observations of the fractional quantized Hall effect[1] show that there exist special stable states of a two-dimensional electron gas, in strong perpendicular magnetic field B, occurring at a set of rational values of ν, the filling factor of the Landau level. Laughlin[2] has constructed an explicit trial wave function (product wave function) to explain the states at $\nu = 1/m$, with m an odd integer, and has argued that the elementary excitations from the stable states are quasiparticles with fractional electric charge. Among the proposals to explain the other observed fractional Hall steps are hierarchical schemes, in which higher-order stable states ν_{s+1} are built up by adding quasiparticles to a stable state ν_s of smaller numerator and denominator.[3–5]

In the present note, we observe that the quantization rules which determine the allowed quasiparticle spacings are just those that would be expected for a set of identical charged particles that obey *fractional statistics*—i.e., such that the wave function changes by a complex phase factor when two particles are interchanged. Moreover, by assuming that the dominant interaction between quasiparticles is just the Coulomb interaction between the quasiparticle charges, we are led to a natural set of approximations for the ground-state energies and energy gaps at all levels of the hierarchy.

The appearance of fractional statistics in the present context is strongly reminiscent of the fractional statistics introduced by Wilczek to describe charged particles tied to "magnetic flux tubes" in two dimensions.[6] As in Ref. 6, the quasiparticles can *also* be described by wave functions obeying Bose or Fermi statistics, the various representations being related by a "singular gauge transformation." The boson description was, in fact, used in Refs. 3 and 4 and the fermion description in Ref. 5. However, the boson or fermion descriptions require, in effect, a long-range interaction between quasiparticles which alters the usual quantization rules. The transformation between representations is analogous to the well-known transformation between impenetrable bosons and fermions in one dimension.

As in previous discussions of the fractional quantized Hall effect, we consider a two-dimensional system of electrons in the lowest Landau level, with a uniform positive background. The filling factor ν is defined by $\nu = n/2\pi l_0^2$, where n is the density of electrons, and $l_0 = |Be/\hbar c|^{1/2}$ is the magnetic length; hence ν is the number of electrons per quantum of flux.

Let ν_s be a stable rational filling factor obtained at level s of the hierarchy. I assert that the low-lying energy states for filling factors near to ν_s can be described by the addition of a small density of quasiparticle excitations to the ground state at ν_s. The elementary quasiparticle excitations are of two types—particlelike "p excitations" and holelike "h excitations"—having charges $q_s e$ and $-q_s e$, respectively, according to a sign convention described below. For the present purposes we need only consider states with one type of excitation present. We shall describe these states by a pseudo wave function Ψ, which is a function of the coordinates $\vec{R}_k$ of the N_s quasiparticles present. I assert that the the allowed pseudo wave functions can be written in the form

$$\Psi[\vec{R}_k] = P[Z_k]Q_s[Z_k]\prod_{k=1}^{N_s}\exp(-|q_s||Z_k|^2/4l_0^2), \tag{1}$$

where $Z_k = X_k \mp iY_k$ is the position in complex notation, with the sign depending on the sign of the charge of the quasiparticle, $P[Z_k]$ is a *symmetric polynomial* in the variables Z_k, and

$$Q_s = \prod_{k<l}|Z_k - Z_l|^{-\alpha/m_s}. \tag{2}$$

In Eq. (2), $\alpha = \pm 1$, according to whether we are dealing with particle- or hole-type excitations, and m_s is a rational $\geqslant 1$, to be specified by an iterative equation below. We may interpret $|\Psi[\vec{R}_k]|^2$ as the probability density for finding a quasiparticle at each of the positions $\vec{R}_1, \ldots, \vec{R}_{N_s}$, at least in the case that the $\vec{R}_k$ are not too close to each other. Since the quasiparticles have a finite size (of order l_0), however, there is no direct significance to the behavior of $|\Psi|^2$ when two positions R_k and R_l come very close together. The wave function is normalized if $\int|\Psi|^2=1$, and two wave functions Ψ and Ψ' are orthogonal if $\int\Psi^*\Psi'=0$.

The pseudo wave function (1)–(2) can be derived in different ways, starting from various microscopic descriptions that have been proposed[2–5] for the electronic state with quasiparticle or quasihole excitations. I shall give below a derivation for p excitations using the *pair model* proposed in Ref. 3.

Because there is no direct physical significance to the phase of the pseudo wave function, it is permissible to *redefine* the factor Q in Eq. (1) by *removing the absolute value sign* in Eq. (2). (This operation may be described as a singular gauge transformation.)[6] If $m_s \neq 1$, the new wave function is a multivalued function of the positions $\{\vec{R}_k\}$, and one should consider it as a function defined on the appropriate Riemann surface for $\{Z_k\}$. [Alternatively one could use a single-valued definition and specify discontinuities along cuts in the variable $(Z_k - Z_l)$.] Now if we continuously interchange the positions of two quasiparticles, the wave function will change by a complex phase factor $(-1)^{\pm 1/m_s}$, with the sign depending on the sense of rotation as the quasiparticles pass by each other. Although the extra phase factor is perhaps a complication, the pseudo wave function now has the esthetically pleasing property that it is an eigenstate of the differential operator $[\nabla_k \mp iq_s e\vec{A}(\vec{R}_k)/\hbar c]^2$ with special boundary conditions at the points $Z_k = Z_l$, where $\vec{A}$ is the vector potential in the symmetric gauge. Then Eq. (1) may be described as a general wave function appropriate to a collection of particles of charge $\pm q_s e$ obeying fractional statistics, all in the lowest Landau level. Of course, in the special case $m_s = 1$, the quasiparticles are ordinary fermions.

In order to find the ground-state configuration for a given density n_s of quasiparticles, we must find the symmetric polynomial $P[Z]$ which leads to the minimum expectation value of the repulsive interaction between the quasiparticles. Using the same reasoning as Laughlin in Ref. 2, we expect that certain choices of P can lead to specially low energies, namely,

$$P[Z_k] = \prod_{k<l}(Z_k - Z_l)^{2p_s+1}, \tag{3}$$

where p_{s+1} is a positive integer. The probability distribution $|\Psi|^2$ is then that of a classical one-component plasma[2] with dimensionless inverse temperature $\Gamma = 2m_{2+1}$, where

$$m_{s+1} = 2p_{s+1} - \alpha_{s+1}/m_s, \tag{4}$$

and $\alpha_{s+1} = 1$ or -1 as particlelike or holelike quasiparticles are involved. The density of the plasma is fixed by a charge neutrality condition,[2] so that the number of quasiparticles in an area $2\pi l_0^2$ is just $n_s = |q_s|/m_{s+1}$. Since each quasiparticle has charge $\alpha_{s+1}q_s$, we may readily calculate the electron density in the new stable state, and we find the filling factor

$$\nu_{s+1} = \nu_s + \alpha_{s+1}q_s|q_s|/m_{s+1}. \tag{5}$$

If we multiply the pseudo wave function described above by the factor $\prod_k Z_k$, for $k=1,\ldots,N_s$, we find a deficiency near the origin of $1/m_{s+1}$ quasiparticles of level s. We identify this state as a hole excitation at level $s+1$. Similarly, we may construct a p excitation having an *excess* of $1/m_{s+1}$ quasiparticles at the origin. The iterative equation for q_s is thus

$$q_{s+1} = \alpha_{s+1}q_s/m_{s+1}. \tag{6}$$

Together with the starting conditions $\nu_0 = 0$, $q_0 = m_0 = \alpha_1 = 1$, the iterative equations (4)–(6) give a sequence of rational filling factors ν_s for any choice of the sequence $\{\alpha_s, p_s\}$. At the level $s=1$, we recover Laughlin's states with $\nu_1 = 1/m_1 = 1, \frac{1}{3}, \frac{1}{5}, \ldots$ for various choices of p_1. If we add holes to the state $\nu_1 = 1$, we find at level $s=2$, the complements to the Laughlin states, $\nu_2 = \frac{2}{3}, \frac{4}{5}, \frac{6}{7}, \ldots$ (In order to stay in the lowest Laudau level, we impose the restriction $\alpha_2 = -1$, if $\nu_1 = 1$.) From the state $\nu_1 = \frac{1}{3}$, we achieve such states as $\nu_2 = \frac{2}{5}$ or $\frac{4}{11}$, with p excitations, and $\nu_2 = \frac{2}{7}$ or $\frac{4}{13}$, with h excitations.

It can be shown, after some algebra, that the allowed values of ν_s may be expressed as continued fractions in terms of the finite sequences $\{\alpha_s, p_s\}$ and that they are identical to those of Haldane.[4] (I have used the opposite sign for α, however, and here p is one-half of Haldane's.) As noted by Haldane, every rational value of ν with odd denominator, with $0 < \nu \leq 1$, is obtained once in this way. There will *not* be a quantized Hall step at *every* such rational ν, however. We know that there exists a maximum allowed value m_c for the parameter m_s, such that if at any stage of the hierarchy the calculated m_s is greater than m_c, then the quasiparticles at the density n_s will form a Wigner crystal rather than a quantum-liquid state.[2] There is then no stabilization of the electron density at the corresponding ν_s, and there will be no meaning to any further states in the hierarchy constructed from this ν_s.

The pseudo wave function (1)–(3) leads to a natural estimate of the potential energy of the system, if we assume that the dependence on the positions of the quasiparticles can be approximated by the pairwise Coulomb interaction between point particles of charge $q_s e$, in the background dielectric constant ϵ. If $E(\nu)$ is the energy per quantum of magnetic flux, we have

$$E(\nu_{s+1}) \cong E(\nu_s) + n_s \epsilon_s^{\pm} + n_s |q_s|^{5/2} u_{\rm pl}(m_s), \tag{7}$$

where $\epsilon_s^{\pm}$ is the energy to add one particlelike excitation or one holelike excitation, together with neutralizing uniform background, to the state ν_s, and $u_{\rm pl}$ is a smooth function of m_s, given (approximately) by Laughlin's interpolation formula[5]

$$u_{\rm pl}(m) = \frac{-0.814}{m^{1/2}}\left(1 - \frac{0.230}{m^{0.64}}\right)\left(\frac{e^2}{\epsilon l_0}\right). \tag{8}$$

We recall that $u_{\rm pl}(m)$ is the potential energy per particle that one would find for a system of electrons at filling factor $\nu = 1/m$ if one approximates the pair correlation function $g(r)$ for the electrons by the pair correlation function $g_{\rm pl}(r)$ for a one-component plasma at inverse temperature $\Gamma = 2m$; the factor $|q_s|^{5/2}$ in the last term of (7) reflects the smaller charge and larger magnetic length for our quasiparticles.

In order to use Eq. (7), we need an iterative formula for the quasiparticle energies $\epsilon_s^{\pm}$. It is convenient to write

$$\epsilon_s^{\pm} = \tilde{\epsilon}_s^{\pm} \pm m_s^{-1}[\epsilon_{s-1} + \tfrac{3}{2}|q_{s-1}|^{5/2} u_{\rm pl}(m_s)]. \tag{9}$$

The quantity in square brackets is the energy it would take to add one quasiparticle or quasihole of level $s-1$, if one could keep the Laughlin product form (3) for the polynomial P, and simply increase the density n_{s-1} by means of a reduction, of order $1/N$, in the magnetic length l_0 which controls the distance scale in Eq. (1).[7] The term $\tilde{\epsilon}_s^{\pm}$ in (9) may be called the *proper* excitation energy; it is relatively small, but is presumably positive for both quasiparticles and holes. For the proper hole energy, we use the approximate formula

$$\tilde{\epsilon}_s^{-} = 0.313|q_{s-1}|^{5/2} m_s^{-9/4}(e^2/\epsilon l_0). \tag{10}$$

This form has the correct dependence on the charge q_{s-1}; it passes through the exact value $0.313(e^2/\epsilon l_0)$, for $q_0 = 1$, $m_1 = 1$, and it yields $\tilde{\epsilon}_1^{-} = 0.264$, $\tilde{\epsilon}_1^{-} = 0.0837$, for $m_1 = 3$ and $m_1 = 5$, in close agreement with the values obtained by Laughlin.[5,7]

Unfortunately, there does not exist at the present time any reliable calculation of the quasiparticle excitation energy. Therefore, *for purposes of illustration*, I have made the arbitrary approximation $\tilde{\epsilon}_s^{+} = \lambda \tilde{\epsilon}_s^{-}$, where λ is a constant independent of m_s. The resulting curve for $E(\nu)$ is plotted in Fig. 1, for the choice $\lambda = 3$, after subtraction of the "plasma approximation" $E_{\rm pl} = \nu u_{\rm pl}(\nu^{-1})$, which is a smooth function of ν. We can see that there are downward pointing cusps in the energy visible at the low-order rational ν with odd denominators. The approximation also gives *upward*-pointing cusps at all rational ν with even denominators; in fact, I find small discontinuities in E, not visible on the scale of the figure, at all these even points except for $\nu = \frac{1}{2}$, where continuity is guaranteed by the particle-hole symmetry of the cohesive energy, which is respected exactly by the present approximation.[7] Clearly the upward-pointing cusps are unphysical; the system could always lower its energy by breaking up into small regions of larger and smaller density; alternatively there may be a different type of ground state with still lower energy at these values of ν. The behavior of the approximate energy curve near the low-order rationals of odd denominator should be qualitatively and semiquan-

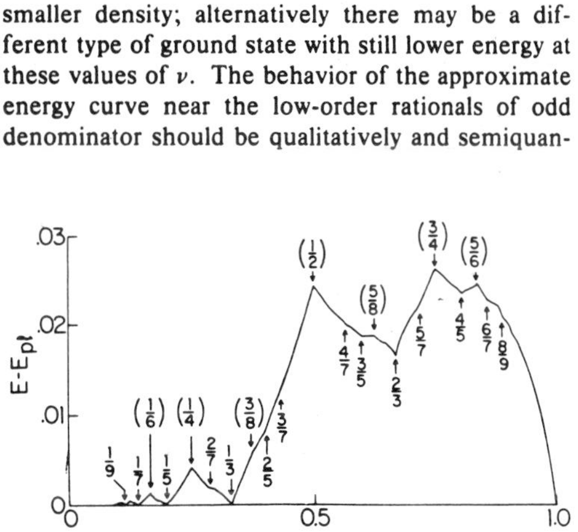

FIG. 1. Potential energy per quantum of magnetic flux, in units of $e^2/\epsilon l_0$, as a function of filling factor ν of the first Landau level, from approximate formulas (7)–(10). Smooth function $E_{\rm pl}(\nu) = \nu u_{\rm pl}(\nu^{-1})$ has been subtracted off.

titatively correct, however. More reliable estimates will be possible when p-excitation energies have been properly calculated, and when corrections are included such as the finite quasiparticle size and effects of virtual excitations of particle-hole pairs.

With the approximation described above, the energy gap $\tilde{\epsilon}_s^+ + \tilde{\epsilon}_s^-$ is equal to

$$0.313(1+\lambda)|q_s|^{5/2} m_s^{1/4}(e^2/\epsilon l_0)$$

[cf. (6) and (10)]. Except for the rather weak factor $m_s^{1/4}$, the gap is determined by the value of $|q_s|^{-1}$, which is the *denominator* of the fraction ν_s. This is in qualitative agreement with reported experimental observations on GaAs samples.[1]

Finally, we derive by induction the starting equation (1). For $s=0$, the Z_k are positions of bare electrons, and Eqs. (1) and (2) are correct, with $q_0 = m_0 = \alpha_1 = 1$. We assume that the p excitations of level $s=1$ can be formed out of *pairs* of electrons, by a generalization of Eq. (23) of Ref. 3. A system containing N_1 pairs of electrons, together with $N_0 - 2N_1$ unpaired electrons, is then described by choosing the polynomial in (1) to have the (schematic) form

$$P[Z_k] = \mathcal{S}\bar{P}[z_i]\prod_{i<j}(z_i - z_j)^{8p_1-4}\prod_{i,\gamma}(z_i - \tilde{Z}_\gamma)^{4p_1-1}\prod_{\gamma<\delta}(\tilde{Z}_\gamma - \tilde{Z}_\delta)^{2p_1}, \tag{11}$$

where z_i are the positions of the centers of gravity of the bound pairs, $\tilde{Z}_\gamma$ are the positions of the *unpaired* electrons, $\bar{P}$ is a symmetric polynomial, and $\mathcal{S}$ is an operator which symmetrizes with respect to the positions of all N_0 electrons. I have assumed that the separation between two members of a pair is small, and have dropped the variables describing these separations. To calculate the probability distribution of the pairs, we ignore the symmetrizer $\mathcal{S}$, and take the trace of $|\Psi[Z_k]|^2$ over the unpaired electron positions $\tilde{Z}_\gamma$. The result can be expressed in the form $|\bar{\Psi}[z_i]|^2\Phi[z_i]$, where $\bar{\Psi}$ has again the form of (1) and (2), with P replaced by $\bar{P}$, and with $m_1 = 2p_1 - 1/m_0$, $\alpha = 1$, and $q_1 = q_0/m_1$, while the remaining factor Φ is the partition function of a classical one-component plasma with sources of strength $2 - m_1^{-1}$, located at the positions z_i. Now Φ will be independent of the positions z_i, provided that the sources are sufficiently separated so that their screening clouds do not overlap. Thus it is consistent to interpret $\bar{\Psi}$ as a pseudo wave function for the positions of the pairs. Higher levels may be obtained iteratively.

Derivation of the pseudo wave function for hole excitations is more complicated because of the necessity to use an integral representation, such as Eq. (25) of Ref. 3.[7]

The author is grateful for helpful discussions with R. B. Laughlin, R. Morf, H. Stormer, P. A. Lee, D. Yoshioka, S. Girvin, and P. Ginsparg. This work was supported in part by the National Science Foundation under Grant No. DMR-82-07431.

[1]See H. L. Stormer *et al.*, Phys. Rev. Lett. **50**, 1953 (1983).

[2]R. B. Laughlin, Phys. Rev. Lett. **50**, 1395 (1983).

[3]B. I. Halperin, Helv. Phys. Acta **56**, 75 (1983).

[4]F. D. M. Haldane, Phys. Rev. Lett. **51**, 605 (1983).

[5]R. B. Laughlin, in Proceedings of the Conference on Electronic Properties of Two-Dimensional Systems, Oxford, 1983 (to be published).

[6]F. Wilczek, Phys. Rev. Lett. **49**, 957 (1982).

[7]Details will be given elsewhere.

Magneto-roton theory of collective excitations in the fractional quantum Hall effect

S. M. Girvin
Surface Science Division, National Bureau of Standards, Gaithersburg, Maryland 20899

A. H. MacDonald
National Research Council of Canada, Ottawa, Canada, K1A 0R6

P. M. Platzman
AT&T Bell Laboratories, Murray Hill, New Jersey 07974
(Received 16 September 1985)

We present a theory of the collective excitation spectrum in the fractional quantum Hall effect which is closely analogous to Feynman's theory of superfluid helium. The predicted spectrum has a large gap at $k=0$ and a deep magneto-roton minimum at finite wave vector, in excellent quantitative agreement with recent numerical calculations. We demonstrate that the magneto-roton minimum is a precursor to the gap collapse associated with the Wigner crystal instability occurring near $\nu=\frac{1}{7}$. In addition to providing a simple physical picture of the collective excitation modes, this theory allows one to compute rather easily and accurately experimentally relevant quantities such as the susceptibility and the ac conductivity.

I. INTRODUCTION

The quantum Hall effect[1–4] is a remarkable macroscopic quantum phenomenon occuring in the two-dimensional electron gas (inversion layer) at high magnetic fields and low temperatures. The Hall resistivity is found to be quantized with extreme accuracy[5] in the form

$$\rho_{xy}=h/e^2 i \ . \tag{1.1}$$

In the integral Hall effect the quantum number i take on integral values. In the fractional quantum Hall effect (FQHE) the values of i are rational fractions with odd denominators.

Associated with and of central importance to this quantization of ρ_{xy} is the appearance of exponentially small dissipation,

$$\rho_{xx}\sim e^{-\Delta/2T} \ . \tag{1.2}$$

The activation energy $\Delta\approx 100$ K for the integral case is associated with disorder and the mobility gap between Landau levels and is thus primarily a single-particle effect. The fractional case occurs in low-disorder, high-mobility samples with partially filled Landau levels for which there is no single-particle gap. In this case the excitation gap is a collective effect arising from many-body correlations due to the Coulomb interaction and is therefore smaller in magnitude ($\Delta_{1/3}\approx 6$ K). Considerable progress has recently been made toward understanding the nature of the many-body ground state, which at least for Landau-level–filling factors of the form $\nu=1/m$, where m is an odd integer, appears to be well described by Laughlin's variational wave function.[6] We have recently reported a theory[7] of the collective excitation spectrum which is closely analogous to Feynman's theory of superfluid helium.[8] The spectrum has an excitation gap which is found to be relatively large at zero wave vector but at finite wave vector exhibits a deep "magneto-roton minimum" quite analogous to the roton minimum in helium. The purpose of the present paper is to present a more complete description and derivation of this theory and to use it to make specific experimental predictions.

The outline of the paper is as follows. Section II contains a review of Feynman's arguments for superfluid ^{4}He. In Sec. III we discuss the application of these ideas to fermion systems in a magnetic field and in Sec. IV we specialize to the case of the lowest Landau level. In Secs. V–VII we discuss evaluation of the static structure factor, the collective mode dispersion, and the role of backflow corrections. In Sec. VIII we consider the large-wave-vector limit. In Secs. IX–XI we discuss finite-thickness effects, disorder, and linear response, and in Sec. XII we present a summary of our conclusions.

II. REVIEW OF FEYNMAN'S THEORY OF ^{4}He

In a beautiful series of papers[8] Feynman has laid out an elegantly simple theory of the collective excitation spectrum in superfluid ^{4}He. Even though the underlying ideas were developed for a neutral Bose system in three dimensions, they are sufficiently general that they can be applied *mutatis mutandis* to a charged Fermi system in two dimensions in a high magnetic field. Let us therefore now review Feynman's arguments.

Because ^{4}He is a Bose system, the ground-state wave function is symmetric under particle exchange and has no nodes. Using these facts Feynman argues that there can be no low-lying single-particle excitations, so that the only low-energy excitations are long-wavelength density oscillations (phonons). Now suppose that somehow one knew the exact ground-state wave function, ψ. Then one could write the following variational wave function to describe a density-wave-excited state which still contains most of the favorable correlations occurring in the ground state:

$$\phi_{\mathbf{k}}(\mathbf{r}_1, \ldots, \mathbf{r}_n) = N^{-1/2}\rho_{\mathbf{k}}\psi(\mathbf{r}_1, \ldots, \mathbf{r}_N) , \tag{2.1}$$

where $\rho_{\mathbf{k}}$ is the Fourier transform of the density

$$\rho_{\mathbf{k}} = \int d^2\mathbf{R} \exp(-i\mathbf{k}\cdot\mathbf{R})\rho(\mathbf{R}) , \tag{2.2}$$

$$\rho(\mathbf{R}) = \sum_{j=1}^{N} \delta^2(\mathbf{R}-\mathbf{r}_j) , \tag{2.3}$$

so that

$$\rho_{\mathbf{k}} = \sum_{j=1}^{N} \exp(-i\mathbf{k}\cdot\mathbf{r}_j) . \tag{2.4}$$

Note that since the ansatz wave function contains the ground-state wave function as a factor, favorable correlations are automatically built in. Nevertheless, the excited state is orthogonal to the ground state as required. This may be seen by writing

$$\langle \psi | \phi_{\mathbf{k}} \rangle = N^{-1/2} \int d^2\mathbf{R} \exp(-i\mathbf{k}\cdot\mathbf{R}) \langle \psi | \rho(\mathbf{R}) | \psi \rangle . \tag{2.5}$$

By hypothesis is the ground state $|\psi\rangle$ is a liquid with a homogeneous density. Hence (for $k \neq 0$) the overlap integral in Eq. (2.5) vanishes.

To see that $\phi_{\mathbf{k}}$ represents a density wave, consider the following. $\phi_{\mathbf{k}}$ is a function of the particle positions. For configurations in which the particles are more or less uniformly distributed, $\rho_{\mathbf{k}}$, and therefore $\phi_{\mathbf{k}}$, will be close to zero. Hence such configurations will have a low probability of occurrence in the excited state. On the other hand, configurations with some degree of density modulation at wave vector k will have a finite value of $\rho_{\mathbf{k}}$, and $\phi_{\mathbf{k}}$, will be proportionately large, making the probability of such configurations greater. Hence this represents a density wave. Note, however, that the phase of $\rho_{\mathbf{k}}$, and thus $\phi_{\mathbf{k}}$, will match the phase of the density wave. All phases are equally likely if ψ corresponds to a liquid. Hence the average density is still uniform in the excited state. A simple way of interpreting all of this is to view $\rho_{\mathbf{k}}$ as one of a set of collective coordinates describing the particle configuration. We can then make an analogy with the simple harmonic oscillator in which the exact excited-state wave function ψ_1 is obtained from the ground state ψ_0 by multiplication by the coordinate

$$\psi_1(x) = x\psi_0(x) . \tag{2.6}$$

ψ_1 is orthogonal to ψ_0, and ψ_1 vanishes for the configuration ($x=0$) at which ψ_0 is peaked ("uniform density"). ψ_1 is nonzero when the coordinate is displaced ("density wave"), but has a phase that varies with displacement so that the average value of the coordinate is zero ("uniform average density") even in the excited state.

Having established an ansatz variational wave function, we need to evaluate the energy. This requires a knowledge of the norm of the wave function

$$s(k) = \langle \phi_{\mathbf{k}} | \phi_{\mathbf{k}} \rangle = N^{-1} \langle \psi | \rho_{\mathbf{k}}^{\dagger}\rho_{\mathbf{k}} | \psi \rangle , \tag{2.7}$$

but this is nothing more than the static structure factor for the ground state, a quantity which can be directly measured using neutron scattering. For later purposes it will be convenient to note that $s(k)$ is also related to the Fourier transform of the radial distribution function for the ground state, $g(r)$:

$$s(k) = 1 + \rho \int d^2\mathbf{R} \exp(-i\mathbf{k}\cdot\mathbf{R})[g(\mathbf{R})-1] + \rho(2\pi)^2\delta^2(\mathbf{k}) , \tag{2.8}$$

where ρ is the average density.

The variational estimate for the excitation energy is the usual expression

$$\Delta(k) = f(k)/s(k) , \tag{2.9}$$

where

$$f(k) = \langle \phi_{\mathbf{k}} | H - E_0 | \phi_{\mathbf{k}} \rangle , \tag{2.10}$$

and H is the Hamiltonian and E_0 is the ground-state energy. Using Eq. (2.1) we may rewrite Eq. (2.10) as

$$f(k) = N^{-1} \langle \Psi | \rho_{\mathbf{k}}^{\dagger}[H,\rho_{\mathbf{k}}] | \Psi \rangle , \tag{2.11}$$

which will be recognized as the oscillator strength.[9] Because the potential energy and the density are both simply functions of position, they commute with each other. The kinetic energy, on the other hand, contains derivatives which do not commute with the density. This yields the universal result[9]

$$f(k) = \frac{\hbar^2 k^2}{2m} , \tag{2.12}$$

making the oscillator-strength sum independent of the interaction potential. We emphasize this point because a rather different result will be obtained for the case of the FQHE.

Combining Eqs. (2.12) and (2.9) yields the Feynman-Bijl formula[8,9] for the excitation energy,

$$\Delta(k) = \frac{\hbar^2 k^2}{2ms(k)} . \tag{2.13}$$

We can interpret this as saying that the collective-mode energy is just the single-particle energy $\hbar^2k^2/2m$ renormalized by the factor $1/s(k)$ which represents the effect of correlated motion of the particles. We emphasize that since we have invoked the variational principle, $\Delta(k)$ is a rigorous upper bound to the lowest excitation energy at wave vector k.

In order to gain a better understanding of the meaning of $\Delta(k)$ and the underlying assumptions that have been used, let us rederive Eq. (2.9) by a different method. Consider the dynamic structure factor defined by ($\hbar = 1$ throughout)

$$S(k,\omega) = N^{-1} \sum_n \langle 0 | \rho_{\mathbf{k}}^{\dagger} | n \rangle \delta(\omega - E_n + E_0) \langle n | \rho_{\mathbf{k}} | 0 \rangle , \tag{2.14}$$

where the sum is over the complete set of exact eigenstates. Using Eq. (2.7) the static structure factor is related to $S(k,\omega)$ by

$$s(k) = \int_0^{\infty} d\omega\, S(k,\omega) , \tag{2.15}$$

and using Eq. (2.11) we see that the oscillator strength is related to $S(k,\omega)$ by

$$f(k)=\int_0^\infty d\omega\,\omega\,S(k,\omega)\ . \tag{2.16}$$

Substitution of these results into Eq. (2.9) shows that the Feynman-Bijl expression for $\Delta(k)$ is actually the exact first moment of the dynamic structure factor. That is, $\Delta(k)$ represents the average energy of the excitations which couple to the ground state through the density. This is consistent with the idea that $\Delta(k)$ is a variational bound on the collective-mode energy, since the average energy necessarily exceeds the minimum excitation energy. Note that if the oscillator-strength sum is saturated by a single mode, then the mean excitation energy and the minimum will be the same and the Feynman-Bijl expression will be exact. This idea is consistent with Feynman's argument that there are no low-lying single-particle excitations. The assumption that $\phi_{\mathbf{k}}$ in Eq. (2.1) is a good variational wave function is equivalent to assuming that the density-wave saturates the oscillator-strength sum rule.

How well does the Feynman-Bijl formula work? Evaluating $s(k)$ in Eq. (2.13) from the experimental neutron-scattering cross section and speed-of-sound data yields[8,9] a collective-mode frequency which vanishes linearly at small k (with a slope corresponding exactly to the speed of sound) and exhibits the famous "roton minimum" near $k=2\ \text{Å}^{-1}$. The roton minimum is due to a peak in $s(k)$ associated with the short-range order in the liquid. The predicted dispersion curve is in good qualitative agreement with experiment, but the predicted frequency at the roton minimum is about a factor of 2 too high.[8,9] Feynman and Cohen[8] have shown that this problem can be remedied by including back-flow corrections which guarantee that the continuity equation $\mathbf{\Delta}\cdot\langle\mathbf{J}\rangle=0$ is satisfied by the variational wave function. These corrections bring the predicted mode energy into excellent quantitative agreement with experiment.

These considerations of Feynman's arguments lead us to the following conclusions. One expects on quite general grounds that (at least for long wavelengths) the low-lying excited states of any system will include density waves. If because of special circumstances (such as those occurring in superfluid ^{4}He) there are no low-lying single-particle states, then the Feynman-Bijl expression for the collective-mode energy will be valid. The expression is also qualitatively correct even at short wavelengths and yields quantitative agreement with experiment if explicit back-flow corrections are included.

III. APPLICATION TO FERMIONS

The single-mode approximation seems less likely to succeed for Fermi systems. For instance, examination of $S(k,\omega)$ for the three-dimensional electron gas (jellium) shows the existence of not only a collective mode (the plasmon) but also a large continuum of single-particle excitations.[9] The low-lying continuum is due to the small kinetic energy of excitations across the Fermi surface. Despite this problem Lundqvist[10] and Overhauser[11] have shown that very useful results can be obtained from the single-mode approximation (SMA). It is straightforward to prove from the compressibility sum rule[9] that the plasmon mode saturates the oscillator-strength sum rule in the limit $k\to 0$. Hence the SMA gives the exact plasma frequency $\Delta(0)=\hbar\omega_p$. For finite k the SMA breaks down, but as noted above $\Delta(k)$ is the exact first moment of the excitation spectrum. Thus for large k, $\Delta(k)$ lies at the centroid of the continuum. Hence the SMA is exact at long wavelengths and gives a reasonable fit to the single-particle continuum even at short wavelengths. The same statement is true for the two-dimensional case, where the plasmon dispersion is $\omega_p\sim k^{1/2}$ at long wavelengths.

We are now in a position to ask what the effect of a large magnetic field is on the Fermi system. It is important to note that the neutral excitations are still characterized[12] by a conserved wave vector $\mathbf{k}$, but that in two dimensions the magnetic field quenches the single-particle continuum of kinetic energy, leaving a series of discrete, highly degenerate Landau levels evenly spaced in energy at intervals of $\hbar\omega_c$, where ω_c is the classical cyclotron frequency. Consider first the case of the filled Landau level ($\nu=1$). Because of Pauli exclusion the lowest excitation is necessarily the cyclotron mode in which particles are excited into the next Landau level. Furthermore, from Kohn's theorem[13] we know that in the limit of zero wave vector, the cyclotron mode occurs at precisely $\hbar\omega_c$ and saturates the oscillation-strength sum rule. Hence, once again the SMA is exact at long wavelengths and yields $\Delta(0)=\hbar\omega_c$. To see explicitly that this is so, note that for $\nu=1$ the radial distribution function is known exactly[6,14] (neglecting mixing of Landau levels in the ground state),

$$g(r)=1-\exp(-r^2/2l^2)\ , \tag{3.1}$$

where $l=(eB/\hbar c)^{-1/2}$ is the magnetic length. Using (2.8) and $\rho=\nu/(2\pi l^2)$, we have, for $k\neq 0$,

$$s(k)=1-\exp(-k^2l^2/2)\ . \tag{3.2}$$

Using (2.9) the predicted mode energy is

$$\Delta(k)=\frac{\hbar^2k^2}{2m[1-\exp(-k^2l^2/2)]}\ , \tag{3.3}$$

but $\hbar^2/ml^2=\hbar\omega_c$, so that we finally obtain

$$\Delta(k)=\hbar\omega_c\frac{(kl)^2/2}{1-\exp(-k^2l^2/2)}\ , \tag{3.4}$$

which has the correct limit $\Delta(0)=\hbar\omega_c$. We emphasize that this was derived with a ground-state structure factor calculated by neglecting Landau-level mixing, but that Kohn's theorem[13] requires that the same result be obtained (for $k\to 0$) for the exact ground state.

For the FQHE we are interested not in the case $\nu=1$, but rather we need to consider the fractionally filled Landay level. The Pauli principle now no longer excludes low-energy intra-Landau-level excitations. It is these low-lying excitations rather than the high-energy inter-Landau-level cyclotron modes which are of primary importance to the FQHE. Let us therefore consider what the SMA yields for the case $\nu<1$ by recalling the argument leading to Kohn's theorem.[13] At asymptotically long wavelengths an external perturbation couples only to the center-of-mass (c.m.) motion. The c.m. degree of freedom has the excitation spectrum of a free particle in the magnetic field and is unaffected by the correlations and interactions among the individual particles. Hence once again the cyclotron mode saturates the oscillator-strength sum rule and the SMA yields $\Delta(0)=\hbar\omega_c$ independent of the filling factor and the interaction potential. The SMA thus tells us nothing about the low-energy modes of interest.

The root of this difficulty can be traced back to Eq. (2.1). For small k the variational excited state has most of its weight in the next-higher Landau level. We can greatly improve the variational bound on the excitation energy by insisting that the excited state lie entirely within the lowest Landau level. This can be enforced by replacing Eq. (2.1) by

$$\phi_{\mathbf{k}}=N^{-1/2}\bar{\rho}_{\mathbf{k}}\psi \, , \tag{3.5}$$

where $\bar{\rho}_{\mathbf{k}}$ is the projection of the density operator onto the subspace of the lowest Landau level. Providing that this projection can be explicitly carried out, we may derive a new approximation (the projected SMA) for the low-lying collective-mode frequency,

$$\Delta(k)=\bar{f}(k)/\bar{s}(k) \, , \tag{3.6}$$

where $\bar{f}$ and $\bar{s}$ are, respectively, the projected analogs of the oscillator strength and the static structure factor. Formulation of the projection scheme and the derivation of (3.6) are carried out in the next section.

IV. PROJECTION ONTO THE LOWEST LANDAU LEVEL

The formal development of quantum mechanics within the subspace of the lowest Landau level has been presented elsewhere.[15] We briefly review here the pertinent results. Taking the magnetic length to be unity ($l=1$) and adopting the symmetric gauge (with $\mathbf{B}=-B\hat{\mathbf{z}}$), the single-particle eigenfunctions of kinetic energy and angular momentum in the lowest Landau level are[6,15]

$$\Phi_m(z)=\frac{1}{(2\pi 2^m m!)^{1/2}}z^m \exp(-|z|^2 4) \, , \tag{4.1}$$

where m is a non-negative integer and $z=x+iy$ is the complex representation of the particle position. We wish to focus on the analytic part of the wave function and ignore as much as possible the ubiquitous Gaussian factor. Hence we define, following Bargmann,[15,16] a Hilbert space of analytic functions with an inner product

$$(f,g)=\int d\mu(z) f^*(z) g(z) \, , \tag{4.2}$$

where the Gaussian factors from (4.1) have been absorbed into the measure:

$$d\mu(z)=(2\pi)^{-1}dx\,dy\,\exp(-|z|^2/2) \, . \tag{4.3}$$

Operators on this space must take analytic functions into analytic functions. Hence a natural pair to consider is

$$a^\dagger=z \, , \tag{4.4a}$$

$$a=2\frac{d}{dz} \, . \tag{4.4b}$$

These are mutually adjoint with respect to the measure defined in (4.3) and represent ladder operators for the angular momentum.[15]

At this point it is useful to note that the adjoint of z is not the same as the Hermitian conjugate of z, which is z^*. z^* is not analytic in z and hence takes functions out of the Hilbert space (it mixes Landau levels). However, for any two states f and g in the Hilbert space,

$$(f,z^*g)=(zf,g)=(f,z^\dagger g)=\left(f,2\frac{d}{dz}g\right) \, . \tag{4.5}$$

Thus the projection of z^* onto the lowest Landau level is

$$\overline{z^*}=z^\dagger=2\frac{d}{dz} \, . \tag{4.6}$$

Since $\overline{z^*}$ and z do not commute, one has to normal order the derivatives to the left.[15] These results are trivially extended to the many-particle case and the projection of the density operator is easily accomplished as follows. Writing the dot product in complex notation, the density operator from Eq. (2.4) becomes

$$\rho_{\mathbf{k}}=\sum_{j=1}^{N}\exp\left(-\frac{ik}{2}z_j^*-\frac{ik^*}{2}z_j\right) \, . \tag{4.7}$$

The projected version is simply (note the normal ordering)

$$\bar{\rho}_k=\sum_{j=1}^{N}\exp\left(-ik\frac{\partial}{\partial z_j}\right)\exp\left(-\frac{ik^*}{2}z_j\right) \, . \tag{4.8}$$

We are now in a position to project the Hamiltonian. The kinetic energy can be ignored since it is an irrelevant constant $N\hbar\omega_c/2$ in the lowest Landau level. The potential energy is

$$V=\frac{1}{2}\int\frac{d^2q}{(2\pi)^2}v(q)\sum_{i\neq j}\exp[i\mathbf{q}\cdot(\mathbf{r}_i-\mathbf{r}_j)] \, , \tag{4.9}$$

where $v(q)$ is the Fourier transform of the interaction potential. The projection of V is

$$\bar{V}=\frac{1}{2}\int\frac{d^2q}{(2\pi)^2}v(q)(\bar{\rho}_{\mathbf{q}}\bar{\rho}_{\mathbf{q}}-\rho e^{-q^2/2}) \, . \tag{4.10}$$

Note that, just as in the usual case, $\bar{\rho}_{\mathbf{q}}^\dagger=\bar{\rho}_{-\mathbf{q}}$.

In analogy with Eq. (2.11), the projected oscillator strength is

$$\bar{f}(k)=N^{-1}\langle 0|\bar{\rho}_{\mathbf{k}}^\dagger[\bar{H},\bar{\rho}_{\mathbf{k}}]|0\rangle \, , \tag{4.11}$$

where $|0\rangle$ is the ground state (represented as a member of the Hilbert space of analytic functions). Note that previously it was the kinetic energy which contained derivatives and hence failed to commute with the density. Now the kinetic energy is an irrelevant constant, but both the potential-energy and density operators contain derivatives. Thus Eq. (4.11) becomes

$$\bar{f}(k)=N^{-1}\langle 0|\bar{\rho}_{\mathbf{k}}^\dagger[\bar{V},\bar{\rho}_{\mathbf{k}}]|0\rangle \, . \tag{4.12}$$

The meaning of this is that since the kinetic energy has been quenched by the magnetic field, the scale of the collective-mode energy is set solely by the scale of the interaction potential, which is, of course, as it should be.

In order to evaluate (4.12) it is convenient to note that the projected density operators obey a closed Lie algebra defined by

$$[\bar{\rho}_{\mathbf{k}},\bar{\rho}_{\mathbf{q}}]=(e^{k^*q/2}-e^{kq^*/2})\bar{\rho}_{\mathbf{k}+\mathbf{q}} \; . \tag{4.13}$$

It is convenient to use parity symmetry in k to rewrite Eq. (4.12) as a double commutator,

$$\bar{f}(k)=\frac{1}{2N}\langle 0\,|\,[\bar{\rho}_{\mathbf{k}}^{\dagger},[\bar{V},\bar{\rho}_{\mathbf{k}}]]\,|\,0\rangle \; . \tag{4.14}$$

Using Eq. (4.10), Eq. (4.14) is readily evaluated with the commutation relation given in (4.13),

$$\bar{f}(k)=\frac{1}{2}\sum_q v(q)(e^{q^*k/2}-e^{qk^*/2}) \times[\bar{s}(q)e^{-k^2/2}(e^{-k^*q/2}-e^{-kq^*/2}) +\bar{s}(k+q)(e^{k^*q/2}-e^{kq^*/2})] \; , \tag{4.15}$$

where $\bar{s}(q)$ is the projected static structure factor,

$$\bar{s}(q)=N^{-1}\langle 0\,|\,\bar{\rho}_{\mathbf{q}}^{\dagger}\bar{\rho}_{\mathbf{q}}\,|\,0\rangle \; . \tag{4.16}$$

Using

$$\overline{\rho_{\mathbf{q}}^{\dagger}\rho_{\mathbf{q}}}=\bar{\rho}_{\mathbf{q}}^{\dagger}\bar{\rho}_{\mathbf{q}}+(1-e^{-|q|^2/2}) \; , \tag{4.17}$$

one obtains the relation

$$\bar{s}(q)=s(q)-(1-e^{-|q|^2/2}) \; , \tag{4.18}$$

where $s(q)$ is the ordinary static structure factor given in Eqs. (2.7) and (2.8). From Eq. (3.2) we see that $\bar{s}(q)$ vanishes identically for the filled Landau level. This is simply a reflection of the fact that it is not possible to create any excitations within the lowest Landau level when it is completely filled.

Clearly, Eq. (4.15) is more complicated than its analog, (2.12). Nevertheless, it is still true that knowledge of the static structure factor is all that is required to evaluate (4.15) and (4.16) and hence obtain the projected SMA mode energy:

$$\Delta(k)=\bar{f}(k)/\bar{s}(k) \; . \tag{4.19}$$

The essence of the Feynman-Bijl result (2.13) is still maintained—namely that one can express a dynamical quantity, the collective-mode energy, solely in terms of static properties of the ground state.

Let us begin our evaluation of (4.19) by consideration of the small-k limit. We assume throughout that the ground state is an isotropic and homogeneous liquid. Direct expansion of Eq. (4.15) shows that, for small k, $\bar{f}(k)$ vanishes like $|k|^4$. Indeed, one can show that this is true, in general, because Kohn's theorem tells us that the total oscillator-strength sum $f(k)=\hbar^2k^2/2m$ is saturated by the cyclotron mode (to leading order in $|k|^2$). Hence the intra-Landau-level contribution [which is $\bar{f}(k)$] must quite generally vanish faster than $|k|^2$ for both solid and liquid ground states. Given that

$$\bar{f}(k)\sim|k|^4 \; , \tag{4.20}$$

it follows from (4.19) that a necessary (but not sufficient) condition for the existence of a finite direct ($k=0$) gap is

$$\bar{s}(k)\sim|k|^4 \; . \tag{4.21}$$

Equation (4.21) is a *sufficient* condition only within the SMA, but as the following argument shows, Eq. (4.21) is always a *necessary* condition for a gap. Equation (4.19) gives the exact first moment of the (intra-Landau-level) excitation spectrum. If $\bar{s}(k)$ vanishes slower than $|k|^4$, then the mean excitation energy vanishes as k approaches zero and there can be no gap. If $\bar{s}(k)\sim|k|^4$, then the mean excitation energy is finite for small k. This does not prove that there is a gap; however, it seems plausible that in this system (unlike ordinary jellium) there can be no low-lying single-particle excitations to invalidate the SMA and defeat the gap since the kinetic energy necessary to produce such a continuum has been quenched by the magnetic field.

Having established the importance of the condition $\bar{s}(k)\sim|k|^4$, let us investigate whether or not this condition obtains. Using Eq. (4.18) and expanding Eq. (2.8) for small k shows that $\bar{s}(k)\sim|k|^4$ if and only if $M_0=M_1=-1$, where

$$M_n\equiv\rho\int d^2r(r^2/2)^n[g(r)-1] \; . \tag{4.22}$$

In general, for a liquid ground state one can express the two-point correlation function in terms of the occupation of the single-particle angular-momentum eigenstates of Eq. (4.1),

$$g(r)=\rho^{-2}\sum_{\alpha,\beta,\gamma,\delta}\phi_\alpha(0)\phi_\beta(r)\phi_\gamma^*(r)\phi_\delta^*(0)\langle c_\alpha^\dagger c_\beta^\dagger c_\gamma c_\delta\rangle \; , \tag{4.23}$$

where $c_\alpha^\dagger$ is the creation operator for state α. Using Eq. (4.1) and conservation of angular momentum, we have

$$\rho[g(r)-1]=(2\pi\nu)^{-1}\sum_{m=0}^{\infty}\frac{(r^2/2)^m}{m!}\exp(-r^2/2) \times(\langle n_m n_0\rangle-\langle n_m\rangle\langle n_0\rangle -\nu\delta_{m0}) \; , \tag{4.24}$$

where $n_m=c_m^\dagger c_m$ is the occupation number for state m. Inserting Eq. (4.24) into Eq. (4.22) yields

$$M_0=\nu^{-1}(\langle Nn_0\rangle-\langle N\rangle\langle n_0\rangle)-1 \; , \tag{4.25}$$

$$M_1=\nu^{-1}[\langle(L+N)n_0\rangle-\langle L+N\rangle\langle n_0\rangle]-1 \; , \tag{4.26}$$

where

$$N\equiv\sum_{m=0}^{\infty}n_m \tag{4.27}$$

is the total particle number and

$$L\equiv\sum_{m=0}^{\infty}mn_m \tag{4.28}$$

is the total angular momentum. Since L and N are constants of the motion, their fluctuations vanish, leaving $M_0 = M_1 = -1$.

This general result implies that, for any liquid ground state in the lowest Landau level, $\bar{s}(k) \sim |k|^4$. Hence any liquid state automatically satisfies the SMA gap condition $\Delta(0) > 0$ discussed above. Interpreting $\bar{s}(k)$ as the mean-square density fluctuation at wave vector k, the condition $\bar{s}(k) \sim |k|^4$ is a statement of the lack of density fluctuations or the incompressibility of the quantum system at long wavelengths. This is the source of the finite gap.

Within the SMA the existence of a gap for liquid ground states appears to be the rule rather than the exception. The interesting question of whether or not liquid ground states must have a rational filling factor is an entirely separate issue, about which nothing has been proved by these arguments.

Within the SMA, gapless excitations can occur only as Goldstone modes in systems with broken translational symmetry (which therefore violate our assumption of a liquid ground state). It is worth noting in this connection that the SMA analog of Eq. (4.19) yields the correct transverse magneto-phonon dispersion curve for the Wigner crystal.

We can shed additional light on the meaning of Eq. (4.21) and (4.22) by considering the specific case of the Laughlin ground state.[6] Invoking the analogy with the two-dimensional one-component plasma (2DOCP),[6,17,18] we see that $M_0 = -1$ is the charge-neutrality sum rule and $M_1 = -1$ is the perfect screening sum rule for the 2DOCP.[17,18] Making use of the 2DOCP compressibility sum rule,[17,18] we obtain M_2 and hence the exact leading term in $\bar{s}(k)$,

$$\bar{s}(k) = \frac{1-\nu}{8\nu} |k|^4 + \cdots . \tag{4.29}$$

This result emphasizes the profound importance of the existence of long-range forces in the 2DOCP analog system. These long-range forces are responsible for the charge-neutrality and perfect screening sum rules. From these it follows that there is (within the SMA) a finite excitation gap at $k=0$ and from these also follows the exactness of the fractional charge $\pm\nu$ of the Laughlin quasiparticles.[6]

V. STATIC STRUCTURE FACTOR

In order to go beyond the small-k limit in evaluating Eq. (4.19), we need to have $\bar{s}(k)$ for finite k. Lacking the experimental structure factor that was available for the case of ^{4}He, we are forced to adopt a specific model for the ground state. We have chosen to use the Laughlin ground-state wave function[6] since it appears to be quite accurate[19,20] and because the static structure factor is available through the 2DOCP analogy.[6,21,22]

The static structure factor for the 2DOCP has been computed by both Monte Carlo[17,21] (MC) and hypernetted-chain[18,23] (HNC) methods. The MC results for $g(r)$ used in Ref. 7 are shown in Fig. 1. Recall from Eq. (2.8) that we need to Fourier-transform $g(r)$ to obtain $s(k)$. This is most easily accomplished by transforming an analytic function which has been fitted to the MC data. Fortunately, we can take advantage of the known

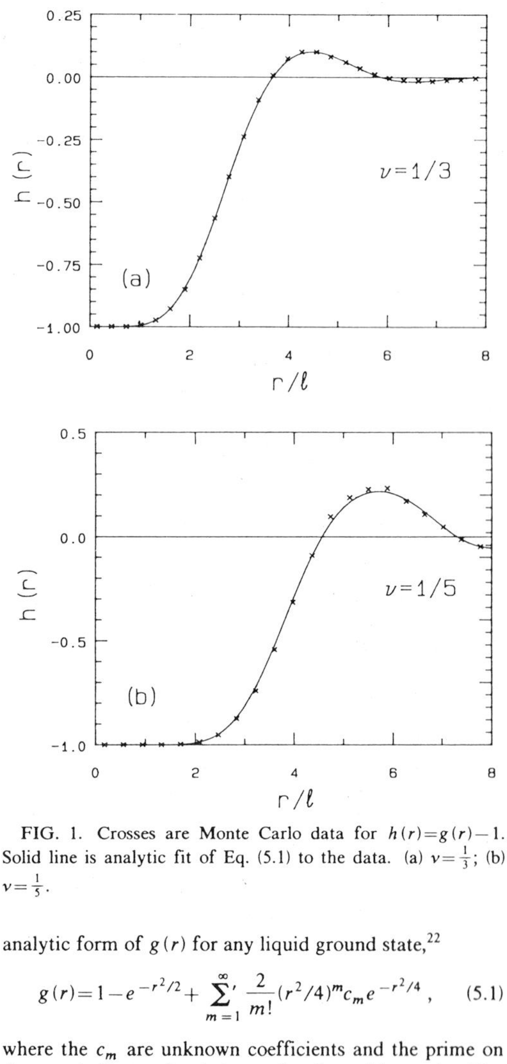

FIG. 1. Crosses are Monte Carlo data for $h(r) = g(r) - 1$. Solid line is analytic fit of Eq. (5.1) to the data. (a) $\nu = \frac{1}{3}$; (b) $\nu = \frac{1}{5}$.

analytic form of $g(r)$ for any liquid ground state,[22]

$$g(r) = 1 - e^{-r^2/2} + \sum_{m=1}^{\infty}{}' \frac{2}{m!}(r^2/4)^m c_m e^{-r^2/4} , \tag{5.1}$$

where the c_m are unknown coefficients and the prime on the sum indicates that it is restricted to odd m only. The latter is a reflection of the Fermi statistics, which requires that pairs of (spin-polarized) particles have odd relative angular momentum. For the Mth Laughlin state ($\nu = 1/M$) the coefficients c_m are constrained by the 2DOCP charge-neutrality, perfect screening, and compressibility sum rules[17,18] to obey[22]

$$\sum_{m=1}^{\infty}{}' c_m = (1-M)/4 , \tag{5.2}$$

$$\sum_{m=1}^{\infty}{}' (m+1)c_m = (1-M)/8 , \tag{5.3}$$

$$\sum_{m=1}^{\infty}{}' (m+2)(m+1)c_m = (1-M)^2/8 . \tag{5.4}$$

We fit a finite number (27) of the coefficients to the MC data subject to the constraints (5.2)–(5.4). The best-fit

TABLE I. Coefficients c_m obtained from fitting Eq. (5.1) to the MC data for $g(r)$ subject to the constraints (5.2)–(5.4). The $\nu=\frac{1}{5}$ fit is somewhat less reliable than the $\nu=\frac{1}{3}$ case (see text).

m	$c_m(\nu=\frac{1}{3})$	$c_m(\nu=\frac{1}{5})$
1	$-1.000\,00$	-1.0000
3	$+0.510\,53$	-1.0000
5	$-0.020\,56$	$+0.6765$
7	$+0.310\,03$	$+0.3130$
9	$-0.490\,50$	-0.1055
11	$+0.201\,02$	$+0.8910$
13	$-0.009\,04$	-0.3750
15	$-0.001\,48$	-0.7750
17	$+0.000\,00$	$+0.3700$
19	$+0.001\,20$	$+0.0100$
21	$+0.000\,60$	-0.0050
23	$-0.001\,80$	-0.0000
25	$+0.000\,00$	-0.1000
27	$+0.000\,00$	$+0.1000$

values for $\nu=\frac{1}{3}$ and $\frac{1}{5}$ are displayed in Table I and the resultant analytic $g(r)$ is shown in Fig. 1 along with the MC data. Having obtained an analytic form, the required Fourier transform is readily computed.

An alternative method of obtaining $s(k)$ is to use a modified hypernetted-chain (MHNC) approximation[23] which guarantees that the sum rules[17,18] on $s(k)$ are satisfied. This method gives a value for the energy in the $\nu=\frac{1}{3}$ Laughlin state of $E_{1/3}=-0.4092$, which is quite close to the value of $E_{1/3}=-0.4100\pm0.0001$) from the essentially exact MC method. Figure 2 displays $s(k)$ computed by the MHNC and MC methods.

Having obtained $s(k)$ we compute $\bar{s}(k)$ from Eq. (4.18) and then use this in Eq. (4.15). We also require the interaction potential $v(q)$. Taking the unit of energy to be $(e^2/\epsilon l)$, where ϵ is the dielectric constant of the background medium, the Coulomb potential is $V(r)=1/r$, which has the transform

$$v(q)=\frac{2\pi}{q} . \tag{5.5}$$

Using (5.5), the quadratures in (4.15) were computed numerically to obtain the oscillator strength and hence the gap function $\Delta(k)$.

VI. EVALUATION OF THE GAP

Using the results of the preceding section, we have evaluated the collective-mode dispersion for filling factors $\nu=\frac{1}{3}$, $\frac{1}{5}$, $\frac{1}{7}$, and $\frac{1}{9}$ using the MHNC structure factors. The various gap functions for the case of the pure Coulomb potential are shown in Fig. 3. The MC structure factors for $\nu=\frac{1}{3}$ and $\frac{1}{5}$ yield nearly identical results,[7] except for a small discrepancy in the $\nu=\frac{1}{5}$ curve at small k. We believe that this is due to the difficulty of extracting accurate information on the long-distance behavior of $g(r)$ from the $\nu=\frac{1}{5}$ MC data and we therefore consider the MHNC result more reliable for this case.

Note that, as discussed earlier, the gap is finite at zero

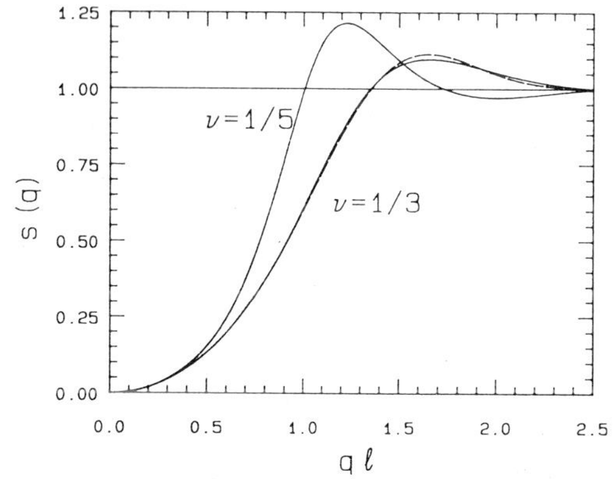

FIG. 2. Static structure factor. Solid line is modified-hypernetted-chain calculation. Dashed line is from fit to Monte Carlo data.

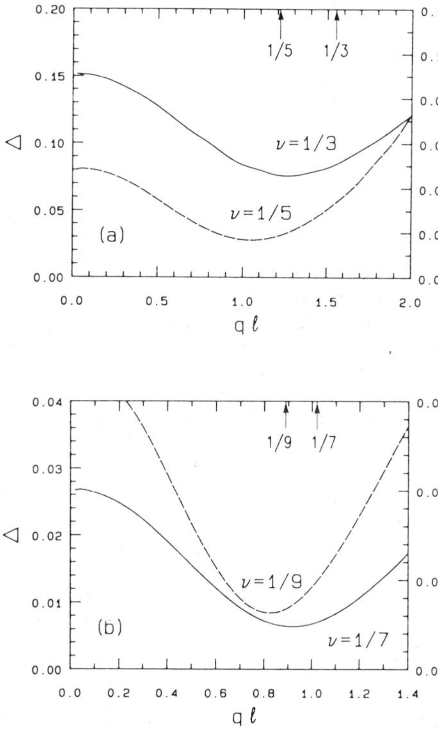

FIG. 3. Collective-mode dispersion. Arrows at the top indicate magnitude of primitive reciprocal-lattice vector of corresponding Wigner crystal. (a) $\nu=\frac{1}{3}$ (scale on left); $\nu=\frac{1}{5}$ (scale on right). (b) $\nu=\frac{1}{7}$ (scale on left); $\nu=\frac{1}{9}$ (scale on right).

wave vector, but exhibits a deep minimum at finite k. This magneto-roton minimum is caused by a peak in $\bar{s}(k)$ and is, in this sense, quite analogous to the roton minimum in helium.[8,9] We interpret the deepening of the minimum in going from $\nu=\frac{1}{3}$ to $\nu=\frac{1}{7}$ to be a precursor of the collapse of the gap which occurs at the critical density ν_c for Wigner crystallization. From Fig. 3 we see that the minimum gap is very small for $\nu<\frac{1}{5}$. This is consistent with a recent estimate[24] of the critical density, $\nu_c=1/(6.5\pm0.5)$. Within mean-field theory, the Wigner crystal transition is weakly first order and hence occurs slightly before the roton mode goes completely soft. Further evidence in favor of this interpretation of the roton minimum is provided by the fact that the magnitude of the primitive reciprocal-lattice vector for the crystal lies close to the position of the magneto-roton minimum, as indicated by the arrows in Fig. 3.

These ideas suggest the physical picture that the liquid is most susceptible to perturbations whose wavelength matches the crystal lattice vector. This will be illustrated in more detail in Sec. XI.

Having provided a physical interpretation of the gap dispersion and the magneto-roton minimum, we now examine how accurate the SMA is. Figure 4 shows the excellent agreement between the SMA prediction for the gap and exact numerical results for small ($N=6,7$) systems recently obtained by Haldane and Rezayi.[20] Those authors have found by direct computation that the single-mode approximation is quite accurate, particularly near the roton minimum, where the lowest excitation absorbs 98% of the oscillator strength.[25] This means that the overlap between our variational state and the exact lowest excited eigenstate *exceeds* 0.98. We believe this agreement confirms the validity of the SMA and the use of the Laughlin-state static structure factor.

Near $k=0$ there is a small ($\sim$20%) discrepancy between $\Delta_{\rm SMA}(0)$ and the numerical calculations.[20] It is interesting to speculate that the lack of dispersion near the roton minimum may combine with residual interactions to produce a strong pairing of rotons of opposite momenta leading to a two-roton bound state of small total momentum. This is known to occur in helium.[26] For the present case $\Delta_{1/3}(0)$ happens to be approximately twice the minimum roton energy. Hence the two-roton bound state which has zero oscillator strength could lie slightly below the one-phonon state which absorbs all of the oscillator strength. For $\nu<\frac{1}{3}$ the two-roton state will definitely be the lowest-energy state at $k=0$. It would be interesting to compare the numerical excitation spectrum with a multiphonon continuum computed using the dispersion curves obtained from the SMA.

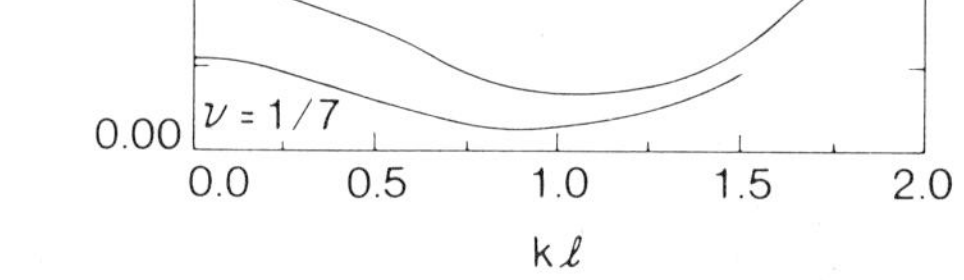

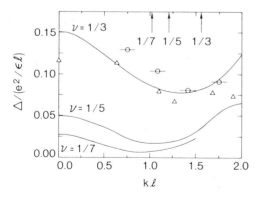

FIG. 4. Comparison of SMA prediction of collective mode energy for $\nu=\frac{1}{3}, \frac{1}{5}, \frac{1}{7}$ with numerical results of Haldane and Rezayi (Ref. 20) for $\nu=\frac{1}{3}$. Circles are from a seven-particle spherical system. Horizontal error bars indicate the uncertainty in converting angular momentum on the sphere to linear momentum. Triangles are from a six-particle system with a hexagonal unit cell. Arrows have same meaning as in Fig. 3.

VII. BACKFLOW CORRECTIONS

It is apparent from Fig. 4 that the SMA works extremely well—better, in fact, than it does for helium.[8,9] Why is this so? Recall that, for the case of helium, the Feynman-Bijl formula overestimates the roton energy by about a factor of 2. Feynman[8] traces this problem to the fact that a roton wave packet made up from the trial wave functions violates the continuity equation

$$\nabla\cdot\langle \mathbf{J}\rangle=0\ . \tag{7.1}$$

To see how this happens, consider a wave packet

$$\phi(r_1,\ldots,r_N)=\int d^2k\,\xi(k)\rho_k\psi(r_1,\ldots,r_N)\ , \tag{7.2}$$

where $\xi(k)$ is some function (say a Gaussian) sharply peaked at a wave vector k located in the roton minimum. It is important to note that this wave packet is quasistationary because the roton group velocity $d\Delta/dk$ vanishes at the roton minimum. Evaluation of the current density gives the result schematically illustrated in Fig. 5(a). The current has a fixed direction and is nonzero only in the region localized around the wave packet. This violates the continuity equation (7.1) since the density is (approximately) time independent for the quasistationary packet. The modified variational wave function of Feynman and Cohen[8] includes the backflow shown in Fig. 5(b). This gives good agreement with the experimental roton energy and shows that the roton can be viewed as a smoke ring (closed vortex loop).

A rather different result is obtained for the case of the quantum Hall effect. The current density operator is

$$\mathbf{J}(\mathbf{R})=\frac{1}{2m}\sum_{j=1}^{N}\left[\delta^2(\mathbf{R}-\mathbf{r}_j)\left[\mathbf{p}_j+\frac{e\mathbf{A}(\mathbf{r}_j)}{c}\right]+\left[\mathbf{p}_j+\frac{e\mathbf{A}(\mathbf{r}_j)}{c}\right]\delta^2(\mathbf{R}-\mathbf{r}_j)\right]\ . \tag{7.3}$$

Taking ϕ and ψ to be any two members of the Hilbert space of analytic functions described in Sec. IV, it is straightforward to show that

$$\langle\Phi|\mathbf{J}(\mathbf{R})|\Psi\rangle=-\tfrac{1}{2}\nabla\times\langle\Phi|\mathbf{M}(\mathbf{R})|\Psi\rangle\ , \tag{7.4}$$

where

$$\mathbf{M}(\mathbf{R})=\rho(\mathbf{R})\hat{\mathbf{z}}\ , \tag{7.5}$$

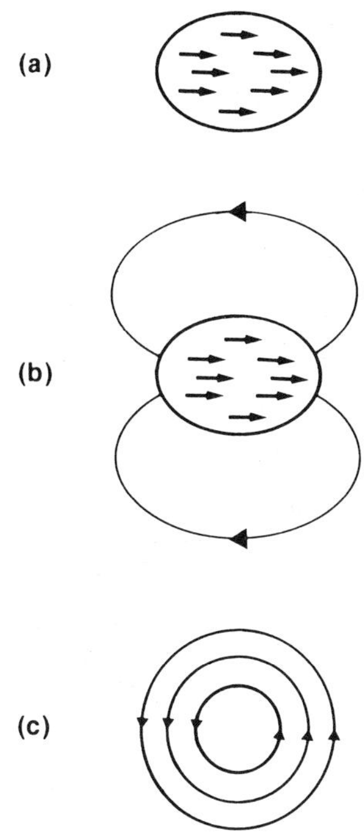

FIG. 5. Schematic illustration of the current distribution in a roton wave packet: (a) helium with no backflow corrections, (b) helium with backflow corrections, and (c) lowest-Landau-level case.

and $\rho(R)$ is the density and $\hat{\mathbf{z}}$ is the unit vector normal to the plane. It follows immediately from (7.4) that

$$\nabla\cdot\langle \mathbf{J}(R)\rangle=0 \tag{7.6}$$

for any state in the lowest Landau level. Hence the backflow condition is automatically satisfied. The current flow for the magneto-roton wave packet calculated from (7.4) is illustrated schematically in Fig. 5(c). We see that the magneto-roton circulation is rather different from the smoke ring in bulk helium shown in Fig. 5(b).

Equation (7.6) is paradoxical in that it implies $\partial\rho/\partial t=0$ for every state in the lowest Landau level. This is merely a reflection of the fact that the kinetic energy has been quenched and perturbations can cause particles to move only be means of (virtual) transitions to higher Landau levels. One can resolve this paradox by noting that there are really two different current operators we can consider. The first is the ordinary (instantaneous) current discussed above. The second is the slow (time-averaged) $\mathbf{E}\times\mathbf{B}$ drift of the particles in the magnetic field. Restriction of the Hilbert space to the lowest Landau level eliminates the fast degrees of freedom associated with the cyclotron motion but retains the slow (drift) coordinates.[15] To illustrate this note that for an external potential projected onto the lowest Landau level

$$V=\sum_{\mathbf{q}} V(q)\bar{\rho}_{\mathbf{q}}\ , \tag{7.7}$$

we have, by virtue of the noncommutivity of the density operators,

$$i\frac{\partial}{\partial t}\bar{\rho}_{\mathbf{k}}=-\sum_{\mathbf{q}} V(q)\bar{\rho}_{\mathbf{k}+\mathbf{q}}(e^{kq^*/2}-e^{k^*q/2})\ . \tag{7.8}$$

Expanding this to lowest order in q (smooth potential) gives

$$i\frac{\partial}{\partial t}\bar{\rho}_{\mathbf{k}}=-\sum_{\mathbf{q}}(\mathbf{q}\times\mathbf{k})_z V(q)\bar{\rho}_{\mathbf{k}+\mathbf{q}}\ . \tag{7.9}$$

Using the continuity equation, this can be rewritten as the usual $\mathbf{E}\times\mathbf{B}$ drift equation

$$\langle \mathbf{J}(\mathbf{r})\rangle=[\nabla V(\mathbf{r})\times\hat{\mathbf{z}}]\rho(\mathbf{r})\ . \tag{7.10}$$

We therefore recover the physically correct result even though the Hilbert space has been restricted to the lowest Landau level.

For the case of the magneto-roton wave packet discussed above, we note that the excess particle density is circularly symmetric. Hence the (mean) electric field is radial and the particle drift is purely circular, as illustrated previously in Fig. 5(c). Hence one is once again led to the conclusion that the continuity condition is automatically satisfied by the magneto-roton wave packet. We believe that this accounts for the excellent results obtained using the SMA.

VIII. SMA AT LARGE WAVE VECTORS

We saw in Sec. VI that the SMA is quantitatively accurate out to the magneto-roton minimum. For larger wave vectors the SMA rapidly breaks down as many different states begin to couple to the density. This is simply because the density wave is not a sensible excitation for wavelengths smaller than the interparticle spacing.

Even though Eq. (4.19) does not give a good variational bound on the energy for large k, it still gives the exact first moment of the excitation spectrum. As we shall see below, it is interesting to consider this first excitation moment in the limit of large k:

$$\Gamma=\lim_{k\to\infty}\bar{f}(k)/\bar{s}(k)\ . \tag{8.1}$$

For large k, Eqs. (2.8), (4.18), and (5.1) yield

$$\bar{s}(k)\sim(1-\nu)e^{-|k|^2/2}\ . \tag{8.2}$$

Using (8.2) to evaluate $\bar{s}(k+q)$ in Eq. (4.15) and taking advantage of rotational symmetry yields

$$\bar{f}(k)\sim\sum_q v(q)e^{-|k|^2/2}[(1-\nu)e^{-|q|^2/2}-\bar{s}(q)]\ . \tag{8.3}$$

Using (4.18) this may be rewritten as

$$\bar{f}(k)=\sum_q v(q)e^{-|k|^2/2}\{[1-s(q)]-\nu[1-s_1(q)]\}\ , \tag{8.4}$$

where

$$s_1(q)\equiv 1-e^{-|q|^2/2} \tag{8.5}$$

is the static structure factor for a filled Landau level ($\nu=1$). The sum in (8.4) is simply related to the ground-state energy *per particle*, $E(\nu)$,

$$\bar{f}(k)=-2e^{-|k|^2/2}[E(\nu)-\nu E(1)]\ , \tag{8.6}$$

or, equivalently,

$$\bar{f}(k)=2e^{-|k|^2/2}E_{\rm coh}(\nu)\ , \tag{8.7}$$

where $E_{\rm coh}(\nu)$ is the cohesive energy *per particle*. Substitution of (8.7) into (8.1) yields

$$\Gamma=\frac{2E_{\rm coh}(\nu)}{1-\nu}\ . \tag{8.8}$$

This is the exact first moment of the (intra-Landau-level) density-fluctuation spectrum at $k=\infty$. Note that because the kinetic energy has been quenched, Γ is finite. Also note that this result does not rely on the validity of the SMA or on any particular assumptions about the structure of the ground state (other than the restriction to the lowest Landau level). For the particular case of Laughlin's ground state, $\Gamma(\nu)=0.603,0.508$ at $\nu=\frac{1}{3},\frac{1}{5}$.

We can gain greater insight into the meaning of (8.8) by considering the form of the variational excited state at large k. From (4.8) we have

$$\bar{\rho}_k\psi=\sum_{j=1}^{N}e^{-|k|^2/2}e^{-ik^*z_j/2}\times\psi(z_1,\ldots,z_{j-1},z_j-ik,z_{j+1},\ldots,z_N)\ . \tag{8.9}$$

We see that $\bar{\rho}_k$ acts, in part, like a translation operator moving particles (transversely) by a distance $|k|$. For large k this corresponds to an incoherent single-particle excitation which leaves a hole and a particle uncorrelated with their immediate surroundings ("undressed"). This explains why the cohesive energy appears in (8.8).

As an aside, we note that Eq. (8.9) can also teach us something about the nature of the collective modes at small k. Recall Halperin's[27] interpretation of the Laughlin wave function which notes that the Laughlin state makes the most efficient use of the zeroes of the wave function by putting them all at the locations of the particles. We see from (8.9) that the density wave is a linear superposition of states with some of their zeroes displaced a distance ik from one of the particles. This accounts for the increased energy of the state.

Equation (8.8) gives the exact first moment Γ of the density-fluctuation excitation spectrum at $k=\infty$. It would also be useful to find not just Γ but $\Delta(\infty)$, the minimum excitation energy at very large wave vectors. This may be done as follows. Equations (8.8)–(8.9) show us that the density wave creates a particle-hole excitation. Recall, however, that for the Laughlin state with $\nu=1/m$, a hole (electron) is precisely a superposition of m fractionally charged quasiholes (quasielectrons).[28,29] This suggests that at large wave vectors an improvement on our density-wave variational state can be obtained by producing a single quasihole-quasielectron pair. An exciton[12,30] formed from such quasiparticles would carry (dimensionless) momentum kl if the particles were separated by a (dimensionless) distance mkl. Hence the energy dispersion for large k is[12,30]

$$\Delta_{\rm ex}(k)=\Delta(\infty)-\frac{1}{m^3kl}\ , \tag{8.10}$$

where $\Delta(\infty)$ is the sum of the quasihole and quasielectron energies and the second term on the right-hand side of (8.10) represents the Coulomb attraction between the fractional charges. Equation (8.10) is valid for large k, where the particles are widely separated compared to their intrinsic size ($\sim l$). For smaller k we know that the density-wave state is a good description. Suppose we arbitrarily assume that the collective mode crosses over from being a density wave to being an exciton at the wave vector of the roton minimum, $k_{\rm min}$. Then we can equate the SMA and exciton values of the gap,

$$\Delta_{\rm ex}(k_{\rm min})=\Delta_{\rm SMA}(k_{\rm min})\ . \tag{8.11}$$

Using (8.10) leads to the prediction

$$\Delta(\infty)=\Delta_{\rm SMA}(k_{\rm min})+\frac{1}{m^3k_{\rm min}l}\ . \tag{8.12}$$

From the numerical values of $\Delta_{\rm SMA}$ and $k_{\rm min}$ obtained previously, we find $\Delta_{1/3}(\infty)=0.106$ and $\Delta_{1/5}(\infty)=0.025$. These values lie considerably above the results of HNC-approximation calculations of Laughlin,[6,31] $\Delta_{1/3,1/5}(\infty)=0.057,0.014$, and Chakraborty,[32] $\Delta_{1/3,1/5}(\infty)=0.053,0.014$. However, preliminary Monte Carlo results of Morf and Halperin[33] yield a larger value, $\Delta_{1/3}(\infty)=0.099\pm0.009$. In addition, the small-system calculations of Haldane and Rezayi[20] yield a value (extrapolated to $N=\infty$) of $\Delta_{1/3}(\infty)=0.105\pm0.005$, in excellent agreement with the present result. In summary, while the SMA breaks down for large k, it still yields the exact first moment of the excitation spectrum [Eq. (8.8)] and can be used to obtain a good estimate of $\Delta(\infty)$, the sum of the quasihole and quasielectron energies.

IX. FINITE-THICKNESS EFFECTS

The calculations discussed in the preceding sections have all used the Coulomb interaction given by Eq. (5.5). In order to make a comparison with experiment, it is important to recognize that the finite extent of the electron wave functions perpendicular to the plane cuts off the divergence of the Coulomb interaction at short distances. For both GaAs and Si devices, this thickness is on the order of[34] 100 Å, which exceeds the magnetic length at high fields ($l=66$ Å at $B=15$ T).

We have used the Fang-Howard[35] variational form for the charge distribution normal to the plane,

$$g(z)=\frac{b^3}{2}z^2e^{-bz}\ , \tag{9.1}$$

from which the Stern-Howard[34,36] interaction may be obtained,

$$V(q)=\left[\frac{2\pi}{q}\right]\tfrac{1}{8}(1+q/b)^{-3}[8+9(q/b)+3(q/b)^2]\ . \tag{9.2}$$

This form for $V(q)$ may be found in Eq. (2.52) of Ref. 34. Figure 6 shows the collective-mode dispersion for four

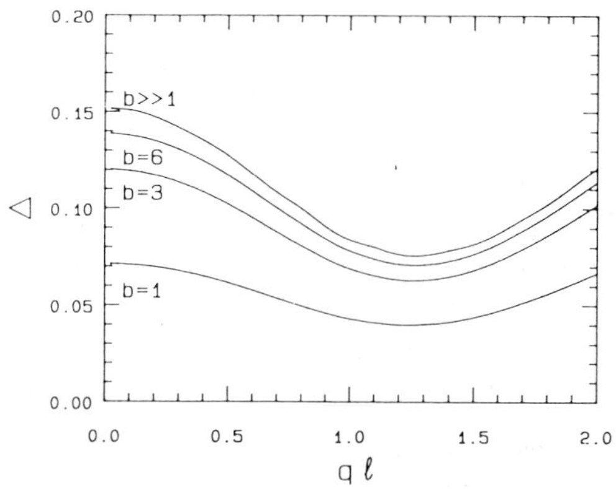

FIG. 6. Effect of finite thickness on collective-mode dispersion. The values of the Fang-Howard parameter bl are ∞, 6, 3, and 1 (going from the top curve to the bottom). All curves are for $\nu=\frac{1}{3}$.

different values of the dimensionless thickness parameter bl. We see that experimentally relevant values of bl cause a significant reduction in the size of the gap. Furthermore, this reduction is B-field dependent. Naively, we expect $\Delta \sim B^{1/2}$ (for fixed ν) since the natural unit of energy is $e^2/\epsilon l$. However, $bl \sim B^{-1/2}$ and therefore Δ does not rise as rapidly as might be expected at large B. This may at least partially (but not fully) account for the saturation in the Δ-versus-B experimental curve recently obtained by Boebinger *et al.*[37]

X. ROLE OF DISORDER

There does not exist at present a good understanding of the temperature and disorder dependence of the conductivity. The experimental activation energy for σ_{xx} is generally believed to be determined by the minimum energy to create charged excitations (quasiparticles). It has been suggested, however, that the scattering of thermally activated rotons may also contribute to the dissipation.[38] The *lowest* excitation energy $\Delta(k)$ occurs at $k=k_{\min}$, corresponding to a neutral density wave with charges of magnitude e separated by a distance $d=kl^2$. For $k>k_{\min}$ and $\nu=1/m$ the excitation is a quasiexciton consisting of charges of magnitude $e^*=e/m$ separated by a distance $d=mkl^2$. Thus the charge-excitation gap corresponds to $\Delta(\infty)$, and from the law of mass action[39] the number of quasiparticles is activated with energy $\Delta(\infty)/2$. The experimentally observed[37,40] activation energies for dissipation are much smaller than the predicted values of $\Delta(\infty)/2$ and $\Delta_{\min}$. Part of the discrepancy is due to finite-thickness effects which (as discussed in Sec. IX) soften the Coulomb repulsion at short distances and hence lower the gap. The remaining discrepancy is presumably due to the disorder in the sample and needs to be further investigated.

It has recently been suggested[41] that, as the disorder is increased, the quasiparticle excitation gap continuously closes, thereby destroying the fractional quantum Hall effect. This picture assumes a liquid state plus independent quasiparticles which interact with the disorder. This may well be correct; however, we wish to point out a possible alternative which may occur for some or even all of the fractional states: The continuous-gap closing transition may be preempted by the formation of a Wigner-glass state[42,43] corresponding to the collapse of the gap near $k_{\min}$ rather than $k=\infty$. If the gap collapses to zero at finite wave vector, then the dissipation is controlled by the generation of rotons as the glassy crystal slides past the impurities, just as occurs for phonon production in charge-density-wave systems.[44] Such a glassy state has been proposed to explain cyclotron-resonance anomalies in Si devices.[43]

The reasons this gap collapse might occur are the following. In the absence of disorder the lowest gap is at $k=k_{\min}$, the wave vector at which the system is already nearly unstable to crystallization. The quasiexciton is made of particles of charge $|e^*|=e/m$ for $\nu=1/m$. Hence it responds much less strongly to the randomly fluctuating Coulomb potential than the roton which consists of charges of magnitude e. Thus the roton gap may collapse first. On the other hand, the charges in the roton are separated by a finite distance $d=k_{\min}l^2$. If the length scale of the random potential is larger than d, then the rotons will respond less strongly to the disorder. These questions will be discussed in greater detail elsewhere.[45]

The lowest filling factor at which evidence for a fractional Hall state has been found is $\nu=\frac{1}{5}$. This state is extremely weak, however, and does not develop a plateau even at very low temperatures.[46] As will be seen in the next section, the $\frac{1}{5}$ state is much more susceptible to external perturbations than the $\frac{1}{3}$ state because of the incipient Wigner crystal instability. Thus the ideas presented here may well explain the weakness of the anomaly at $\nu=\frac{1}{5}$.

XI. STATIC LINEAR RESPONSE

We are now in the position of having rather accurate eigenfunctions $\bar{\rho}_k|0\rangle$ and eigenvalues $\Delta(k)$ for those states which couple to the ground state through the density. The dynamical structure factor is (in the SMA) given by

$$S(q,\omega)=\bar{s}(q)\delta(\omega-\Delta(q)) . \qquad (11.1)$$

From this we can compute the susceptibility $\chi(q,\omega)$, the dielectric function $\epsilon(q,\omega)$ the loss function $\mathrm{Im}[-1/\epsilon(q,\omega)]$, and related quantities such as the coupling between rotons and the substrate phonons. For example, using the memory-function formalism we have computed the ac conductivity in the presence of weak impurity scattering. This work will be reported elsewhere.[47]

As a simple example of what can be done with Eq. (11.1), we consider here the static susceptibility to an external perturbation. Zhang *et al.*[48] and Rezayi and Haldane[49] have recently performed numerical calculations of the charge distribution and relaxation energy of the $\nu=\frac{1}{3}$

state in the presence of a Coulomb impurity of charge Ze. Below we present simple analytical calculations of the same quantities.

The static susceptibility is given by[9]

$$\chi(q) = -2\int_0^\infty \frac{d\omega}{\omega}\, S(q,\omega)\ . \tag{11.2}$$

Using Eq. (11.1) we have

$$\chi(q) = \frac{-2\bar{s}(q)}{\Delta(q)}\ . \tag{11.3}$$

The quantity $\alpha(q) \equiv \bar{s}(q)/\Delta(q)$ is shown plotted in Fig. 7. Note that the susceptibility is sharply peaked at the wave vector corresponding to the roton minimum and that the magnitude of the susceptibility rises very rapidly as the filling factor approaches $\nu = \frac{1}{7}$. This is consistent with our previous discussion in Sec. VI of the Wigner crystal instability near $\nu = \frac{1}{7}$.

The Fourier transform of the perturbed charge density is (within linear response theory)

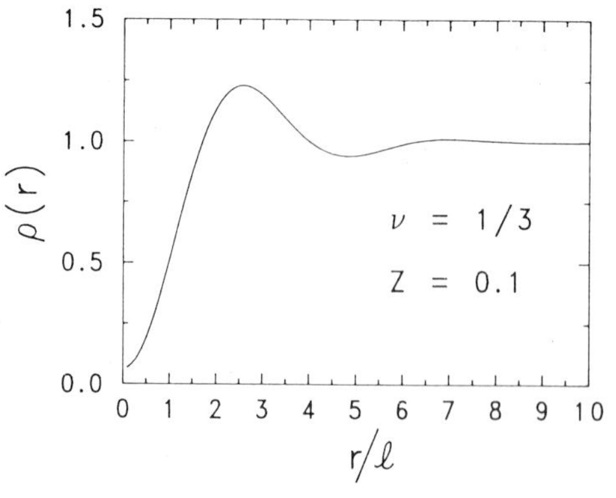

FIG. 8. Normalized charge distribution near a repulsive Coulomb impurity with $Z = 0.1$. Note the oscillations at the wavelength corresponding to the roton minimum.

$$\langle \delta\bar{\rho}_q \rangle = \rho v(q)\chi(q)\ , \tag{11.4}$$

where $\rho = e\nu/2\pi l^2$ is the mean density and $v(q) = 2\pi Z e^2/\epsilon q$ is the transform of the Coulomb potential. The real space-charge distribution is therefore

$$\frac{\langle \rho(r) \rangle}{\rho} = 1 + Z\int_0^\infty dq\, \chi(q) J_0(qr)\ . \tag{11.5}$$

This is plotted in Fig. 8 for the case $Z = 0.1$. Note that because $\chi(q)$ is sharply peaked at the roton (crystallization) wave vector, the spatial distribution of the charge is oscillatory. This provides a simple physical explanation of the charge oscillations first observed numerically by Zhang *et al.*[48]

The impurity relaxation energy in linear response is

$$\Delta E = \frac{1}{2}\int \frac{d^2q}{(2\pi)}\, v(q) \langle \delta\bar{\rho}_q \rangle\ . \tag{11.6}$$

Evaluation of ΔE using Eqs. (11.3) and (11.4) yields $\Delta E = -1.15Z^2$ for $\nu = \frac{1}{3}$. This compares very favorably with the numerical result $\Delta E = -1.2Z^2$ obtained by Zhang *et al.*[48] for a small system (five particles).

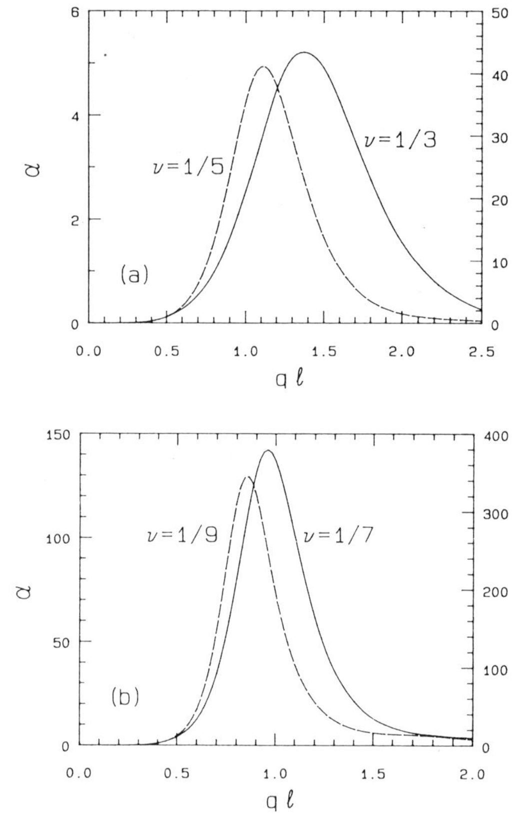

FIG. 7. Susceptibility parameter $\alpha \equiv \bar{s}(q)/\Delta(q)$. (a) $\nu = \frac{1}{3}$ (scale on left), $\nu = \frac{1}{5}$ (scale on right); (b) $\nu = \frac{1}{7}$ (scale on left), $\nu = \frac{1}{9}$ (scale on right).

XII. SUMMARY AND CONCLUSIONS

We have presented here a theory of the collective excitations in the fractional quantum Hall regime which is closely analogous to Feynman's theory of superfluid helium. The elementary collective excitations are density waves which exhibit a finite gap at zero wave vector and have a distinct magneto-roton minimum at finite wave vector. This feature is analogous to the roton minimum in helium and is a precursor to the gap collapse associated with Wigner crystallization which occurs near $\nu = \frac{1}{7}$. For larger wave vectors the lowest-lying mode crosses over from being a density wave to being a quasiparticle exciton.[12,30]

Our theory is based on the single-mode approximation, which assumes that 100% of the (intra-Landau-level) oscillator-strength sum is saturated by a single mode. By direct numerical diagonalization of the Hamiltonian for small systems ($N \leq 8$), Haldane and Rezayi[25] have found that the single-mode approximation is quite accurate, particularly near the roton minimum, where the lowest eigenmode absorbs 98% of the oscillator strength. Another way to state this is that the overlap between our variational excited-state wave function and the exact eigenfunction exceeds 0.98.

We thus have rather accurate eigenvalues and wave functions for those states which couple to the ground state through the density. From these we have calculated the dynamic structure factor and the susceptibility. As an illustration of the utility of this theory we presented a simple calculation of the perturbed charge distribution and relaxation energy due to a Coulomb impurity potential. The results were found to be in good agreement with numerical calculations of Zhang *et al.*[48]

The two main advantages of the present theory are that it gives a clear and useful physical picture of the nature of the collective modes and it allows one to calculate rather easily and accurately experimentally relevant quantities such as the collective-mode dispersion and the susceptibility. There remain, however, many further interesting questions related to the existence of two-roton bound states, the nature of the collective excitations for the compound ground states further down in the hierarchy,[50] the role of higher Landau levels, and so forth.

It would be particularly interesting to see whether states further down in the hierarchy act as multicomponent fluids having more than one collective excitation branch. The projection formalism[15] discussed here can be generalized to discuss fractional states in higher Landau levels[51] and to investigate inter-Landau-level cyclotron-resonance modes[52] as well.

ACKNOWLEDGMENTS

During the course of this work we have benefited from many helpful discussions with numerous colleagues. We would particularly like to thank C. Kallin, B. I. Halperin, R. B. Laughlin, P. A. Lee, F. D. M. Haldane, F. C. Zhang, and S. Das Sarma.

[1]K. von Klitzing, G. Dorda, and M. Pepper, Phys. Rev. Lett. **45**, 494 (1980).

[2]D. C. Tsui, H. L. Störmer, and A. C. Gossard, Phys. Rev. Lett. **48**, 1559 (1982).

[3]H. L. Störmer, A. Chang, D. C. Tsui, J. C. M. Hwang, A. C. Gossard, and W. Wiegmann, Phys. Rev. Lett. **50**, 1953 (1983).

[4]H. L. Störmer, in *Festkörperprobleme (Advances in Solid State Physics)*, edited by P. Grösse (Pergamon/Vieweg, Braunschweig, 1984), Vol. 24, p. 25.

[5]D. C. Tsui, A. C. Gossard, B. F. Field, M. E. Cage, and R. F. Dziuba, Phys. Rev. Lett. **48**, 3 (1982).

[6]R. B. Laughlin, Phys. Rev. Lett. **50**, 1395 (1983).

[7]S. M. Girvin, A. H. MacDonald, and P. M. Platzman, Phys. Rev. Lett. **54**, 581 (1985).

[8]R. P. Feynman, *Statistical Mechanics* (Benjamin, Reading, Mass., 1972), Chap. 11; Phys. Rev. **91**, 1291, 1301 (1953); **94**, 262 (1954); R. P. Feynman and M. Cohen *ibid.* **102**, 1189 (1956).

[9]G. D. Mahan, *Many-Particle Physics* (Plenum, New York, 1981).

[10]B. I. Lundqvist, Phys. Kondens. Mater. **6**, 206 (1967).

[11]A. W. Overhauser, Phys. Rev. B **3**, 1888 (1971).

[12]C. Kallin and B. I. Halperin, Phys. Rev. B **30**, 5655 (1984).

[13]W. Kohn, Phys. Rev. **123**, 1242 (1961).

[14]B. Jancovici, Phys. Rev. Lett. **46**, 386 (1981).

[15]S. M. Girvin and T. Jach, Phys. Rev. B **29**, 5617 (1984).

[16]V. Bargmann, Rev. Mod. Phys. **34**, 829 (1962).

[17]J. M. Caillol, D. Levesque, J. J. Weis, and J. P. Hansen, J. Stat. Phys. **28**, 235 (1982).

[18]Marc Baus and Jean-Pierre Hansen, Phys. Rep. **59**, 1 (1980).

[19]D. Yoshioka, Phys. Rev. B **29**, 6833 (1984); J. Phys. Soc. Jpn. **53**, 3740 (1984).

[20]F. D. M. Haldane and E. H. Rezayi, Phys. Rev. Lett. **54**, 237 (1985).

[21]D. Levesque, J. J. Weiss, and A. H. MacDonald, Phys. Rev. B **30**, 1056 (1984).

[22]S. M. Girvin, Phys. Rev. B **30**, 558 (1984).

[23]J. P. Hansen and D. Levesque, J. Phys. C **14**, L603 (1981).

[24]P. K. Lam and S. M. Girvin, Phys. Rev. B **30**, 473 (1984); **31**, 613(E) (1985).

[25]F. D. M. Haldane and E. H. Rezayi, Bull. Am. Phys. Soc. **30**, 381 (1985).

[26]A. Zawadowski, J. Ruvalds, and J. Solana, Phys. Rev. A **5**, 399 (1972); V. Celli and J. Ruvalds, Phys. Rev. Lett. **28**, 539 (1972).

[27]B. I. Halperin, Helv. Phys. Acta **56**, 75 (1983).

[28]P. W. Anderson, Phys. Rev. B **28**, 2264 (1983).

[29]S. M. Girvin, Phys. Rev. B **29**, 6012 (1984).

[30]R. B. Laughlin, Physica **126B**, 254 (1984).

[31]R. B. Laughlin, Surf. Sci. **142**, 163 (1982).

[32]Tapash Chakraborty, Phys. Rev. B **31**, 4016 (1985).

[33]R. Morf and B. I. Halperin, Phys. Rev. B (to be published).

[34]Tsuneya Ando, Alan B. Fowler, and Frank Stern, Rev. Mod. Phys. **54**, 437 (1982).

[35]F. F. Fang and W. E. Howard, Phys. Rev. Lett. **16**, 797 (1966).

[36]F. Stern and W. E. Howard, Phys. Rev. **163**, 816 (1967).

[37]G. S. Boebinger, A. M. Chang, H. L. Störmer, and D. C. Tsui, Bull. Am. Phys. Soc. **30**, 301 (1985); Phys. Rev. Lett. **55**, 1606 (1985).

[38]P. M. Platzman, S. M. Girvin, and A. H. MacDonald, Phys. Rev. B **32**, 8458 (1985).

[39]Neil W. Ashcroft and N. David Mermin, *Solid State Physics* (Holt, Rinehart and Winston, New York, 1976), p. 574.

[40]A. M. Chang, P. Berglund, D. C. Tsui, H. L. Störmer, and J. C. M. Hwang, Phys. Rev. Lett. **53**, 997 (1984); S. Kawaji, J. Wakabayashi, J. Yoshino, and H. Sakaki, J. Phys. Soc. Jpn. **53**, 1915 (1984).

[41]R. B. Laughlin, M. L. Cohen, J. M. Kosterlitz, H. Levine, S. B. Libby, and A. M. M. Pruisken, Phys. Rev. B **32**, 1311 (1985).

[42]H. Fukuyama and P. A. Lee, Phys. Rev. B **18**, 6245 (1978).

[43]B. A. Wilson, S. J. Allen, Jr., and D. C. Tsui, Phys. Rev. B **24**, 5887 (1981).

[44]H. Fukuyama and P. A. Lee, Phys. Rev. B **17**, 535 (1977).

[45]A. H. MacDonald, K. L. Liu, S. M. Girvin, and P. M. Platz-

man, Bull. Am. Phys. Soc. **30**, 301 (1985); Phys. Rev. B (to be published).

[46]E. E. Mendez, M. Heiblum, L. L. Chang, and L. Esaki, Phys. Rev. B **28**, 4886 (1983).

[47]P. M. Platzman, S. M. Girvin, and A. H. MacDonald, Bull. Am. Phys. Soc. **30**, 301 (1985); Phys. Rev. B **32**, 8458 (1985).

[48]F. C. Zhang, V. Z. Vulovic, Y. Guo, and S. Das Sarma, Phys. Rev. B **32**, 6920 (1985).

[49]E. H. Rezayi and F. D. M. Haldane, Bull. Am. Phys. Soc. **30**, 300 (1985).

[50]F. D. M. Haldane, Phys. Rev. Lett. **51**, 605 (1983).

[51]A. H. MacDonald, Phys. Rev. B **30**, 3550 (1984).

[52]A. H. MacDonald, H. C. A. Oji, and S. M. Girvin, Phys. Rev. Lett. **55**, 2208 (1985).

Off-Diagonal Long-Range Order, Oblique Confinement, and the Fractional Quantum Hall Effect

S. M. Girvin
Surface Science Division, National Bureau of Standards, Gaithersburg, Maryland 20899

and

A. H. MacDonald
National Research Council, Ottawa, Ontario, Canada K1A OR6
(Received 24 November 1986)

We demonstrate the existence of a novel type of off-diagonal long-range order in the fractional-quantum-Hall-effect ground state. This is revealed for the case of fractional filling factor $\nu = 1/m$ by application of Wilczek's "anyon" gauge transformation to attach m quantized flux tubes to each particle. The binding of the zeros of the wave function to the particles in the fractional quantum Hall effect is a (2+1)-dimensional analog of *oblique confinement* in which a condensation occurs, not of ordinary particles, but rather of composite objects consisting of particles and gauge flux tubes.

PACS numbers: 72.20.My, 71.45.Gm, 73.40.Lq

A remarkable amount of progress has recently been made in our understanding of the fractional quantum Hall effect (FQHE)[1] following upon the seminal paper by Laughlin.[2] There remains, however, a major unsolved problem which centers on whether or not there exists an order parameter associated with some type of symmetry breaking.[3-6] The apparent symmetry breaking associated with the discrete degeneracy of the ground state in the Landau gauge[5] is an artifact of the toroidal geometry[6,7] and is not an issue here. Rather, the questions that we are addressing have been motivated by the analogies which have been observed to exist[4,8] between the FQHE and superfluidity and by recent progress towards a phenomenological Ginsburg-Landau picture of the FQHE.[4] Further motivation has come from the development of the correlated-ring-exchange theory of Kivelson *et al.*[9] (see also Chui, Hakim, and Ma,[10] and Chui,[10] and Baskaran[11]). The existence of infinitely large ring exchanges is a signal of broken gauge symmetry in superfluid helium[12] and is suggestive of something similar in the FQHE. However, the concept of ring exchanges on large length scales has not as yet been fully reconciled with Laughlin's (essentially exact[7]) variational wave functions which focus on the short-distance behavior of the two-particle correlation function. Furthermore it is clear that we cannot have an ordinary broken gauge symmetry since the particle density (which is conjugate to the phase) is ever more sharply defined as the length scale increases. The purpose of this Letter is to unify all these points of view by demonstrating the existence of a novel type of off-diagonal long-range order (ODLRO) in the FQHE ground state.

In second quantization the one-body density matrix is given by

$$\rho(z,z') = \sum_{m,n} \varphi_m^*(z)\varphi_n(z')\langle 0|c_n^\dagger c_m|0\rangle, \tag{1}$$

where $\varphi_n(z)$ is the nth lowest-Landau-level single-particle orbital[1] in the symmetric gauge, and z is a complex representation of the particle position vector in units of the magnetic length.[1] It is an unusual feature of this problem that there is a unique single-particle state for each angular momentum. Hence by making only the assumption that the ground state is isotropic and homogeneous we may deduce $\langle 0|c_n^\dagger c_m|0\rangle = \nu\delta_{nm}$, and thereby obtain the powerful identity:

$$\rho(z,z') = \nu g(z,z') = (\nu/2\pi)\exp(-\tfrac{1}{4}|z-z'|^2)\exp[\tfrac{1}{4}(z^*z' - zz'^*)], \tag{2}$$

where $g(z,z')$ is the ordinary single-particle Green's function.[13]

We see from (2) that the density matrix is short ranged with a characteristic scale given by the magnetic length, just as occurs in superconducting films in a magnetic field.[14] The same result can be obtained within first quantization via the expression

$$\rho(z,z') = \frac{N}{Z}\int d^2z_2 \cdots d^2z_N \Psi^*(z,z_2,\ldots,z_N)\Psi(z',z_2,\ldots,z_N), \tag{3}$$

where Z is the norm of Ψ.

If the lowest Landau level has filling factor $\nu = 1/m$ and the interaction is a short-ranged repulsion, then in the low-electron mass limit,[7] the *exact* ground-state wave function is given by Laughlin's expression:

$$\Psi(z_1,\ldots,z_N) = \prod_{i<j}(z_i - z_j)^m \exp\left(-\tfrac{1}{4}\sum_k |z_k|^2\right). \tag{4}$$

Laughlin's plasma analogy[2,15] proves that the ground state is a liquid of uniform density so that Eq. (2) is valid. The rapid phase oscillations of the integrand in (3) cause ρ to be short ranged. There is, nevertheless, a peculiar type of long-range order hidden in the ground state. For reasons which will become clear below, this order is revealed by considering the singular gauge field used in the study of "anyons"[16,17]:

$$\mathcal{A}_j(z_j) = \frac{\lambda\Phi_0}{2\pi}\sum_{i\neq j}\nabla_j \mathrm{Im}\ln(z_j - z_i), \tag{5}$$

where $\Phi_0 = hc/e$ is the quantum of flux and λ is a constant. The addition of this vector potential to the Hamiltonian is not a true gauge transformation since a flux tube is attached to each particle. If, however, $\lambda = m$ where m is an integer, the net effect is just a change in the phase of the wave function:

$$\Psi_{\mathrm{new}} = \exp\left(-im\sum_{i<j}\mathrm{Im}\ln(z_i - z_j)\right)\Psi_{\mathrm{old}}. \tag{6}$$

Application of (6) to the Laughlin wave function (4) yields

$$\Psi(z_1,\ldots,z_N) = \prod_{i<j}|z_i - z_j|^m \exp\left(-\tfrac{1}{4}\sum_k |z_k|^2\right), \tag{7}$$

which is purely real and is symmetric under particle exchange for both even and odd m. Hence we have the remarkable result that both fermion and boson systems map into bosons in this singular gauge.

Substituting (7) into (3) and using Laughlins's plasma analogy,[2,15] a little algebra shows that the singular-gauge density matrix $\tilde\rho$ can be expressed as

$$\tilde\rho(z,z') = (\nu/2\pi)\exp[-\beta\Delta f(z,z')]\,|z - z'|^{-m/2}, \tag{8}$$

where $\beta \equiv 2/m$, and $\Delta f(z,z')$ is the difference in free energy between two impurities of charge $m/2$ (located at z and z') and a single impurity of charge m (with arbitrary location). Because of complete screening of the impurities by the plasma, the free-energy difference $\Delta f(z,z')$ rapidly approaches a constant as $|z - z'| \to \infty$. This proves the existence of ODLRO[18] characterized by an exponent $\beta^{-1} = m/2$ equal to the plasma "temperature." For $m = 1$ the asymptotic value of Δf can be found exactly: $\beta\Delta f_\infty = -0.03942$. For other values of m, $\beta\Delta f(z,z')$ can be estimated either by use of the ion-disk approximation[2,15] or the static (linear response) susceptibility of the (classical) plasma calculated from the known static structure factor[8] (see Fig. 1).

The rigorous and quantitative results we have obtained above are valid for the case of short-range repulsive interactions for which Laughlin's wave function is exact. We now wish to use these results for a qualitative examination of more general cases and to deepen our understanding of the ODLRO. We begin by noting that $\tilde\rho$ can be rewritten in the ordinary gauge as

$$\tilde\rho(z,z') = \frac{N}{Z}\int d^2z_2\cdots d^2z_N \exp\left(-i\frac{e}{\hbar c}\int_{\mathbf{z}}^{\mathbf{z}'} d\mathbf{r}\cdot\mathcal{A}_1\right)\Psi^*(z,z_2,\ldots,z_N)\Psi(z',z_2,\ldots,z_N), \tag{9}$$

where $\mathbf{z}$ and $\mathbf{z}'$ are vector representations of z and z'. The line integral in (9) is multiple valued but its exponential is single valued because the flux tubes are quantized. The additional phases introduced by the singular gauge transformation will cancel the phases in Ψ nearly everywhere, and produce ODLRO in $\tilde\rho$ if and only if the zeros of Ψ (which must necessarily be present because of the magnetic field[19]) are bound to the particles. Thus ODLRO in $\tilde\rho$ *always* signals a condensation of the zeros onto the particles (independent of whether or not the composite-particle occupation of the lowest momentum state diverges[18]). Because the gauge field $\mathcal{A}_1$ depends on the positions of *all* the particles, $\tilde\rho$ differs not just in the phase but in *magnitude* from ρ. This multiparticle object, which explicitly exhibits ODLRO, is very reminiscent of the topological order parameter in the *XY* model[20] and related gauge models[21,22] and is intimately connected with the frustrated *XY* model which arises in the correlated-ring-exchange theory.[9]

For short-range interactions, the zeros of Ψ are directly on the particles and the associated phase factors are exactly canceled by the gauge term in (9) [see Eq. (7)]. As the range of the interaction increases, $m-1$ of the zeros move away from the particles but remain nearby

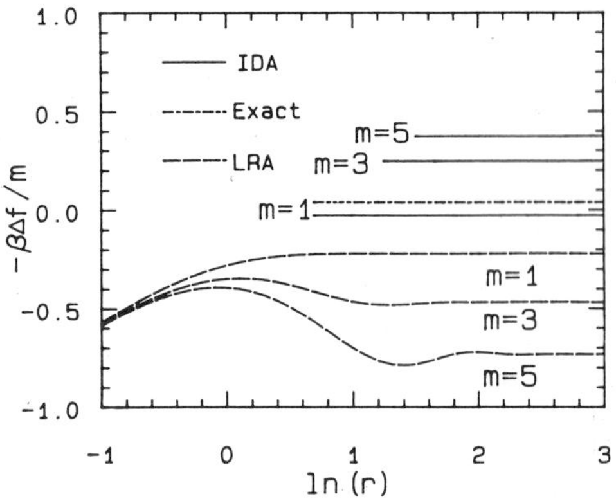

FIG. 1. Plot of $-\beta\Delta f(z,z')/m$ vs $r \equiv |z - z'|$ for filling factor $\nu = 1/m$. LRA is linear-response approximation, IDA is ion-disk approximation (shown only for separations exceeding the sum of the ion-disk radii). Because the plasma is strongly coupled, the IDA is quite accurate at $m = 1$ and improves further with increasing m. The LRA is less accurate at $m = 1$ and worsens with increasing m.

and bound to them.[7,23] The gauge and wave function phase factors in (9) now appear in the form of the bound vortex-antivortex pairs. We expect such bound pairs *not* to destroy the ODLRO and speculate (based on our understanding of the Kosterlitz-Thouless transition[20]) that the effect is at most to renormalize the exponent of the power law in (8). As the range of the potential is increased still further, numerical computations[7] indicate that a critical point is reached at which the gap rather suddenly collapses and the overlap between Laughlin's state and the true ground state drops quickly to zero. We propose that this gap collapse corresponds to the unbinding of the vortices and hence to the loss of ODLRO and the onset of short-range behavior of $\tilde{\rho}(z,z')$. Recall that the distinguishing feature of the FQHE state is its long wavelength excitation gap. At least within the single-mode approximation,[8] this gap can only exist when the ground state is homogeneous and the two-point correlation function exhibits perfect screening:

$$M_1 \equiv (v/2\pi)\int d^2r(r^2/2)[g(r)-1] = -1.$$

In the analog plasma problem, the zeros of Ψ act like point charges seen by each particle and the M_1 sum rule implies that electrons see each other as charge-$(m = 1/v)$ objects; i.e., that m zeros are bound to each electron. Thus (within the single-mode approximation) there is a one-to-one correspondence between the existence of ODLRO and the occurrence of the FQHE.

The exact nature of the gap-collapse transition, which occurs when the range of the potential is increased,[7] is not understood at present. However, it has been proven[8] that the M_1 sum rule is satisfied by every homogeneous and isotropic state in the lowest Landau level. Hence the vortex unbinding should be a first-order transition to a state which breaks rotation symmetry (like the Tao-Thouless state[24]) and/or translation symmetry (like the Wigner crystal[4,8]). We know that as a function of *temperature* (for fixed interaction potential) there can be no Kosterlitz-Thouless transition[20] since isolated vortices (quasiparticles) cost only a finite energy in this system[4,25] (see, however, Ref. 10).

Further insight into the gap collapse can be obtained by considering the exceptional case of Laughlin's wave function with $m > 70$. In this case the zeros are still rigorously bound to the particles so that the analog plasma contains long-range forces (and $\tilde{\rho}$ exhibits ODLRO), but the plasma "temperature" has dropped below the freezing point.[2,15] If such a state exhibits (sufficiently[10]) long-range positional correlations, the FQHE would be destroyed by a gapless Goldstone mode associated with the broken translation symmetry. Hence in this exceptional case the normal connection between ODLRO and the FQHE would be broken by a gap collapse due to positional order at a finite wave vector.

The existence of ODLRO in $\tilde{\rho}$ is the type of infrared property which suggests that a field-theoretic approach to the FQHE would be viable. It is clear from the results presented here that the binding of the zeros of Ψ to the particles can be viewed as a condensation,[18] not of ordinary particles, but rather of composite objects consisting of a particle and m flux tubes. (We emphasize that these are *not* real flux tubes, but merely consequences of the singular gauge. The assumption that electrons can bind real flux tubes[26] is easily shown to be unphysical.[27]) The analog of this result for hierarchical daughter states of the Laughlin states[7,15] would be a condensation of composite objects consisting of n particles and m flux tubes (cf. Halperin's "pair" wave functions[19]). This seems closely analogous to the phenomenon of *oblique confinement*[22] and it ought to be possible to derive the appropriate field theory from first principles by use of this idea.

Since the singular gauge maps the problem onto interacting bosons, coherent-state path integration[28] may prove useful. A step in this direction has been taken recently in the form of a Landau-Ginsburg theory which was developed on phenomenological grounds.[4] In the static limit, the action has the "θ vacuum" form

$$S = \int d^2r\,|(-i\nabla+\mathbf{a})\psi(\mathbf{r})|^2 + i\phi(\mathbf{r})(\psi^*\psi-1) - i(\theta/8\pi^2)(\phi\nabla\times\mathbf{a}+\mathbf{a}\times\nabla\phi), \tag{10}$$

where $\mathbf{a}$ is not the physical vector potential but an effective gauge field[4] representing frustration due to density deviations away from the quantized Laughlin density and ϕ is a scalar potential which couples both to the charge density and the "flux" density. From (10) the equation of motion for $\mathbf{a}$ is (in the static case):

$$\theta\nabla\times\mathbf{a} = (\psi^*\psi - 1). \tag{11}$$

This equation and the parameter θ, which determines the charge carried by an isolated vortex, originally had to be chosen phenomenologically.[4] Now, however, it can be justified by examination of Eq. (5) which shows that the curl of $\mathcal{A}_j$ is proportional to the density of particles. If we identify $\mathbf{a}$ in (10) and (11) as

$$\mathbf{a} = \mathcal{A}_1 + \mathbf{A}, \tag{12}$$

where $\mathbf{A}$ is the physical vector potential and we take $\psi^*\psi$ as the particle density relative to the density in the Laughlin state, then Eq. (11) follows from (5) with the θ angle being given by $\theta = 2\pi/m$. This yields[4] the correct charge of an isolated vortex (Laughlin quasiparticle) of $q^* = 1/m$. The connection between this result and the Berry phase argument of Arovas *et al.*[29] should be noted (see also Semenoff and Sodano[30]). To summarize, it is the strong phase fluctuations induced by the frustration associated with density variations [Eq. (11)] which pin the density at rational fractional values and account for the differences between the FQHE and ordinary superfluidity.[4]

We believe that these results shed considerable light on the FQHE, unify the different pictures of the effect, and emphasize the topological nature of the order in the zero-temperature state of the FQHE. The present picture leads to several predictions which can be tested by numerical computations by use of methods very similar to those now in use.[31] ODLRO will be found only in states exhibiting an excitation gap. The decay of the singular-gauge density matrix will be controlled by the distribution of distances of the zeros of the wave function from the particles. This distribution, which can be artificially varied by changing the model interaction, directly determines the short-range behavior of the density-density correlation function and hence the ground-state energy.[7,23]

The authors would like to express their thanks to C. Kallin, S. Kivelson, and R. Morf for useful conversations and suggestions.

[1] *The Quantum Hall Effect*, edited by R. E. Prange and S. M. Girvin (Springer-Verlag, New York, 1986).

[2] R. B. Laughlin, Phys. Rev. Lett. **50**, 1395 (1983).

[3] P. W. Anderson, Phys. Rev. B **28**, 2264 (1983).

[4] S. M. Girvin, in Chap. 10 of Ref. 1.

[5] R. Tao and Yong-Shi Wu, Phys. Rev. B **30**, 1097 (1984).

[6] D. J. Thouless, Phys. Rev. B **31**, 8305 (1985).

[7] F. D. M. Haldane, in Chap. 8 of Ref. 1.

[8] S. M. Girvin, A. H. MacDonald, and P. M. Platzman, Phys. Rev. Lett. **54**, 581 (1985), and Phys. Rev. B **33**, 2481 (1986); S. M. Girvin in Chap. 9 of Ref. 1.

[9] S. Kivelson, C. Kallin, D. P. Arovas, and J. R. Schrieffer, Phys. Rev. Lett. **56**, 873 (1986).

[10] S. T. Chui, T. M. Hakim, and K. B. Ma, Phys. Rev. B **33**, 7110 (1986); S. T. Chui, unpublished.

[11] G. Baskaran, Phys. Rev. Lett. **56**, 2716 (1986), and unpublished.

[12] R. P. Feynman, Phys. Rev. **91**, 1291 (1953).

[13] S. M. Girvin and T. Jach, Phys. Rev. B **29**, 5617 (1984).

[14] E. Brézin, D. R. Nelson, and A. Thiaville, Phys. Rev. B **31**, 7124 (1985).

[15] R. B. Laughlin, in Chap. 7 of Ref. 1.

[16] F. Wilczek, Phys. Rev. Lett. **49**, 957 (1982).

[17] D. P. Arovas, J. R. Schrieffer, F. Wilczek, and A. Zee, Nucl. Phys. B **251**, 117 (1985).

[18] We refer to this as ODLRO or condensation because of the slow power-law decay even though the largest eigenvalue $\lambda \equiv \int d^2 z \, \tilde{\rho}(z,z')$ of the density matrix diverges only for $m \leq 4$ [see C. N. Yang, Rev. Mod. Phys. **34**, 694 (1962)].

[19] B. I. Halperin, Helv. Phys. Acta **56**, 75 (1983).

[20] J. V. José, L. P. Kadanoff, S. Kirkpatrick, and D. R. Nelson, Phys. Rev. B **16**, 1217 (1977).

[21] J. B. Kogut, Rev. Mod. Phys. **51**, 659 (1979).

[22] J. L. Cardy and E. Rabinovici, Nucl. Phys. B **205**, 1 (1982); J. L. Cardy, Nucl. Phys. B **205**, 17 (1982).

[23] D. J. Yoshioka, Phys. Rev. B **29**, 6833 (1984).

[24] R. Tao and D. J. Thouless, Phys. Rev. B **28**, 1142 (1983). The symmetric gauge version of this state exhibits threefold rotational symmetry.

[25] A. M. Chang, in Chap. 6 in Ref. 1.

[26] M. H. Friedman, J. B. Sokoloff, A. Widom, and Y. N. Srivastava, Phys. Rev. Lett. **52**, 1587 (1984), and **53**, 2592 (1984).

[27] F. D. M. Haldane and L. Chen, Phys. Rev. Lett. **53**, 2591 (1984).

[28] L. S. Schulman, *Techniques and Applications of Path Integration* (Wiley, New York, 1981).

[29] D. Arovas, J. R. Schrieffer, and F. Wilczek, Phys. Rev. Lett. **53**, 722 (1984).

[30] G. Semenoff and P. Sodano, Phys. Rev. Lett. **57**, 1195 (1986).

[31] F. D. M. Haldane and E. H. Rezayi, Phys. Rev. Lett. **54**, 237 (1985), and Phys. Rev. B **31**, 2529 (1985); F. C. Zhang, V. Z. Vulovic, Y. Guo, and S. Das Sarma, Phys. Rev. B **32**, 6920 (1985); G. Fano, F. Ortolani, and E. Colombo, Phys. Rev. B **34**, 2670 (1986).

SUBJECT INDEX

AUTHOR INDEX

PERSPECTIVES IN CONDENSED MATTER PHYSICS

Published Volumes

1. G. Margaritondo, *Electronic Structure of Semiconductor Heterojunctions*
2. A. H. MacDonald, *Quantum Hall Effect: a Perspective*

Forthcoming Volumes

G. Jacucci, *Simulation Approach to Solids*
G.J. Iafrate, *Quantum Transport and New Semiconductor Devices*
G.M. Kalvius and W. Zinn, *Magnetic Materials*

Executive Editor: L. Miglio

Dipartimento di Fisica dell'Università di Milano
Via Celoria, 16 I-20133 MILANO
Fax + 39/2/2366583; **Telex** 334687 INFNMI; **Tel.** + 39/2/2392.408

finito di stampare nel mese
di ottobre 1989
dalla Nuova Timec s.r.l.
Albairate (MI)

Editoriale Jaca Book spa
Via Aurelio Saffi 19, 20123 Milano

spedizione in abbonamento
postale TR editoriale
aut. D/162247/PI/3
direzione PT Milano